OIL FROM SHALE AND TAR SANDS

OIL FROM SHALE AND TAR SANDS

Edward M. Perrini

NOYES DATA CORPORATION
Park Ridge, New Jersey London, England
1975

Library of Congress Catalog Card Number: 75-10357
ISBN: 0-8155-0583-3
Printed in the United States

Published in the United States of America by
Noyes Data Corporation
Noyes Building, Park Ridge, New Jersey 07656

FOREWORD

The detailed, descriptive information in this book is based on U.S. patents issued since 1960 that deal with obtaining oil from shale and tar sands. Where it was necessary to round out the complete technological picture, even earlier, but very relevant patents were included.

This book serves a double purpose in that it supplies detailed technical information and can be used as a guide to the U.S. patent literature in this field. By indicating all the information that is significant, and eliminating legal jargon and juristic phraseology, this book presents an advanced, commercially oriented review of modern methods of oil shale and tar sands processing as depicted in U.S. patents.

The U.S. patent literature is the largest and most comprehensive collection of technical information in the world. There is more practical, commercial, timely process information assembled here than is available from any other source. The technical information obtained from a patent is extremely reliable and comprehensive; sufficient information must be included to avoid rejection for "insufficient disclosure." These patents include practically all of those issued on the subject in the United States during the period under review, there has been no bias in the selection of patents for inclusion.

The patent literature covers a substantial amount of information not available in the journal literature. The patent literature is a prime source of basic commercially useful information. This information is overlooked by those who rely primarily on the periodical journal literature. It is realized that there is a lag between a patent application on a new process development and the granting of a patent, but it is felt that this may roughly parallel or even anticipate the lag in putting that development into commercial practice.

Many of these patents are being utilized commercially. Whether used or not, they offer opportunities for technological transfer. Also, a major purpose of this book is to describe the number of technical possibilities available, which may open up profitable areas of research and development. The information contained in this book will allow you to establish a sound background before launching into research in this field.

Advanced composition and production methods developed by Noyes Data are employed to bring our new durably bound books to you in a minimum of time. Special techniques are used to close the gap between "manuscript" and "completed book." Industrial technology is progressing so rapidly that time-honored, conventional typesetting, binding and shipping methods are no longer suitable. We have bypassed the delays in the conventional book publishing cycle and provide the user with an effective and convenient means of reviewing up-to-date information in depth.

The Table of Contents is organized in such a way as to serve as a subject index. Other indexes by company, inventor and patent number help in providing easy access to the information contained in this book.

15 Reasons Why the U.S. Patent Office Literature Is Important to You —

1. The U.S. patent literature is the largest and most comprehensive collection of technical information in the world. There is more practical commercial process information assembled here than is available from any other source.

2. The technical information obtained from the patent literature is extremely comprehensive; sufficient information must be included to avoid rejection for "insufficient disclosure."

3. The patent literature is a prime source of basic commercially utilizable information. This information is overlooked by those who rely primarily on the periodical journal literature.

4. An important feature of the patent literature is that it can serve to avoid duplication of research and development.

5. Patents, unlike periodical literature, are bound by definition to contain new information, data and ideas.

6. It can serve as a source of new ideas in a different but related field, and may be outside the patent protection offered the original invention.

7. Since claims are narrowly defined, much valuable information is included that may be outside the legal protection afforded by the claims.

8. Patents discuss the difficulties associated with previous research, development or production techniques, and offer a specific method of overcoming problems. This gives clues to current process information that has not been published in periodicals or books.

9. Can aid in process design by providing a selection of alternate techniques. A powerful research and engineering tool.

10. Obtain licenses — many U.S. chemical patents have not been developed commercially.

11. Patents provide an excellent starting point for the next investigator.

12. Frequently, innovations derived from research are first disclosed in the patent literature, prior to coverage in the periodical literature.

13. Patents offer a most valuable method of keeping abreast of latest technologies, serving an individual's own "current awareness" program.

14. Copies of U.S. patents are easily obtained from the U.S. Patent Office at 50¢ a copy.

15. It is a creative source of ideas for those with imagination.

CONTENTS AND SUBJECT INDEX

INTRODUCTION

According to many geologists, oil shale may well be petroleum which nature never allowed to completely develop. As left by nature, oil shale is actually a substance called kerogen, made up of approximately 80% carbon, 10% hydrogen, and varying lesser amounts of nitrogen, oxygen and sulfur. Mixed with sediment, it forms a tough, dense rock ranging in color from light tan to black. The Ute Indians, upon observing outcroppings burst into flames after being hit by lightening, referred to it as "the rock that burns."

As early as the 1850s, shale oil was being touted as a replacement for the wood that America depended on for its energy. As Colonel Drake brought in his first oil well in Titusville, Pa., shale oil was generally pushed into the background and forgotten about. Scotland had an operating shale industry from 1850 to 1864 when the low price of foreign crude forced it to close. Interestingly, British Petroleum, one of the giants of the oil world and a large producer of Middle East crude was actually founded as a shale oil company. Oil shale from Estonia once supplied fuel gas for nearby Leningrad.

Over the years, most booms in shale were ended by declines in the price of conventional crude oil. During the 1920s, for example, some 200 companies were formed to mine the shale, but the 1930 discovery of the great East Texas field dropped the price of conventional crude to ten cents a barrel. In the 1950s, in an area of Colorado where Parachute Creek has dug a canyon into the dry earth, Union Oil retorted shale recovering some 300 barrels a day from its operation. The plant was eventually closed as imported oil was flooding the country and the low depletion allowance for shale (10% vs 22% for conventional oil) provided little incentive.

A large number of companies have an interest in finding, mining and making a commercial success out of shale oil. The list includes many experienced veterans of the oil exploration business, among them Shell, Mobil, Superior, Sun, American Petrofina, Atlantic Richfield, Standard of California, Ohio Standard and Texaco. Other smaller companies, attracted by the potential bonanza of these oil reserves, include Equity Oil, Austral Oil, Western Oil Shale, Skyline Oil and the Oil Shale Corp.

While many events in the private, public and international sectors still tend to give a confused picture of the role that shale will play in the very complex and competitive worldwide oil picture, events continue to point in the favor of shale as a future source of energy. While shale is found in such diverse corners of the world as Scotland, Russia, China, Germany, Australia and Thailand, the richest and most extensive deposit is in the Green River formation in the Colorado-Wyoming-Utah triangle where test drillings are extensively conducted. The shale is about 2,300 feet thick at the center of the basin (Piceance Creek Basin, Colorado) and is located some 1,000 feet beneath the surface. According to surveys, the Basin contains 480 to 500 billion barrels of oil in deposits with at least 25 gallons of oil per ton of shale. Another 800 billion barrels rest in deposits with a richness of 10 to 25 gallons a ton. Most of this land is owned by the U.S. Government.

One of the significant factors behind shale's slow development relates to technology. The processing of rock is substantially different than producing conventional crude. Two major processes have been developed to date. With the in-situ method, oil is obtained from the shale while it is still in the ground by injecting a hot gas or other heated substance into the structure, then recovering the oil via producing wells which are connected to the injecting wells. This method has been used in Sweden with limited success. A modified approach involves a system of fracturing the oil-bearing rock using hydraulic techniques, then igniting the rubble and driving the oil off through holes drilled into the fractured area. Various techniques have also been developed to use an atomic device to loosen the shale.

In the U.S.S.R., where domestic political restraints are largely nonexistent, nonmilitary applications of nuclear energy are widely tested. The Soviet plowshare projects include detonations at moderate depths within clay, shale, sandstone and limestone. Soviet oil stimulation projects resulted in immediate production increases stabilized in experimental fields at 30 to 60% above previous rates, with at least a 10% increase in ultimate recovery.

Although the in-plant fracturing would save on cost, most beds of shale have virtually no permeability, or porosity, and make poor conductors of heat. Thus Oil Shale Corp., and others, have developed a process termed the hot balls system. With this method, the rock is mined and brought to the surface using conventional mining techniques. It is then put into an enclosed chamber where it is mixed with small, preheated balls. Heat from the balls retorts the shale and oil vapors are drawn off and condensed in another chamber.

Vast reserves of oil are also found in tar sands. One of the largest of these reserves is the Athabasca deposit in Canada. Oil from these tar sands is generally obtained by mining with huge shovels and the use of aqueous separation techniques to remove the oil from the sand.

At a recent conference in Edmonton, a proposal to use nuclear energy as a heavy oil recovery technique was discussed. The project involves the use of a 10 kiloton device releasing both thermal and shock energy to convert the otherwise immobile viscous oil deposits into conventionally recoverable hydrocarbons. The resulting subterranean detonation cavern would also facilitate secondary and tertiary in situ recovery methods such as steam injection and fireflood.

Today, as the OPEC countries, and others, are demanding more and more profits and asking for increased equity participation in the operation of the oil companies and downstream chemicals production, the economics and the recovery of oil from shale and tar sands are becoming increasingly attractive. Indeed, in the past year it has been clearly demonstrated that the political stability and long range viability of many industrial nations is largely dependent on developing alternate energy sources.

This book describes over 215 processes related to the retorting and refining of oil shale and the separation of oil from tar sands. Many of these processes are now in the pilot plant stage and will form the basis for large scale commercialization over the next decade.

OIL SHALE RETORTING

Shale oil is not a naturally occurring product, but it is formed by the pyrolysis or distillation of organic matter (commonly called kerogen) found in certain shale-like rock. The organic material has limited solubility in ordinary solvents, and therefore, cannot be recovered by extraction. Upon strong heating, the organic material of some of these materials decomposes into gas, condensible liquids and residual carbonaceous material remaining on the spent shale.

In its basic aspects, the retorting of oil shale, and also other similar solid, hydrocarbonaceous materials, is a simple operation. The major steps involve the heating of the solid material to the proper temperature and the recovery of the vapor evolved. However, for a commercially feasible process it is necessary to consider and properly choose one of the many of the possible methods of physically moving the solid material through a vessel in which the retorting is to be carried out, as well as many other variants and operating parameters, all of which are interrelated.

The choice of a particular method of moving the solids through the vessel must include a consideration of mechanical aspects as well as the chemistry and the processes involved. Further, it is necessary to consider the many possible sources of heat that may be used for the pyrolysis, or destructive distillation, including the most economical and advantageous method of utilizing such heat to produce the desired products. In order for a retorting operation to be economically attractive, it must be self-supporting as far as energy requirements are concerned.

Generally the most feasible process is to derive the heating requirements from some of the various products of the pyrolysis using, however, a minimum of the valuable products of the process. Another area of consideration is the possible method of transferring heat to the solid material and the recovery of heat not actually utilized in the process itself, utilizing the principles of countercurrent heat exchange between gas and solids.

In order to achieve a retorting process that is economically attractive, the various operating parameters must be controlled so that the overall process is economical, continuous and highly reliable. Any equipment usable in the process must permit a high unit throughput of material since enormous quantities of the oil shale must be handled for a relatively nominal recovery of shale oil. The largest deposits of oil shale found in the United States lie in the semiarid locations of Colorado, Utah and Wyoming. Process equipment for these shale reserves must have a high thermal efficiency which requires little if any cooling water. As in the case of all mechanical devices, the retorting equipment should be as simple as possible so that relatively proven and economically attractive mechanical devices may be utilized in the operation of the retort. In an effort to provide an economic commercial process, literally hundreds of retorting processes have been proposed, each of which offers a somewhat different choice and/or combination of the many possible operating conditions.

GAS COMBUSTION

Multiprocessing Zones in Kiln

J.B. Jones, Jr. and A.A. Reeves; U.S. Patent 3,736,247; May 29, 1973; assigned to Paraho Corporation describe a process for retorting solid hydrocarbonaceous material in a continuously moving vertical column and to release heat in the column by the preferential oxidation of the carbonaceous residue on the solid material, and to provide a heat release which is controlled to minimize the endothermic decomposition of carbonates in the solid matter.

In the process, oil shale is treated in a vertical shaft vessel or kiln. Pulverulent oil shale is fed into the top of the kiln at a sufficient rate to maintain the vessel completely filled and to replace spent shale withdrawn from the bottom. A series of eight processing zones are established in the column including from top to bottom zones of solids preheat, oil mist formation, pyrolysis, residue stripping and water-gas shift, two zones for partial combustion, combustion, and heat recovery. These zones are contiguous in the column and are not physically separated. The pulverulent solids provide interstitial spaces between the solids where gas to gas reactions take place. The gas to solids reactions in the transfer of heat initiate at the solids-gas interfaces throughout the entire column.

To establish the desired conditions in each processing zone, it is necessary to inject fluids at different locations in the column; the fluids compositions and rates are preferably selected from the different processing zones. The principle of lateral injection and distribution of fluids into shaft vessels are described in U.S. Patent 3,432,348. The patent describes the equipment necessary for the lateral distribution of fluids into a bed of pulverulent solids for uniform distribution across the lateral extent of the bed. According to this process there is defined a range of ratios of fluids to solids to be established by injection through the suitable fluid distribution devices into the various zones mentioned above.

Of particular importance in the process is to interpose a residual stripping and water-gas shift zone between the pyrolysis zone and a partial combustion zone for distributing water or water vapor into the bed of solids. The method also contemplates the injection of air, oxygen enriched air, or other oxygen containing gas at more than one vertical location in the shaft vessel. The percentage of oxygen in the injected mixture is established for each level so that controlled reactions and controlled vertical temperature profiles may be established optimally.

An important aspect of the process is the control of the solids in the column which includes feeding the pulverulent solids at a constant and controlled weight rate into the fixed processing volume and to withdraw the residue after processing at a rate to maintain the volume in the kiln relatively constant. Suitable withdrawing devices are described in U.S. Patent 3,401,992 and U.S. Patent 3,373,892. Various sensing devices and controllers for controlling the level of the solids in the volume may be used since many are known in the art.

In one form of the process control is provided by adjusting the withdrawal rate of one element to maintain a relatively constant temperature in one of the processing zones, and the off-gas temperature or the solid exit temperature are relatively easy to measure. Thus, the temperature of these two elements may be used for the control in combination with the level control. By establishing the abovementioned zones, injected air combines preferentially with the carbonaceous residue from the pyrolysis of the oil shale. This differentiates from the prior art processes wherein internal combustion retorting is accomplished by the burning of a major portion of recycled product gas resulting from the pyrolysis combined with injected air. To utilize the combustion of residue carbonaceous matter injected mixtures of air and gas are selected so that they are outside the limits of flammability for the temperature of the product gas in the various zones where the product gas is cycled or recycled. The retort gas should contain no oxygen but CO, CH_4, H_2. Allowing for variations in composition and effective temperature (higher temperature increases the range of flammability) injected mixtures should be less than about 25% gas or greater than about 75% gas.

Based on the flammability limits of the recycled gas, a large portion of the process air requirements are mixed with the recycled gas which is used as a coolant for the spent shale. For example, 2,000 scf air per ton mixed with 13,000 scf recycled gas per ton is an 86.6% gas mixture, too rich to burn, thus, this mixture, containing about 2.8% oxygen, is well mixed externally of the shaft vessel and is diffused throughout the cross section of the column of pulverulent material and just above the grate where the spent shale is removed. This gaseous mixture cools the shale residue and the mixture becomes preheated as it rises through the shaft vessel.

In a zone several feet above the grate, the oxygen combines preferentially with the carbon residue inside and outside of the retorted shale particles. The heat release takes place over a large cross-sectional area in the presence of a large amount of gas and shale residue. Therefore, the temperature level at this carbon combustion area remains relatively low (about 1100°F) and inorganic carbonate decomposition is minimized. In the slow oxidation of this residual carbon, complete combustion to carbon dioxide is achieved, which is especially so in this zone which is relatively poor in carbon content.

The above process describes the lowest zone in the shaft kiln where preheating of recycled gas-air mixture is accomplished and the cooling of the spent shale is accomplished. The next zone described is the carbon combustion zone. Just above the lowest level of distributors **45** in zone **C**, Figure 1.1b, the desired combustion reaction is the partial combustion of carbon residue to produce carbon monoxide, leaving residual carbon on the spent shale for combustion in the next lower zone. With a portion of the total process air requirements injected in this level, 500 scf air per ton of raw shale, there is relative excess of carbon, thereby favoring the partial combustion reaction (Reaction 2 which is $2C + O_2 = 2CO$). A diluent may be added to the air injection at this point to assist in diffusing the oxygen into a broader zone. The injected mixture should be too rich to burn at the injection temperature.

FIGURE 1.1: RETORTING PROCESS

a.

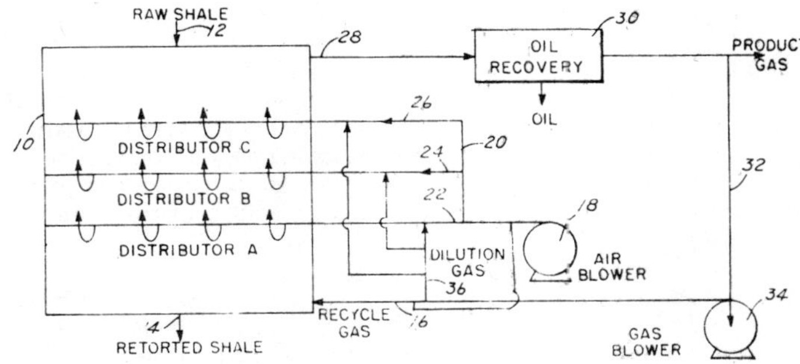

Schematic Flow Sheet of the Process

(continued)

FIGURE 1.1: (continued)

b.

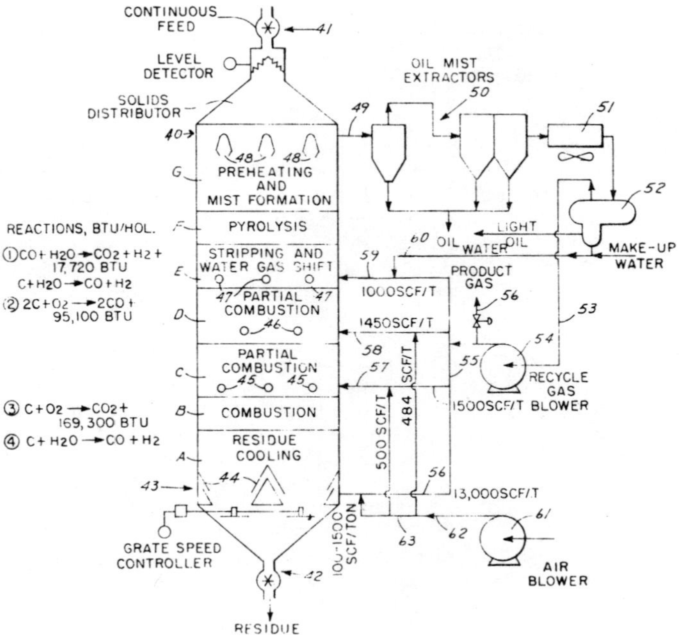

Schematic Materials Flow and Reaction Sheet of the Process

c.

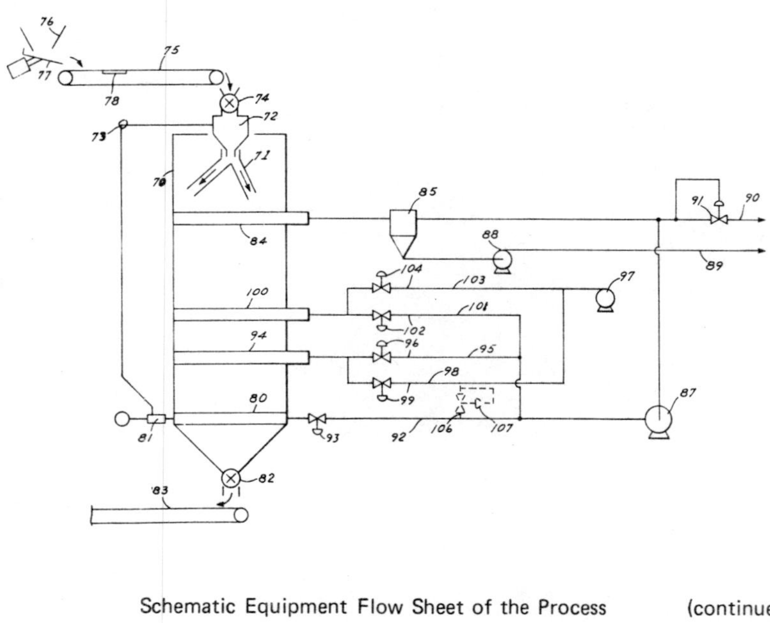

Schematic Equipment Flow Sheet of the Process (continued)

FIGURE 1.1: (continued)

d.

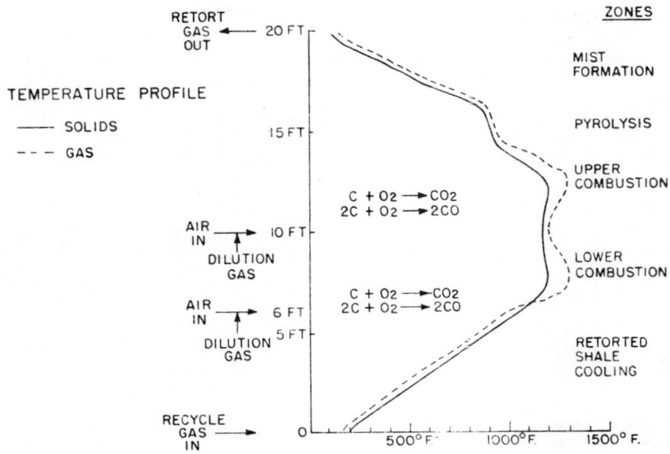

Temperature Profile Chart of the Reaction

e.

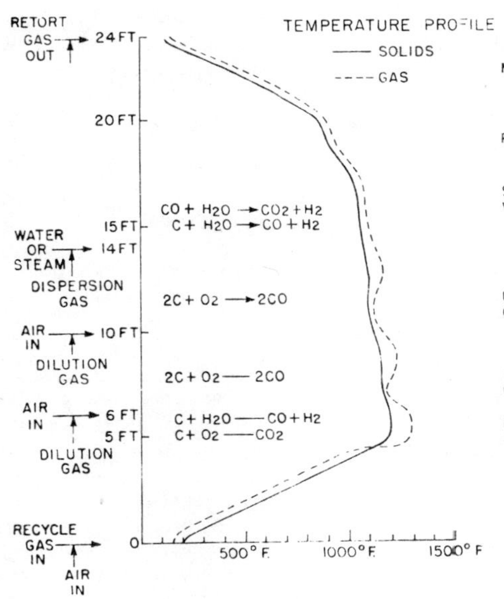

Modified Temperature Profile Chart of the Reaction Zones in a Particular Vertical Kiln

Source: J.B. Jones, Jr. and A.A. Reeves; U.S. Patent 3,736,247; May 29, 1973

Recycled gas is used as the diluent (in one process) with the air and it should exceed 1,500 scf per ton of the raw shale. Recovered CO_2 may be used as a portion of the diluent gas, maintaining it below about 25%. The next higher level of distributors **46** are used in the same manner for partial combustion of carbon. The distributors of this level may be staggered laterally and are spaced above the lower set, the distance between the two being at a greater distance than the horizontal spacing of the distributors of the lower set.

The staggered configuration of both sets gives maximum exposure of the gas-air mixtures to the shale residue and at the same time results in minimum resistance to the downward descent of the solids. The balance of the total air input is injected with enough recycled gas diluent to produce a too rich mixture. The rates of air and gas shown in Figure 1.1b are 484 scf air and at least 1,450 scf of recycled gas respectively.

The temperature levels will remain relatively low in the two partial combustion zones since the heat release is throughout the large volume of solids and in the presence of relatively large volume of gases flowing upwardly. Carbonate decomposition in the shale should be quite low since both of these zones are at relatively low temperatures and there is a relatively high partial pressure effect from the carbon dioxide present in the fluids rising from below.

A stripping and water-gas shift zone is established above the top level of the distributors **47** by entraining a water fog in a portion of recycled gas. The reaction is mildly exothermic but some heat will be absorbed in vaporizing the water. This zone **E** separates the highest temperature combustion zone from the pyrolysis zone and helps to prevent oxygen from entering the pyrolysis zone. The water vapor formed is a good stripping medium for the carbonaceous residue leaving the pyrolysis zone thereby increasing the liquid oil yield from the process.

The hydrogen formed by the water-gas shift reaction, $CO + H_2O = H_2 + CO_2 + 17{,}720$ Btu will increase the partial pressure of hydrogen in the heated gas entering the pyrolysis zone. Another method of producing H_2 is the water-gas reaction $C + H_2O = CO + H_2$, (–70,000 British thermal units). While the effect is relatively minor, pyrolysis will be more selective in increasing the liquid oil yield of the process.

Another important feature of the process is a controlling of the temperature gradient in the mist formation zone which is controlled independently of the residue cooling and the combustion processes by injecting a controlled quantity of recycled gas above the combustion zones. This concept produces high efficiency in the distillation. It is noted that in the process the amount of recycled gas required for cooling the shale residue has been decreased by adding a portion of the air necessary for the process to the shale residue cooling medium at the grate **43**.

Temperature levels in the combustion zones **B, C** and **D** have been decreased, resulting in less carbon dioxide production from carbonate (in the shale) decomposition, while on the other hand, carbon gasification to carbon monoxide in the partial combustion zones increases the gas quantity in the mist formation zone.

Another important feature of the process is the ability to control the yield distribution from pyrolysis of hydrocarbonaceous materials. As the relative amounts of gas, oil, and carbonaceous residues are varied due to changes in the process parameters, the properties of the resultant gas and oil, also, change. The carbon to hydrogen ratio of the oil product is an important consideration in subsequent processing of the oil to finished products.

The utilization of the product gas will depend on the unit heating value of that gas (Btu/ft^3) and the total heating value of the gas (Btu/ton of raw shale). The use of multiple level air injection sites at selected oxygen to combustible gas ratios at the different zones, along with the injection of diluent gas and entrained water (or steam) at a predetermined rate permits the control of selective yield distribution. The flow sheet of Figure 1.1a schematically shows the process. A vertical shaft vessel **10** is provided with a feed of raw shale **12** and discharge of spent shale **14** at the bottom. The process requires that the kiln be maintained completely filled of particulate shale, usually in the particle size range of ½" to 4". Air and recycled gas enters the bottom of the vessel through line **16** which includes recycled gas and some air. An air blower **18** supplies air to a manifold **20**, into a lower level distribution system **22** including distributor **A**, an intermediate distribution system **24** including distributor **B**, and an upper distribution system **26** including distributor **C**. Off-gas is removed from the kiln by an off-gas line **28** which passes through a recovery system **30** and a portion of the product gas is recycled along line **32** into a gas blower **34**. Part of the recycled gas is transferred by line **36** into the distribution systems **22, 24** and **26**. This provides the concept of at least three levels of fluid injection into the system, providing unique control of the combustion zones.

In the specific application of the process, Figure 1.1b, a vertical kiln **40** is provided with a continuous feed mechanism for pulverulent oil shale, shown in general by numeral **41**, and a spent shale discharge mechanism, shown in general **43**, is similar to the grate mechanisms referred to above. A gas injection system **44** is provided above or in the grates for injecting recycled gas and air into the spent shale; this cools the shale and heats the incoming fluids. A lower distributor system **45** provides means for introducing recycled gas and air into a partial combustion zone immediately thereabove and above a combustion zone **A** for the gasification of carbon residue on the shale.

Distributors **46**, also, provide for the introduction of air and recycled gas for a second partial combustion zone immediately above the distributors **46**. The combustion zone **B** and the partial combustion zones are established and are controlled by the quantity and composition of the fluids in the zone. Upper distributors **47** provide means for introducing recycled gas and water vapor into the kiln to produce a stripping and water-gas shift zone **E** immediately above that zone provides for the destructive distillation of the hydrocarbonaceous material in the kiln.

The last zone **G** is a preheating and mist (liquid product) formation zone wherein the feed is produced. Off-gas collectors **48** permit withdrawal of the gas and oil mist from the shale bed. The gas and mist is withdrawn from the kiln by line **49** which passes through a series of oil mist extractors, shown in general at **50**, which may be cyclones, wire mesh pads, electrostatic precipitators, etc., followed by an air cooled (or other type) condensor **51**. The stream from the air cooled condensor passes through a separator **52** for removing light oil from product gas which is removed from the separator and a line **53** which enters a recycled gas blower **54**.

A dryer may be installed on line **56** or any other line as desired. From the outlet of the gas blower the recycled gas goes into a manifold **55** or alternatively to product gas line **56**. The manifold **55** feeds lower line **56** which enters the distribution system **44** just above the grate **43** in the kiln, manifold **57** which feeds the distributors **45**, and manifold **58** which feeds the distributors **46**, and another manifold **59** which feeds the distributors **47**. A water makeup line **60** provides means for injecting water into the upper distributor system **47** for producing the stripping and water-gas shift. Air from a blower **61** provides air through an outlet line **62** into a manifold **63** which feeds the air to the manifolds **56, 57** and **58** for producing some combustion in the kiln.

The various zones in the kiln are identified as zone **A** where spent shale is cooled, a combustion zone **B**, a partial combustion zone **C**, a second partial combustion zone **D**, a stripping and water-gas shift zone **E**, a pyrolysis zone **F**, and a preheating and mist formation zone **G** which also includes the off-gas collectors. The reactions which occur in the various zones, based on Btu per mol include $C + O_2 = CO_2$ + 169,300 Btu, $2C + O_2 = 2CO$ + 95,000 Btu and $C + H_2O = CO + H_2$ (-70,000 Btu) in combustion zone **B**.

The first partial combustion zone **C** provides a reaction of $2C + O_2 = 2CO$ + 95,100 Btu and $C + H_2O = CO + H_2$, while the second partial combustion zone **D** also includes the same reaction of $C + O_2 = CO_2$ with the same Btu production. The first combustion zone **B** may include the $C + H_2O = CO + H_2$ due to residual moisture in the recycle gas and/or in the air.

The stripping and water-gas shift, zone **E**, combines carbon monoxide and water according to the following formula: $CO + H_2O$ goes to $CO_2 + H_2 + 17{,}720$ Btu. In the pyrolysis section the reaction is endothermic utilizing some of the heat produced by the combustion and partial combustions below the pyrolysis zone.

The amount of recycled gas per ton of shale is shown in Figure 1.1b, where 13,000 scf per ton (shale) is injected into the bottom along with 100 to 1,500 scf per ton of air. Into the first level of distributors **45**, 500 scf air per ton is injected with recycled gas which is greater than about 1,500 scf per ton. In the intermediate distributors **46**, 484 scf of air is injected with recycled gas which is greater than about 1,450 scf per ton. In the upper distributors 1,000 scf of recycled gas per ton of shale is injected with water and no air.

As known from the prior art, injection of air alone into the combustion zone produces high temperatures and burning a part of the gas product as well as part of the produced oil product. Therefore, it is important that all the oxygen be consumed below the pyrolysis or retorting zone so that little or no oil is burned, when the oil is the principle valuable product of the retorting process. It is also important that the gas burning in the combustion zone **B** be maintained at a minimum, particularly where the gas product is required to have some value.

As explained above, combustible gas and air mixtures have upper and lower limits of flammability (in reference to the percentage of combustible gas in the air-gas mixture). The spread between the upper and lower limits decreases as a diluent (recycle gas and/or CO_2) is added, and the addition of CO_2 finally extinguishes the flame at a concentration of about 24% CO_2 in the mixture. Lower CO_2 concentrations will permit ignition to take place, but the limits of flammability are quite narrow.

The process provides means to prevent the burning of gas in the kiln or permits control combustion by controlling the amount of gas for the burning. In the lower levels of the kiln, oxygen in the injected air combines with the carbonaceous residue on the shale which amounts to 2 to 6% by weight of the retorted shale. This carbonaceous residue is of doubtful economic value since it is mixed with such large amounts of the spent shale. The residue will burn, however, at the temperature levels and the oxygen levels which may be maintained in the various zone in the retort.

In addition to the reaction of the burning of carbon with oxygen as shown in the combustion zone above, the parameters in the combustion zone may be controlled so as to produce the following reaction: $CO_2 + C = 2CO$ (endothermic). By the law of mass action increasing the CO_2 concentration favors the formation of carbon monoxide, a gas having moderate heating value. Thus in the combustion zone the main reactions are the combining of carbon with oxygen to produce carbon dioxide and the combining of carbon and oxygen to produce carbon monoxide. The exothermic reactions to produce carbon dioxide and carbon monoxide is at a relatively low temperature level and is controlled to produce a combustible gas having some economic value. The heat liberated is used to supply the heat requirements for pyrolysis in the zone above the stripping and water-gas shift.

Carbon dioxide necessary to produce the carbon monoxide may be obtained by scrubbing the retort product gas with one of the ethanolamines using well-known methods. In the operation of the shaft vessel, some carbon dioxide is produced from the combustion of organic matter in the shale, some will be produced by the decomposition of some of the magnesium carbonate in the shale, and some is produced by the actual burning of the carbon after pyrolysis.

A side benefit of the use of ethanolamines scrubbing is to remove hydrogen sulfide and sulfur dioxide and thereby permit an economic recovery of sulfur from the product gas. The introduction of carbon dioxide as a diluent in the product gas provides an additional control on the flammability of the injected cool gaseous mixture, which, also, incidentally suppresses combustion near the point of injection into the shale bed. As the gaseous mixture becomes heated by the hot shale it reaches a temperature where the mixture is flammable and the injected gas mixture rises above the distributors into the partial combustion zones where the controlled oxidation takes place. Thus, by controlling the carbon dioxide content and the recycled gas content of the injected gas-air mixture, the distance of the flame from the point of injection may be accurately controlled. By providing the two partial combustion zones above the combustion zone the residence time of the shale in the higher temperature level zones in the kiln may be greatly extended without increasing the shale temperature above the decomposition temperatures of the carbonates in the shale.

Specific controls of the process are shown in Figure 1.1c where a kiln **70** is provided with a feed distributor **71** to provide uniform distribution of shale into the kiln, and the distributor is fed from a constant level hopper **72** controlled by means of a level controller **73.** A star feeder **74** fed from a belt conveyor **75** maintains the hopper full. A feed bin **76** feeds the belt conveyor by means of a loader **77** controlled by a belt scale **78.**

This arrangement provides for a constant feed into the kiln through the feed distributor. The level controller controls a discharge grate mechanism **80** at the lower part of the kiln, for example, the linear grate explained above, through a control mechanism **81** to thereby provide for a constant level of shale in the kiln. A star feeder **82** feeds a spent shale discharge belt mechanism **83.** Off-gas is collected from the shale bed by means of an off-gas collector **84** passing through a mist separator **85** and the product gas passing through a line **86** to a recycle gas pump **87.**

Liquid from the mist separator passes through a pump **88** to a liquid product discharge line **89.** Product gas is discharged through line **90** which is controlled by a control valve **91.** Recycled gas is injected into the grate mechanism through a line **92** controlled by a valve controller **93.** The first level of distributors **94** is provided with recycled gas through line **95** from the recycle pump **87** controlled by a valve controller **96**, and air from an air pump **97** is also injected into the distributor level **94** through a line **98** controlled by valve controller **99.**

The next level of distributor **100** is provided with recycled gas through line **101** controlled by line controller **102** and air is provided into the distributor through line **103** which is controlled by a valve controller **104.** Where desired, air can be introduced into the recycled gas into line **92** by means of line **106** controlled by line controller **107**, in the manner described above.

The two or more levels of distributors in the bed distribute an oxygen containing gas which may be accurately controlled by injecting a diluent gas with the air. As pointed out, the diluent gas may be recycled product gas, CO_2 or other gas in which to control the atmosphere within the combustion and/or partial combustion zones. The multiple level distributors increases the resident time of the shale at high temperatures in the kiln to produce a lower combustion zone temperature for the release of the heat required for the retorting of the oil shale.

It also permits the introduction of oxygen containing gas below the pyrolysis zone in sufficient quantities to restore the stone temperature to desirable levels to compensate for the endothermic heat of reaction in the pyrolysis zone; the lower combustion zone temperatures will cause fewer carbonates to be decomposed which will produce less carbon dioxide dilution in the product gas. The decrease in carbonate decomposition decreases the heat requirements which will, therefore, decrease the air requirements, and thereby decrease the nitrogen dilution in the product gas.

The net effect of the decrease in carbon dioxide and nitrogen dilution will improve the heating value of the product gas which may be recovered as a product of the pyrolysis. The deepening of the combustion zone, which, also, includes the partial combustion zones, will permit the use of higher solids throughput rates before contacting any oxygen in the carbonaceous distillation zone and the resulting burning of some of the oil values. The temperature profile of Figure 1.1d illustrates injection of fluids into two intermediate levels in the vertical vessel as well as some recycle gas in the bottom of the vessel.

This is for countercurrent flow of solids from top to bottom, and fluids from bottom to top. Shale is introduced into the top of the vessel at the 20 foot level at ambient temperature and rising gas raises the shale to pyrolysis temperature of about 900°F in about 5 feet, or at the 15 foot level. Heat for the pyrolysis is provided by combustion in the upper combustion zone which results from the oxidation of the residual carbon into CO and CO_2. The lower combustion zone starts at about the 6 foot level and carbon is oxidized to CO and CO_2.

The high temperatures (1000° to 1200°F) extends through both combustion zones over about 5 to 14 feet of the vessel, or just under half of the vessel. This provides a long residence time for the solids at the higher temperatures insuring efficient utilization of the residual carbon on the shale following pyrolysis. The incoming gas is heated by the spent shale, which in turn is cooled. The volume of gas induced into the vessel by forced draft is sufficient to entrain the mist which forms by condensation of vaporized liquid (from pyrolysis) contacting the relatively cooler shale. The entrained mist is then carried out for recovery by the off-gas.

The specific example shown by the flow sheet of Figure 1.1b has a temperature profile as shown in Figure 1.1e. This includes three upper levels of fluid injection and three levels for oxygen containing gas injection. The ambient temperature shale enters the vessel, is heated in passing through the mixed formation and pyrolysis zones, which are maintained essentially nonoxygeneous. The heated shale from the pyrolysis zone, at about 1000°F, is subjected to the stripping and water shift zone where both CO and C react with water to form CO_2 and hydrogen and CO and hydrogen, respectively.

The shale then passes into a partial combustion zone where the oxygen content is maintained low to produce a partial oxidation of some of the residual carbon to carbon monoxide. The shale passed to middle partial combustion zone where a controlled oxygeneous atmosphere permits some additional residual carbon to partially oxidize to CO. The temperature of the shale rises somewhat through these two partial combustion zones.

The shale then passes into the lower combustion zone where the remainder of residual carbon is utilized by oxidation and a water-gas shift (due to minor amounts of water entering via the air and recycle gas). The desired oxidation here is the complete oxidation of carbon to carbon dioxide. The spent shale heats the gas and is simultaneously cooled. The residence time of the shale above a 1000°F is from 4 to 17 feet in a 24 foot vessel, or well over half the vessel.

In related work *J.B. Jones, Jr. and A.A. Reeves; U.S. Patent 3,841,992; October 15, 1974; assigned to Paraho Corporation* describe a process for the destructive distillation of the organic matter in hydrocarbonaceous solids using shaft vessels in which pulverulent solids are moved continuously downwardly in a continuous vertical column through successive heating zones and then into a residual cooling zone, thereby volatilizing some of the hydrocarbonaceous material included in the solids.

The distillation is conducted by using produced gases from the distillation, externally heated and recycled through the column of material, in addition to a portion of the product gases which are cooled and recycled into a lower portion of a bed, cooling the solid residue and being heated during passage upwardly through the column of material, maintaining an efficient heat balance for the process. The process includes multiple injection levels in the column of material for a controlled exposure time of the solid material to gases of controlled composition at predetermined temperatures for an efficient and economical process for recovery of volatile organic material from the solids.

Multistage Thermal Processes

N.P. Peet; U.S. Patent 3,384,569; May 21, 1968; assigned to Esso Research and Engineering Company describes a method for retorting oil shale in a plural stage thermal operation in which shale oil is recovered from crushed oil shale.

The plural stage comprises preheating, retorting, burning and cooling zones in which the crushed shale moves serially through the zones and in which shale fines, flue gas and shale oil are formed. In the method, the shale fines are separately removed from each of the zones with the fines removed from the burning and cooling zones being substantially oil-free and discarded.

The fines removed from the preheating and retorting zones are substantially oil rich and are introduced into the burning zone to provide a source of fuel. Also introduced into the burning zone is a free oxygen-containing gas such as air. The oil rich fines may be mixed with the free oxygen-containing gas and the free oxygen-containing gas thus serves as a carrier for the oil rich fines.

Introduced into the retorting and cooling zones is a combustible gas, such as a high Btu content gas, which is preferably separated from the shale oil in a later stage in the process, but which may originate outside the system. Natural gas may be used as the combustible gas having a high Btu content. High Btu content as used means a Btu value of about 500 to 2,500. Preferably, a Btu value for the gas may range from about 800 to 1,200 Btu.

Gas from the cooling zone is introduced into the preheating zone as a preheating fluid while flue gas from the burning zone is introduced into the retorting zone into indirect heat exchange with oil shale therein. Shale oil is recovered from the preheating and retorting zone by virtue of an operation such as briefly described. High overall thermal efficiency is achieved by recovery of heat, and carbonaceous material in the oil shale is substantially completely utilized in the operation. The shale oil is recovered in vaporous form and cooled exteriorly of the preheating and retorting zones and thus, is not available for recycling in the several stages which ordinarily in the prior art processes results in consumption of oil and lowering of yield.

The high Btu gas is separated from the shale oil after cooling and condensing of the vaporous shale oil. The fines removed from the preheating and retorting zones are separated from vaporous shale oil before being introduced into the burning zone, and the fines removed from the burning and cooling zones are separated from flue gas before being discarded. The flue gas which is introduced into the retorting zone in indirect heat exchange is withdrawn from the retorting zone and may be discarded such as through a waste heat boiler for recovery of heat or may be reheated either in a separate vessel or other heating means and reintroduced for indirect heat exchange into the retorting zone.

Referring to Figure 1.2, **11** indicates a charge line or conduit by way of which fresh crushed oil shale is introduced into the system from a source not shown. Line **11** leads into a preheat vessel or zone **12** provided with a distribution means or spider **13** connected to line **10** which introduces hot gas at a temperature of about 1000°F. The oil shale and hot gas flow countercurrently to each other in zone **12** with the preheated shale discharging by line **14** into a vessel comprising a retorting zone **15**.

The vaporous and gaseous materials from zone **12** are discharged by line **16** into a separator **17** which may suitably be a cyclone separator from where the gaseous material and vaporous material may leave by line **18**. The vaporous and gaseous materials in lines **16** and **18** are at a temperature of 250°F and are therefore routed by line **19** containing a cooler-condenser **20** into a separator **21** wherein the condensed hydrocarbons shown as body **22** are separated from water **23** in leg **24** of separator **21**, the water being discharged by line **25** and the hydrocarbons by line **26**.

High Btu content gas is withdrawn from separator **21** by line **27a** part of which is recycled, in a manner which will be described, by line **28** while the remainder may be recovered by line **29** for further use as may be desired. Returning to the retorting zone **15**, this zone is provided with internal tubes **27** for indirect heat exchange with the preheated shale flowing downward through zone **15** from zone **12**. In zone **15**, the preheated shale contacts the high Btu content gas from line **29** which is introduced into retorting zone **15** by branch line **30** connecting to distribution means such as spider **31**.

FIGURE 1.2: MULTISTAGE THERMAL PROCESS

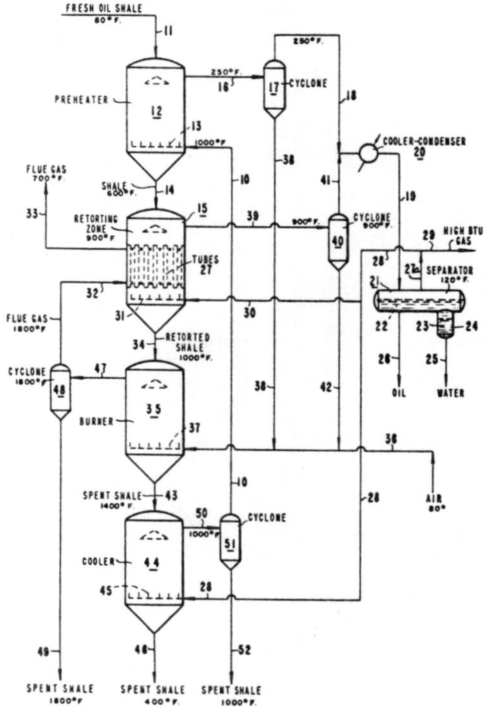

Source: N.P. Peet; U.S. Patent 3,384,569; May 21, 1968

The preheated shale in zone **15** is retorted by heat supplied indirectly to the internal tubes **27** by line **32**, with the flue gas leaving the tubes **27** by line **33**. This flue gas which is at a temperature of 700°F may be reheated by circulation through a furnace or other heating means or by heat exchange with material in the process at a higher temperature. The flue gas introduced in line **32** may be at a temperature of about 1800°F. The shale after being retorted in zone **15** leaves zone **15** by line **34** and is introduced into a vessel providing a burning zone **35**.

The retorted shale introduced into burning zone **35** may be at a temperature of about 1000°F and is contacted with a free oxygen-containing gas such as air introduced by line **36** at a temperature of 80°F from a source not shown, through a distribution means such as spider **37**. Mixed with the air in line **36** for introduction into line **38** are oil rich fines recovered from preheating zone **12** in cyclone separator **17**. These recovered oil rich fines are introduced into line **36** by line **38**.

The vaporous shale oil which may also be in the form of a mist with entrained liquid shale oil is withdrawn from zone **15** by line **39** and discharged at a temperature of 900°F into separator **40** which may be a cyclone separator. The products separated from the fines in separator **40** are introduced by line **41** into branch line **19** for recovery of hydrocarbons and high Btu gas. The oil rich fines separated from the mist or vaporous shale oil are withdrawn from separator **40** by line **42** for introduction into line **36** in a mixture with the air for ultimate injection into burning zone **35**. The spent shale from burning zone **35** is withdrawn by line **43** at a temperature of 1400°F, and introduced thereby into a vessel defining a cooling zone **44** into which there is introduced by line **28**, which connects to distribution means such as spider **45**, the high Btu content gas recovered from separator **21**. The spent shale at a temperature of 400°F is discharged by line **46** from the system.

The flue gas from burning zone **35** discharges by line **47** into a separator such as a cyclone **48** from where the flue gas is withdrawn by line **32** for use in zone **15** with the spent shale fines, being substantially oil-free, being discharged by line **49** from the system. Likewise, the flue gas from zone **44** is discharged by line **50** into a separator **51** which may be a cyclone separator from where the gases by line **10** are introduced into zone **12**.

Substantially oil-free fines are discharged from zone **51** from the system by lines **52**. Since the spent shale fines in lines **49** and **52** are at temperatures, respectively 1800° and 1000°F, this material may be passed into heat exchange either with the flue gas in line **33** or the fresh oil shale in line **11**.

In a process described by *C.E. Hemminger; U.S. Patent 3,484,364; December 16, 1969; assigned to Esso Research and Engineering Company* oil shale is contacted with hot hydrogen in a fluidized system under conditions to conserve heat, to give increased yields and to produce products which are susceptible to removal of nitrogen The hot hydrogen containing gas is produced from the cracking of a heavy residue recycle in the presence of the hydrogen containing gas. The cracking takes place in a separate zone and supplies at least a portion of the heat necessary for the retorting of the shale.

B.L. Schulman and H.P. Dengler; U.S. Patent 3,520,795; July 14, 1970; assigned to Esso Research and Engineering Company have found that in retorting of crushed oil shale in a retorting system having in a single vessel preheating, retorting and cooling zones and the crushed shale flows serially through, a temperature below carbonate decomposition and below shale oil cracking temperature can be maintained by introducing vaporous water into the system adjacent or into the retorting zone and liquid water into the cooling zone, whereby introduction of cooling gas into the cooling zone is reduced, oil and gas yields are improved and gas heating value is enhanced.

Isolation of Retort Zone to Utilize Flue Gases

In a process described by *D.D. MacLaren; U.S. Patent 3,475,319; October 28, 1969; assigned to Esso Research and Engineering Company* oil shale is retorted serially downward through a preheating zone, a retorting zone and a combustion zone. Flue gas generated from burning carbonaceous material on the spent oil shale is withdrawn from the combustion zone and introduced into the preheating zone. The sensible heat of the flue gas is further utilized to raise the temperature of product gas which is introduced into the retorting zone as the source of retorting heat, condensibles having been previously removed from the product gas.

In the process, the several zones including the preheating zone, the retorting zone, and the combustion zone are isolated from each other to prevent vapor from the zone below entering the zone above. A sufficient amount of seal gas is employed to displace other gases in the interstices of the oil shale thereby preventing these gases from being carried along with the oil shale such that there is no net flow of seal gas upward into the zone above. Seal gas approximately equal in volume to the interstices flows downward.

The zones may be isolated from each other by providing a restriction between the zones and introducing into the restriction upward a seal gas which may be steam or other gas such as, but not limited to, product gas, and the like. Steam is the preferred seal gas, since it does not dilute the product gas since the steam is condensed with the liquid product. Steam simply adds to the water generated by burning carbon from the spent shale. The amount of seal gas introduced into the restriction area may range from 5 to 20 scf per ton of raw oil shale introduced into the system.

By sealing the zones from each other to prevent commingling of gaseous material in one zone from gaseous material in another zone, it is possible to withdraw product vapors substantially uncontaminated with flue gases within the system, and it is possible thereby to recover a recycle gas which comprises about 50 to 55% hydrogen with the remainder light hydrocarbons such as methane, ethane and propane, and other gaseous and perhaps normally liquid materials. This enhances the calorific value of the gas and provides a gas having a Btu value within the range from 500 to 750 Btu/scf.

Also, by virtue of the high hydrogen content of the recycle product gas, this gas may be used if desired, in hydrogenating the shale oil to remove nitrogen compounds, sulfur, and other deleterious materials. Also, by isolating or sealing the zones from each other, the flue gas may be withdrawn from the combustion zone to bypass the retorting zone and be introduced into the preheating zone to provide heat for use in preheating the raw oil shale introduced into the system. Also, the recycle product gas introduced into the retorting zone may be suitably heated by passage in indirect heat exchange with the withdrawn flue gas, thus conserving heat and allowing the oil shale to be retorted with a mixture of principally hydrogen and hydrocarbons.

In a process described by *J.E. Lawson, Jr.; U.S. Patent 3,440,162; April 22, 1969; assigned to Esso Research and Engineering Company* independent control of the retorting of crushed oil shale is provided in a unitary operation by flowing the crushed shale downward through preheating, retorting and combustion-cooling zones with vaporous product being removed separately and fractionated. Shale oil bottoms are introduced into the retorting zone and inert heated gas is used to preheat the oil shale with the inert gas being recycled to the burning-cooling zone to cool the discarded shale. Air and combustible gas are introduced into the retorting zone while air may be introduced into the burning-cooling zone.

Increased Residence Time in Retort Zone

R.E. Biddick, C.E. Jahnig and C.W. Tyson; U.S. Patent 3,297,562; January 10, 1967; assigned to Esso Research and Engineering Company describe a method for increasing the time of residence of the shale lumps in the retorting zone to improve the recovery of oil. Channels or tubes are provided for bypassing the combustion gases around at least a portion of the solid shale particles in the retorting zone. The tubes or channels may be mounted internally or externally of the retorting zone. The solid shale particles pass down around the tubes when mounted internally but some of the gas and/or vapor bypasses the shale solids and passes up through the tubes in the shale solids moving bed to a higher region.

In this way the heated oil shale particles from the preheat zone are allowed to spend a longer time at desired retorting temperatures of 800° to 1200°F before entering the combustion zone. In other words, a soaking zone is provided between the preheat and combustion zones to assure complete retorting and maximize oil recovery. Without it, independent control of retorting time and temperature is not possible, since the temperature profile in the bed is set by the relative flow rates of gas and solids.

Increased residence time at the selected desired temperature permits heat diffusion to the inner portions of the large lumps of shale which would otherwise pass into the combustion or hot zone before being fully retorted. This unretorted oil would be lost by combustion or be severely cracked to gas and coke. The channels or tubes in effect increase the length of the retorting zone and in this way increase the time of retorting.

Recycle Gas Bypasses Combustion Zone

J.H. Haddad and J.G. Mitchell; U.S. Patent 3,503,869; March 31, 1970; assigned to Mobil Oil Corporation describe a modified gas combustion shale retorting process where direct contact is minimized between gas combustion products from the combustion zone and both kerogen decomposition products in the retorting zone and recycle gas from the shale cooling zone. Shale having coke on the surface from a retorting zone is passed downward to a combustion zone. In the combustion zone, the shale is contacted with free oxygen containing gas to promote combustion of the coke and heat the shale and gas. The combustion gas and shale in the combustion zone are passed concurrently downward through the combustion zone. The shale from the combustion zone is passed over gas disengagement means located within the retort to separate gaseous combustion products from the spent shale prior to contacting recycle gas from the shale cooling zone. The gaseous combustion products are removed from the retort while the separated spent shale is passed downward through a shale cooling zone.

In the shale cooling zone, the spent shale is contacted with upward moving recycle gas whereby heat is transferred from the spent shale to the recycle gas. The recycle gas is removed from the retort prior to entering the combustion zone by gas disengaging means located within the retort. The spent shale is removed from the retort after passing through the shale cooling zone. The heated recycle gas can contain fines and after being removed from the retort is directed to a fines separation zone wherein fines are separated from the heated recycle gas.

The recycle gas from the fines separation zone is directed to the retorting zone located above the combustion zone and is therein passed upward through the downward moving shale. In the retorting zone, kerogen in the raw shale is thermally decomposed to oil vapors. The resultant gasiform material is passed upward through incoming raw shale to transfer heat from the gasiform material to the raw shale.

This results in the formation of an oil mist or fog which is removed from the shale preheating zone and directed to an oil separation zone. In the oil separation zone, liquid oil product is separated from gasiform material. A portion of the gasiform material is recycled to the lower portion of the shale cooling zone in a manner described above.

Thus by the process, transfer of gaseous products from the combustion zone to either the retorting zone or the shale cooling zone is minimized. This results in substantial advantages for the overall gas combustion process. By minimizing contact between gaseous combustion products and oil formed by kerogen decomposition in the retorting zone, the chances of burning shale oil product are minimized.

In addition, because the large proportion of fines found in a retorting process are formed in the combustion zone, the chances for contacting fines and shale oil product with the attendant undesirable results are minimized. The process also provides substantial advantages by removing the recycle gas from the retort prior to contacting gaseous products from the combustion zone. In this manner, selective burning of coke on the shale is accomplished while minimizing burning of gaseous hydrocarbons. This results in more uniform coke burning and prevents substantial loss of shale oil vapors in the combustion zone. Accordingly, higher retort shale oil yields are obtained since condensible oil not previously separated will not be consumed as fuel in the combustion zone.

Referring to Figure 1.3, crushed raw shale particles having a size of from ¼" to 4" mean diameter is introduced into retort **1** through closed conduit **2**. The shale particles are directed through a plurality of conduits **3** to the top of a downward moving compact bed of shale particles in preheating and retorting zone **4**. In this manner, relatively even particle distribution over the bed surface area is effected. The top of the particle bed in preheating and retorting zone **4** is maintained at a desired distance below the top of the retort **1** to form a plenum chamber **5** which facilitates separation of particles and oil mist.

In preheating and retorting zone **4**, the incoming particles are heated by upward moving vapors and reach retorting temperature as the particles move downward. The heat supplying gas is introduced through conduit **6** into header **7** and is introduced into the bed through distributors **8**. When retorting temperature is reached in zone **4**, the kerogen decomposes to form gas, oil vapors and coke residue. The spent shale having coke on the surface is passed downward through standpipes **9** into combustion zone **10**. The standpipes **9** extend from baffle **11** to form a plenum chamber **12** and to provide back pressure means to pre-

FIGURE 1.3: GAS COMBUSTION RETORTING PROCESS

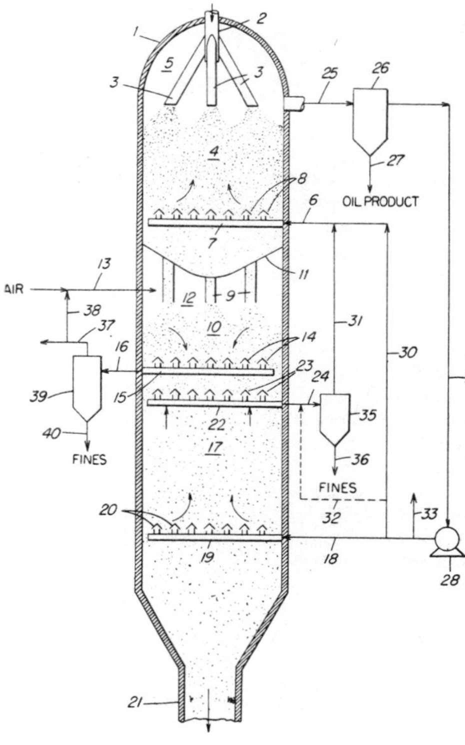

Source: J.H. Haddad and J.G. Mitchell; U.S. Patent 3,503,869; March 31, 1970

vent incoming combustion gas from entering preheating and retorting zone **4**. Provision can be made to introduce seal gas into each standpipe **9**. Combustion gas is introduced into plenum chamber **12** through conduit **13**. The combustion gas is passed downward from the plenum chamber into the combustion zone **10** to effect burning of coke on the shale particles. In combustion zone **10**, gaseous material and shale particles are passed concurrently downward to contact a gas solids disengager comprising a plurality of collectors **14** and a header **15**.

The gaseous combustion products and spent shale fines are removed from the retort through header **15** and conduit **16**. Due to the burning effected in combustion zone **10**, the shale particles become heated. The heated shale particles pass downward from the combustion zone **10** to shale cooling zone **17** where contact is made with relatively cool recycle gas. The recycle gas is introduced into the retort **1** through conduit **18**, header **19** and distributors **20**.

The recycle gas passes upward through the shale cooling zone **17** to contact downward moving shale particles. In this manner, heat is transferred from the shale particles to the recycle gas. The cool shale particles are removed from retort **1** through conduit **21** and discarded. The heated recycle gas is removed from retort **1** by way of gas solids disengager comprising header **22** and collectors **23** and conduit **24**. Means not shown, are provided in conduit **21** for effecting relatively uniform shale flow through the retort **1**.

The oil mist or fog separated from the shale preheating zone **4** is removed from plenum chamber **5** through conduit **25**. The oil mist is directed to a separation step **26** wherein liquid oil product is separated from vaporous material. An electrostatic precipitator and/or cyclone can be employed in the separation step **26**. Oil product is obtained and recovered through conduit **27**. Gaseous material is recovered from separation step **26** and recycled in a manner whereby a portion is directed to the shale cooling zone **17**, another portion can be mixed with the gaseous effluent from the shale cooling zone, and the remainder is vented to be employed in other processes.

The gas recycled to the shale cooling zone **17** is pumped by pump **28** through conduits **29** and **18** into header **19**. The gas to be mixed with hot recycle gas is pumped through conduits **18, 30** and **6** and mixed with hot recycle gas from conduit **31**. Alternately, this cooling gas may pass to conduit **24** through conduit **32**. Excess recycle gas is vented through conduit **33**. Hot recycle gas is removed from the retort through conduit **24** and directed to a fines separation step **35** where fines carried over with the gas are separated and removed through conduit **36**. The hot recycle gas is directed to the preheating and retorting zone **4** through conduits **31** and **6**.

Combustion air is introduced into plenum chamber **12** through conduit **13**. The air can be mixed with combustion products obtained from the retort and directed through conduits **37** and **38** prior to being introduced into the retort. This is an effective means for diluting the combustion air to control burning in the combustion zone **10** and minimize the development of high localized peak temperatures.

The gaseous combustion products are withdrawn from the retort through conduit **16** and directed to a separation zone **39** where carried over fines are separated from the gases and removed through conduit **40**. The combustion gas exits from the retort at relatively high temperatures and the heat can be extracted by heat exchange means not shown. The heat from the gaseous combustion products can be converted to steam, for example, which can then be used to drive the compressors which pump recycle gas and combustion air to and from the retorting process.

The amount of free oxygen which is introduced into the combustion zone is regulated so as to maintain the shale temperature in the combustion zone between 900° and 1400°F, preferably between 950° and 1100°F. By operating in this manner, sufficient heat can be transferred to the recycle gas in the shale cooling zone to support kerogen decomposition in the retorting zone without causing excessive shale disintegration in the combustion zone. The amount of free oxygen in the combustion zone can be regulated by diluting combustion air with flue gas to obtain a free oxygen concentration in the resultant mixture usually between 5 and 15 volume percent.

Since carbon burning is proportional to oxygen partial pressure, the diluted gas will minimize the attainment of high peak temperatures and carbonate decomposition in the combustion zone while the combustion gas is being diffused into the bed. The amount of recycle gas directed to the shale cooling zone is sufficient to reduce the shale temperature to below 400°F and preferably below 300°F.

By operating in this manner, the hot recycle gas from the fines separation step associated with the shale cooling step can have a temperature as high as 1400°F. It therefore may become desirable to reduce the hot recycle gas temperature prior to introducing it into the retorting zone. Thus, the process provides for mixing the hot gas with a portion of the cold recycle gas prior to introducing gas into the retorting zone.

The relative amounts of cold and hot gas mixed is that which will maintain the maximum retorting temperature between 700° and 1100°F, preferably between 850° and 1000°F. In this manner, increased retorting times result which increases oil yield. The oil mist exits from the top of the shale retorting and preheating zone at a temperature below 300°F and preferably between 100° and 200°F. In the process, the hot recycle gas can be introduced by means of a plurality of inlets at different vertical heights.

This permits more effective heating in the retorting zone and permits heat soaking of the shale in the retorting zone. This heat soaking effects increased vaporization of the liquid decomposition products and this provides for increased shale oil yields and flowability of the solids.

J.G. Mitchell and J.H. Haddad; U.S. Patent 3,349,022; October 24, 1967; assigned to Mobil Oil Corporation describes a vertical gas combustion shale retort where a downward moving bed of shale granules is contacted with countercurrently rising gasiform material in a first multichambered open end spent shale cooling zone (recycle gas preheat zone), a combustion zone, a shale retorting zone and a shale preheating zone.

The improvement of the above combinations of steps resides particularly in the design and arrangement of the multichambered shale cooling zone to improve heat transfer and provide for more uniform heat exchange throughout the cross section of the retort between recycle gas and spent shale particles before passing the preheated recycle gas through the combustion section. In addition, the improvement resides in passing all gaseous material recovered from the combustion section of the retort to a zone for removal of solid particle fines before passing hot gases upward through the retorting zone of the retort.

Preheat Requirements for Optimum Operation

A process described by *E.F. Kondis, F.A. Smith and P.W. Snyder, Jr.; U.S. Patent 3,318,798; May 9, 1967; assigned to Mobil Oil Company* involves defining a preferred region of operability of a gas-combustion retort and identifying a correlation of operating conditions which will permit operation in a preferred region of oil shale throughput rate and maximum recovery of desired kerogen decomposition products.

It has been found that the preferred region of operability of a gas combustion retort is confined by (1) a low oil shale flow rate of about 300 lb/hr-ft^2; (2) an upper shale flow rate of about 1,000 lb/hr-ft^2 limit which begins to impose an undesired high pressure drop across the retort thereby requiring usage of expensive compressor equipment to provide adequate gas flow rates; and (3) an upper gas temperature boundary of 1800°F selected to avoid creating excessive costs in structural and material problems; and (4) an upper shale temperature boundary of 1500°F selected to avoid excessive thermal cracking and undesirable high carbonate decomposition temperatures.

In addition, it has been found that employing air rates in excess of 7,000 scf/ton of air falls outside a reasonable operating range since above this air rate the pressure drop in a retort of about 12 feet high would be in excess of about one atmosphere, the shale outlet temperature would exceed a desired upper limit of 500°F and peak shale temperatures would exceed 1500°F leading to excessive thermal cracking and undesired high carbonate decomposition.

The method of operation described permits a preferred gas phase burning to the exclusion of substantial coke burning and little, if any, kerogen burning; a better temperature control within the retort; more kerogen can be decomposed above the combustion zone; a high Btu content off-gas may be obtained and recovered from the retort and more rigid spent shale particles will be present, thereby significantly reducing the dust carry over from the retort.

Although it is possible to operate a gas-combustion retort outside the preferred limits, it is substantially less desirable for many reasons and generally outside the scope of this process for the reasons herein expressed. That is, if shale rates below 300 lb/hr-ft^2 are used, large numbers of retorts are required. On the other hand, employing shale rates above 1,000 and as high as up to 2,000 lb/hr-ft^2 although possible, cause a condition of excessive or large pressure drop in the retort requiring additional compressor equipment and generally undesirable operating conditions leading to inefficient decomposition and recovery of decomposed kerogen. The process also involves maintaining a temperature profile within the gas-combustion retort for preferred shale throughput rates within the range of above 300 to 1,000 lb/hr-ft^2 which will permit decomposition and recovery of desired kerogen product in amounts that may be considered substantially optimum without encountering undesired retorting temperature conditions due to insufficient or uncontrolled burning of available combustible materials in the retort.

Accordingly, in the process, a portion of uncondensible shale gas recovered from the retort and separated from desired kerogen decomposition liquid product is recycled to the retort. A substantial portion of the separated and recycled gas is passed to the lower or bottom portion of the retort for flow upward therethrough under conditions to preheat the recycle gas by countercurrent contact with downward moving hot shale particles.

The remaining recycled portion of recovered shale gas referred to as dilution gas is employed when combined with air as a heat carrying gaseous material to the combustion section of the retort. That is, the portion of dilution gas and air mixture passed to the combustion section of the retort is preheated either directly or indirectly by combustion, for example, or other suitable means to a desired elevated temperature.

The minimum extent of gas heating effected by the methods described is determined in accordance with the formula relationship provided below, so that upon introduction to the retort the temperature profile of the recycled gaseous material having entered the bottom and passed upward through the retort under heat exchange conditions countercurrent to the hot shale particles will not be reduced any substantial amount at the air inlet of the retort combustion section and will be of a heat carrying capacity and combustible material content to substantially limit undesired burning and undesired temperature profiles within the retort.

Other gaseous materials such as steam or flue gas which are considered relatively inert in the process may be combined with the air and/or dilution gas in desired quantities. In addition, the volume of heat carrying gas comprising air and dilution gas passed to the combustion section may be varied considerably depending upon the method of heating employed and within the range of from 3,000 to 7,000 scf/ton of shale. That is, when employing indirect heat exchange means for heating the combustion supporting gas passed to the retort combustion section, use of air volumes less than 3,500 scf/ton and as low as 3,000 scf/ton is sufficient to provide the oxygen combustion supporting requirements of the retort.

On the other hand, when partial combustion means are employed for directly heating the air either with or without the presence of dilution gas, a greater volume of air is generally required which will be at least about 3,500 scf/ton and more usually at least 4,000 scf/ton to provide the oxygen combustion supporting requirements with the retort.

Accordingly, it is important whether direct or indirect heat exchange means are employed to limit the oxygen available for combustion within the retort to substantial combustion of gaseous material while providing kerogen decomposition heat within the retort within the range of from 300,000 to 700,000 Btu/ton. Therefore, the total heat in the retort is the direct or indirect preheat of the combustion supporting gases referred to as the sensible heat plus the potential heat from consuming the oxygen therein and this combined heat input should be controlled within the range of from 300,000 to 700,000 Btu/ton of shale.

Thus, by heating the gas streams in the manner provided, the amount of air employed may vary over a considerable range of from 2,000 to 7,000 scf/ton of shale so that a higher and desired mix gas temperature is thereby obtained when the gas streams in the retort are combined at substantially the gas inlet of the combustion section.

Having a higher gas mix temperature means that the gases have a shorter distance to travel before the oxygen available in the gases is utilized to effect desired limited burning of available combustible materials. Accordingly, more of the heat of retorting is supplied by sensible heat in the gases rather than by heat of combustion on the surface of the shale particles.

Thus, by operating in accordance with the process, significantly more kerogen can be decomposed above the combustion zone and thermal cracking and oxidation of the desired shale oil as it passes out through the surface of the shale particle can be eliminated. Tables 1 and 2 show an order of magnitude of the following improvements which may be effected when operating a gas-combustion oil shale retort in accordance with this process.

A greater portion of the kerogen is decomposed before the shale particles enter the combustion zone.

Operation at much higher shale rates is possible.

Higher gas cooling rates are obtained in the mist forming zone which result in a more desirable mist and less refluxing of the shale oil.

Greater overall heat efficiency, producing lower spent shale temperatures.

High heat value gas is obtained for recycle to the retort.

TABLE 1

	Type of Operation			
	No Preheat		With Preheat	
	Range of Shale Rate			
	Low	High	Low	High
Type of Preheat	None	None	Indirect	Indirect
Operation Conditions:				
Shale Rate, lbs./hr.-ft.²	300	750	300	750
Air Rate, s.c.f./t.	4,000	4,000	3,200	3,200
Air-Dilution Gas Temperature, ° F	133	140	927	1,132
Performance Temperature, ° F.:				
Off-Gas	133	(1)	133	137
Spent Shale	285	(1)	248	279
Peak Gas	1,175	(1)	1,335	1,390
Peak Shale	1,140	(1)	1,201	1,246
Decomposition, Percent:				
Kerogen	100	(1)	100	100
Carbonates	9	(1)	14	14
Kerogen Decomposed Above Combustion Zone, Percent	5	(1)	95	84
Products Burned, Percent:				
Coke	17	(1)	8	6
Gas	12	(1)	21	23
Shale Oil	0.5	(1)	0.1	0.1
Gas Cooling Rate in Mist Forming Zone, ° F./sec	1,245	(1)	1,430	2,600
Heating Value of Dry Gas Make, B.t.u./s.c.f.	139	(1)	172	172

(1) Inoperable.

Retort Length = 12 Feet
Total Recycle Gas Rate = 16,000 s.c.f./ton
Brine = 15 lbs./ton
Shale Richness = 30 Gal./ton
Dilution Gas-to-Air Ratio = 0.4 v./v.

Table 1 shows the effect of preheating the air dilution gas on the performance of the gas-combustion retort for one inch particles and Table 2 shows the effect of preheating the air dilution gas on the performance of the gas-combustion retort for three inch particles. Accordingly, the gas-combustion retort temperature profile throughout the height for a shale flow rate within the range of from 300 to 1,000 lb/hr-ft^2 for air flow rates in the range of from 3,000 to 7,000 scf/ton is very carefully controlled and regulated by preheating of the air dilution gas mixture passed to the combustion section of the gas-combustion retort.

Within these limits, it has been found that the minimum preheat temperature of the air dilution gas mixture or preheated gas passed to the combustion section of the retort for air rates in the range of from 3,000 to 7,000 scf/ton may be determined by the following equation.

$$T = 38.82 - 0.11A + 1.87S - 2.59 \times 10^{-4}AS + 1.63 \times 10^{-5}A^2 + 2.09 \times 10^{-4}S^2$$

In the above equation T is the temperature in °F, S is the shale flow rate in $lb/hr\text{-}ft^2$, and A is the air rate in scf/ton entering the combustion zone. On the other hand, the maximum preheat temperature employed when operating at shale flow rates in the range of 300 to 1,000 $lb/hr\text{-}ft^2$ will be a function of the decomposed shale outlet temperature set not to exceed about 500°F, an upper shale temperature limit of 1500°F, and an upper gas temperature of 1800°F.

TABLE 2

	Type of Operation			
	No Preheat		With Preheat	
	Range of Shale Rate			
	Low	High	Low	High
Type of Preheat	None	None	Direct	Direct
Operating Conditions:				
Shale Rate, lbs./hr.-$ft.^2$	300	750	300	750
Air Rate, s.c.f./t	4,000	4,000	4,000	4,000
Air-Dilution Gas Temperature, ° F	150	150	1,370	1,800
Performance Temperature, ° F.:				
Off-Gas	-------	-------	130	210
Spent Shale	-------	-------	410	500
Peak Gas	-------	-------	1,238	1,340
Peak Shale	-------	-------	1,083	1,134
Decomposition, Percent:				
Kerogen	-------	-------	100	100
Carbonates	(1)	(1)	8	7
Kerogen Decomposed Above Combustion Zone, Percent	(1)	(1)	67	66
Products Burned, Percent:				
Coke	(1)	(1)	4	3
Gas	(1)	(1)	31	32
Shale Oil	(1)	(1)	0.2	0.1
Gas Cooling Rate in Mist Forming Zone, ° F./sec	(1)	(1)	740	1,560
Heating Value of Dry Gas Make, B.t.u./s.c.f	(1)	(1)	139	139

[1] Inoperable.

Retort Length=16 Feet
Total Recycle Gas Rate=16,000 s.c.f./ton
Brine=90 lbs./ton
Shale Richness=30 gal./ton
Dilution Gas-To-Air Ratio=0.4 v./v.

External Indirect Gas Heat Exchange

A process described by *J.H. Smith; U.S. Patent 3,619,405; November 9, 1971; assigned to Continental Oil Company* involves a gas combustion retort modified to withdraw the shale oil product while still predominantly vapor at elevated temperature and condense it externally with a net loss of only about 37% of its contained heat. Roughly one-fourth of the residue gas is burned to preheat the shale to offset this loss.

The process is carried out in a gas retorting system including a vertical retort having a raw shale preheating zone at the top of the retort; an external circuit for cooling and separation of product vapor and gas; and an external circuit for heating cold recycle gas (from the top of the shale preheating zone combined with noncondensable product gasses) and reintroducing it into the bottom of the shale preheating zone.

In the preferred system, the cold recycle gas is indirectly heated first in the product vapor heat exchanger, then indirectly in a fired heater. Cold gas from the final product separator is recycled to the bottom of the retort where it is preheated by the spent shale and then commingled with combustion air to control the flame temperature.

The purpose of preheating the raw shale is to prevent shale oil from condensing on the raw shale and refluxing down into the retorting zone, causing formation of heavy tars and coke which bind the shale into large agglomerates which impair efficiency. Referring to Figure 1.4a, the equipment includes a countercurrent gas combustion retort **1** similar to that developed by the Bureau of Mines at Anvil Points, Col. (see U.S. Patents 2,813,823; 2,901,402).

At the top of the retort is a raw shale preheating zone **2** separated from the retort by two approximately spaced tube sheets **3** through which pass conduits for the gravitational flow of shale from the preheating zone into the retorting zone **7**. These shale conduits extend some distance into the retorting zone to form a void space **4** below the lower tube sheet for the purpose of disengaging the product vapors from the shale and permit their easy withdrawal from the retort. Shielded conduits **5** extend upward from the top tube sheet for conveyance of preheating gas from the plenum chamber **6** between the two tube sheets upward into the raw shale preheating zone **2**.

The raw shale feed pipes **8** extend downward from the shale feed chamber **9** into the raw shale preheating zone a sufficient distance to form a suitable void space **10** for disengaging the cooled preheating gases for easy withdrawal and recycle. The external circuit for the retorting zone **7** consists of an indirect vapor heat exchanger **12**, a vapor cooler **14**, a liquid product separator **16**, a liquid product pump **18**, a recycle compressor **20**, plus appropriate pipes and control instruments. The external circuit for the shale preheating zone **2** consists of a recycle compressor **22** indirect, vapor heat exchanger **12**, a fired gas heater **24** for indirectly heating combined noncondensable product gases with gases recovered from the top of the preheating zone, plus appropriate pipes and control instruments.

FIGURE 1.4: GAS COMBUSTION RETORTING

a.

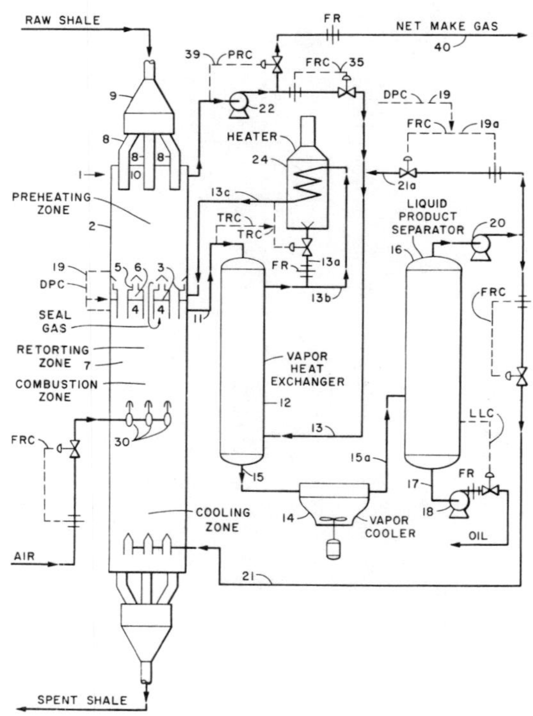

Flow Diagram of Process (continued)

FIGURE 1.4: (continued)

b.

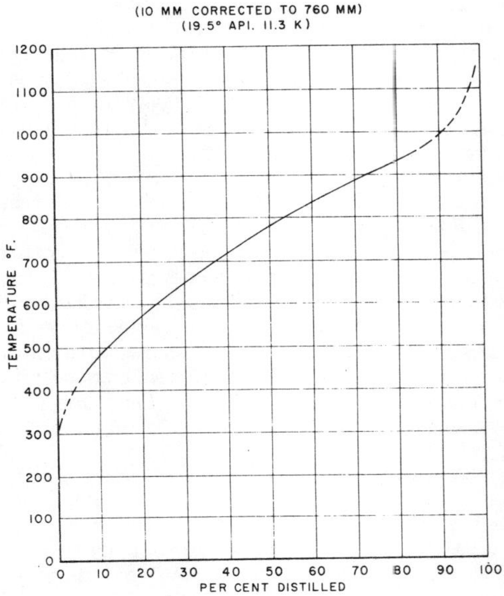

Engler Distillation of Shale Oil

c.

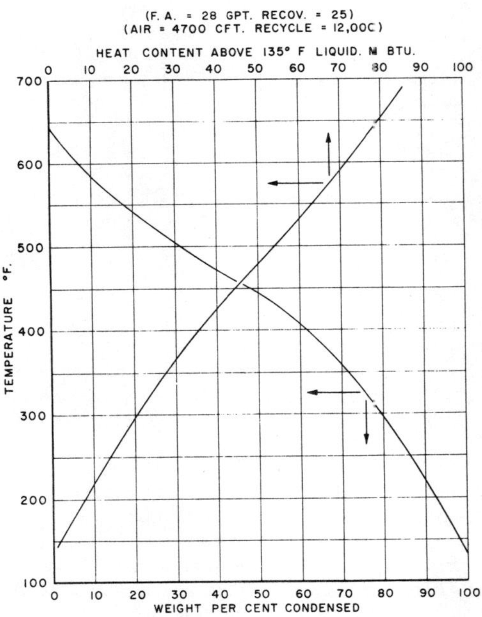

Condensation of Shale Oil Gas Combustion Retort

Source: J.H. Smith; U.S. Patent 3,619,405; November 9, 1971

In operation, raw shale is fed continuously via feed hopper **9** into the top of shale preheating zone **2**. It gravitates through the preheating zone then through the retorting zone **7**. After passing through the cooling zone at the bottom of the retort, the spent shale is discharged at a controlled rate from the bottom of the retort, indirectly controlling the raw shale feed rate.

Air is fed into the middle of the retort proper through a hang of distributors **30** embedded in the shale. Combustion of gases and residual coke on the shale in the vicinity of the air distributors supplies the heat required for retorting. Cold recycle gas is injected into the bottom of the retort proper through line **21** and flows upward countercurrent to descending spent shale recovering heat therefrom.

The preheated recycle gas commingles with the combustion air at the lower edge of the combustion zone to moderate the intensity of combustion and limit the flame temperature to substantially below the spent shale fusion or clinkering temperature (about 2100°F). The commingled recycle and combustion gases (about 1100° to 1400°F) continue to flow upward countercurrent to the descending shale, surrendering heat thereto to effect preheating and retorting of the shale. In the retorting zone (1 to 3 feet above the combustion zone) the kerogen is decomposed and forms shale oil vapors.

The noncondensable gases and shale oil vapors mix thoroughly and flow upward countercurrent to the descending raw shale surrendering heat thereto to bring the shale up to retorting temperature (about 900°F maximum), the gases and vapors themselves being cooled in the process to approximately their dew point, which will vary from about 550° to 750°F depending upon the source of shale and the operating conditions used. By dew point is meant the temperature at which shale oil liquid begins to condense from the vapors.

The thus cooled gases and product vapors are withdrawn from the top of the retorting zone through line **11** at approximately their dew point and are further cooled externally in indirect vapor heat exchanger **12** and vapor cooled **14** to effect condensation of the shale oil vapors. The major part of this cooling is done by exchanging heat (in vapor heat exchanger **12**) with a second recycle gas stream (line **13**) which is used for preheating the raw shale to approximately the dew point of the gas-vapor mixture leaving the retorting zone, before the shale enters the retorting zone.

The cooled mixture of shale oil liquid and noncondensable gases flows from the final cooler through line **15a** into shale oil separator **16** from which the shale oil is pumped into storage on level control. Blower **20** compresses the gases from the shale oil separator delivering a portion as unheated recycle gas to the bottom of the retort on flow rate control through line **21**.

The net product gas (from combustion and retorting) plus seal gas amounting to about 500 ft^3/ton of shale feed is delivered to the shale preheating gas loop via line **21a** which is on flow rate control **19a** reset by a differential pressure control **19** which maintains a very slightly higher pressure (less than 3 inches water column) at the bottom of the shale preheating section than at the top of the retorting zone. This causes a slight flow of seal gas from the preheating section into the retort proper to prevent the flow of shale oil vapors into the shale preheating section.

The gases from the top of the preheating zone are recovered at a temperature approximating their dew point with respect to water, i.e., at a temperature of about 110° to 160°F. A second blower **22** compresses these cold gases from the top of the shale preheating zone delivering a major portion thereof via a flow rate control **35** to combine with the remaining portion of noncondensable gases delivered through loop **21a** to pass through vapor heat exchanger **12** then via line **13b** through heater **24** into the bottom of the raw shale preheating zone **2** at about 600° to 800°F, preferably 700° to 750°F. A portion of this gas is used as fuel for the fired heater via line **13a**. The net make gas is delivered offsite via line **40** on pressure control **39** which maintains essentially atmospheric pressure at the top of the raw shale preheating section. Table 1 lists the compositions and other properties of various gas streams in the retorting zone.

TABLE 1: GAS STREAM PROPERTIES, MODIFIED GAS COMBUSTION RETORT

	Make Gas Retorting Zone	Off Gas Preheating Zone	Furnace Fuel Gas
Component, scf/ton			
H_2	306	231	75
CO	216	163	53
C_1 to C_4	172	130	42
C_5+	72	54	18
N_2	3,722	2,810	912
CO_2	1,594	1,203	391
H_2O	476	1,257	309
Total	6,558	5,848	1,800
Pounds	527	440	138
Pounds/mol	30.5	28.5	29.1
LHV/Btu/scf	102.0	86.3	91.0
Heat content, Btu/mscf			
135°F	0		0
300°F	3,560		3,530
500°F	8,140		8,070
700°F	13,160		13,010

Table 2 lists the fractional composition and ideal K values (vapor-liquid equilibrium constants) for a typical shale oil with gravity and distillation shown in Figure 1.4a. Table 2 also lists K values for fractions of the retort off gas. These K values were used for calculating the shale oil condensation curve shown in Figure 1.4c.

TABLE 2: PROPERTIES OF TYPICAL SHALE OIL

Fractional composition (Per ton of shale)

Fraction	MBP	Gal.	#/Gal.	Pounds	#/Mol	Mol
A	370	1.25	6.83	8.53	145	.0588
B	450	1.25	7.076	8.85	174	.0509
C	525	2.50	7.296	18.25	205	.0890
D	610	2.50	7.505	18.77	248	.0757
E	715	5.00	7.752	38.76	305	.1270
F	830	5.00	7.989	39.95	380	.1052
G	930	5.00	8.212	41.06	456	.0902
H	1,020	1.25	8.386	10.49	520	.0198
I	1,120	1.25	8.570	10.72	620	.0173
Total	775	25.00	7.815	195.38	308	.6339

Ideal V–L equilibrium constants at 12.5 p.s.i.a.

	Temperature, °F.				
Fraction	250	350	450	550	650
A	.16	.88	3.2	8.8	20.
B	.038	.26	1.2	3.8	10.
C	.0088	.080	.42	1.6	3.6
D	.0011	.015	.11	.52	1.8
E	.00010	.0020	.020	.12	.56
F	.00002	.00020	.0020	.018	.10
G		.00003	.0003	.0032	.024
H			.00004	.0004	.0044
I				.00005	.0005
C_4 and L	Inf.	Inf.	Inf.	Inf.	Inf.
C_5–C_7	4.8	14.	34.	Inf.	Inf.
C_8+	.75	3.1	9.2	22.	42.

The following example shows typical specifications, operating conditions, and results.

Example:

	Pounds/Ton	Flow Rate/Ton	Temperature °F
Raw shale to hopper 9	2,000		60
Spent shale	1,587		400

(continued)

	Pounds/Ton	Flow Rate/Ton	Temperature °F
Shale oil product (line **17**)	195	25 gal	135
Air rate to **30**	360	4,700 scf	120
Line **40**	440	5,848 scf	150
Line **13**	1,695	22,058 scf	150
Line **13a**	138	1,800 scf	500
Line **13c**	1,557	20,258 scf	717
Line **21**	964	12,000 scf	150
Line **31**	1,404	20,848 scf	135
Line **31a**	1,128	15,000 scf	150
Line **21a**	567	7,058 scf	150
Line **15**	1,726		322
Line **15a**	1,726		135
Line **11**	1,726	19,298 scf	650
Seal gas: 500 scf			

Basis: 1 ton 28 Fischer assay shale.

The only drawback of the system of Figure 1.4a is the extra cost over the simple Bureau of Mines version. To help offset the higher cost, this modified design will give a higher yield from elimination of refluxing and permitting a higher recycle gas rate. Also, this modified version is capable of retorting rich shale, assaying up to 40 gal/ton or more and particles smaller than one-fourth inch which cannot be processed in the Bureau of Mines retort.

The improved control of the temperature pattern in the retorting zone will permit a reduction in air rate which enhances the yield. The combination of less mineral carbonate decomposition plus the recycling of a drier cooling gas results in less hydration of spent shale. This, together with higher recycle gas rate, lowers spent shale temperature to lessen disposal problems.

Control of Fines

T.C. Lyons; U.S. Patent 3,501,394; March 17, 1970; assigned to Mobil Oil Corporation describes a process for retorting fine particulate solids such as shale or Athabasca tar sands containing hydrocarbonaceous material to recover oil. The fine solids are mixed with a hot noncombustion supporting lift gas in a lift pot and the resultant mixture is passed upward through a riser as a dilute phase. Contact time and temperature in the riser are regulated to effect thermal decomposition of the hydrocarbonaceous material to oil vapor which is subsequently recovered. Spent fines can be burned in a separate combustion zone and at least a portion of the hot gas is recycled to the lift pot.

By the process, an effective means is provided for retorting raw fines containing thermally decomposable hydrocarbonaceous material. When retorting raw fines in dilute phase, in accordance with the process, the random movement of fines is greatly reduced so that the major movement of fines is in an upward direction.

This permits increased control of fines residence time within the riser reactor so that more uniform residence time in the reactor for individual fine particles is attained. In addition, removal of spent fines from the riser reactor promotes process thermal efficiency in that heat is not expended in maintaining spent fines at reaction temperature.

Referring to Figure 1.5a, raw fines are directed to lift pot **1** as compact beds through conduit **2**. This internal portion of the lift pot is structured to provide ease in controlling the amount and velocity of the lift gas entering riser reactor **3**. Baffles **4** and **5** are provided so as to permit secondary gas entering conduits **6** to contact a relatively large portion of the compact bed surface area located in the bottom portion of the lift pot. Primary lift gas is supplied to the interior of the riser **3** through conduit **7**. This conduit has a conical top portion **8** provided with perforations. The primary lift gas contacts the raw fines introduced through conduit **2** to provide a dilute phase within riser **3**.

FIGURE 1.5: GAS LIFT RETORTING PROCESS

a.

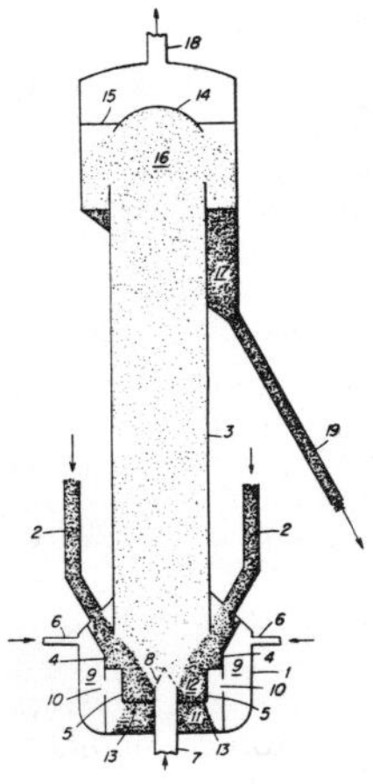

Riser Reactor

b.

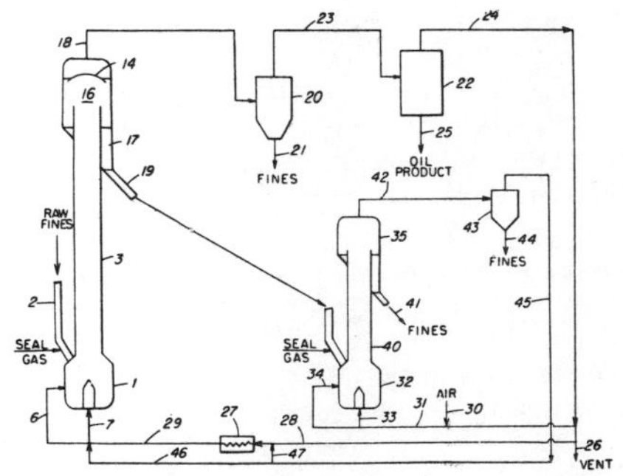

Process Flow Arrangement

Source: T.C. Lyons; U.S. Patent 3,501,394; March 17, 1970

The secondary lift gas stream **6** provides means for controlling the velocity and the amount of lift gas provided to the riser **3**. The secondary lift gas stream is provided with means, not shown, for controlling the amount of gas supplied to the riser. The secondary lift gas is directed to plenum zones **9** and through openings **10** to contact a compact bed of fines in zones **11** and **12**. Baffles **5** are held to conduit **7** by ribs **13**. The secondary lift gas and primary lift gas are supplied at elevated temperatures to effect the thermal decomposition desired.

The mixture of gas and fines pass upward through the riser as a dilute phase and contact deflector **14** which is held by ribs **15** to the interior walls of the separation zone **16**. After contacting the deflector, the spent fines from the riser are caused to pass downward to a compact bed of fines **17**. The lift gas and the vaporized product pass upward through and out of the separation zone through conduit **18**. From the bottom of separator zone **17**, the fines are removed through conduit **19**. The particular type of deflector employed in the separation zone above the riser is not critical. Any type of deflector which effects adequate separation of vapors from fine particles can be employed.

Referring to Figure 1.5b, raw fines are directed as a compact bed to a lift pot **1** through conduit **2**. In the lift pot, the raw fines are contacted with a secondary gas stream from conduit **7**. The fines-gas mixture passes upward through riser reactor **3** as a dilute phase in a manner described previously for Figure 1.5a. In the riser reactor, the hydrocarbonaceous portion of the raw fines is thermally decomposed to produce oil vapors. The oil vapors and spent fines exit the riser reactor and contact deflector **14** in the upper portion of separation zone **16**.

The spent fines are directed downward to a compact bed of spent fines located in chamber **17**. The bottom portion of the chamber slopes down at an angle greater than the angle of repose of the spent fines. The chamber can be provided with stripping means to remove entrained vapors from spent fines. The vaporous material, including oil vapor and lift gas is removed from separation zone **16** through conduit **18** at a temperature from 700° to 1000°F and is directed to cyclone **20**.

In cyclone **20**, vaporous material is separated from any fines carried over from separation zone **16**. The fines are removed from this cyclone through conduit **21** and are discarded. It is desirable to maintain the vaporous material in the cyclone at a high temperature of above 700°F. This is because partial condensation of oil at a lower temperature will result in oil yield loss caused by absorption by the carried over fines. Vapors from the cyclone are directed to a condensation step **22** through conduit **23**. In the condensation step, the vapors are cooled to obtain a liquid oil product and a gas overhead.

The condensation step is carried out in a manner to recover the hydrocarbon product boiling in the naphtha boiling range and above. Hydrocarbons boiling below the naphtha boiling range in addition to the lift gas employed in the riser **3** are withdrawn from the condensation step through conduit **24**. Oil product is withdrawn from the condensation step through conduit **25**. A portion of the gas from the condensation step is vented through conduit **26** while another portion of the gas is employed to support spent fines combustion in a manner described below. The remainder of the gas from the condensation step is directed to heater **27** through conduit **28** to be heated to a temperature which supports conversion. The heated gas is directed from the heater through conduits **29, 7** and **6** to lift pot **1**.

The gas from the condensation step which is used to support spent fines combustion is mixed with air from conduit **30**. The mixture of air and gas is directed to lift pot **32** through conduits **31, 33** and **34**. The lift gas contacts hot spent fines containing unconverted hydrocarbonaceous material and coke in lift pot **32**. The spent fines are obtained from chamber **17** and are directed to the lift pot through conduit **19**. The ratio of combustion air to recycle gas in the lift pot is regulated so that the vaporous combustion products from separation zone **35** are oxygen free. That is, the oxygen entering the lift pot is totally consumed in combustion.

In addition, the amount of gas entering lift pot **32** is regulated so that a dilute phase of fines is maintained in riser **40** without an undesirably high pressure drop across this riser. The hot spent fines and gas are passed upward as a dilute phase in the riser to effect combustion of carbonaceous material in the fines. The fines and vapors are passed upward to separation zone **35** where fines are separated from vapors.

The fines are removed from separation zone **35** by being passed downward through conduit **41**. The vapors from separation zone **35** are passed overhead through conduit **42** to cyclone **43** wherein carried over fines are separated from vapor. The fines are removed from cyclone **43** through conduit **44**. The hot vapors from cyclone **43** are recycled to lift pot **1** through conduits **45, 46, 6** and **7**. A portion of the recycled vapor from cyclone **43** can be directed to heater **27** through conduit **47** to provide additional heat of conversion in riser **3**.

Multistage Retort Process

According to a process described by *R.B. Needham; U.S. Patent 3,487,001; December 30, 1969; assigned to Phillips Petroleum Company* oil yielding solids in particulate form are introduced into a retorting zone, fuel and oxygen-containing gas are introduced into a combustion region within the retorting zone, a condensible gas is introduced into a diffuse region within the retorting zone, noncondensible combustion gases are removed from the retorting zone, a portion of condensible gas and noncondensible combustion gases mixture is removed from the diffuse region, dust is removed from the gas mixture and the gas mixture is recycled to the diffuse region, hydrocarbons are educed from the solids in a pyrolysis region within the retorting zone, the hydrocarbons are recovered from the retorting zone, and particulate solids are removed from the retorting zone.

Referring to Figure 1.6a, particulate oil yielding solids are moved through line **11** and are heated to a desired temperature, e.g., 300°F in preheater **12**. From the preheater, a compacted bed of solids is introduced through line **13** into the retorting column **14**. The compact bed of particulate solids flows into pyrolysis region **15** and educed hydrocarbon products are removed via line **16**.

A condensible gas, e.g., superheated steam, is introduced into the upper portion of diffuse region **17** through line **18**. Noncondensible combustion gases are removed from the lower portion of diffuse region **17** via line **19**. A mixture of steam, fines and small amounts of combustion gases is removed from diffuse region **17** via line **20** to separator **21**. After removal of a substantial portion of the fines, the gases are recycled to diffuse region **17** via line **22**. Fines are removed from separator **21** through line **23**.

FIGURE 1.6: RETORTING PROCESS

a.

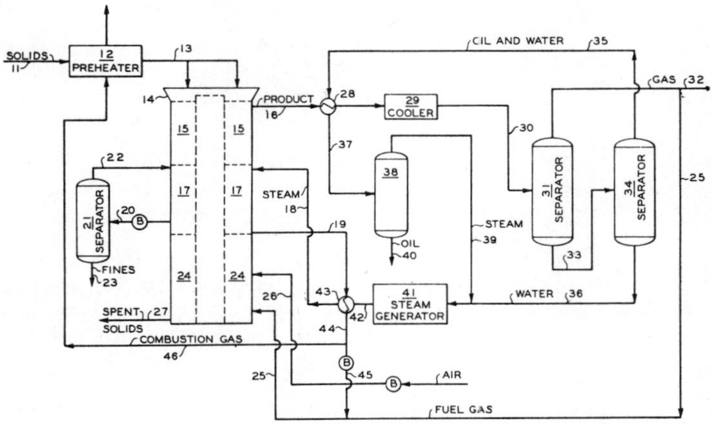

Process Schematic

b.

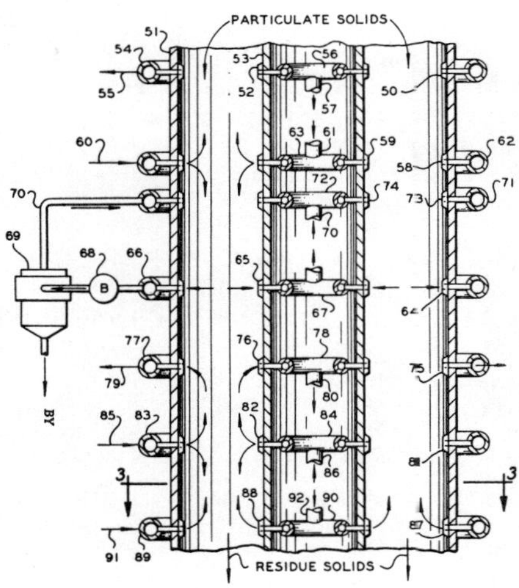

Vertical Cross Section of Retorting Vessel (continued)

FIGURE 1.6: (continued)

c.

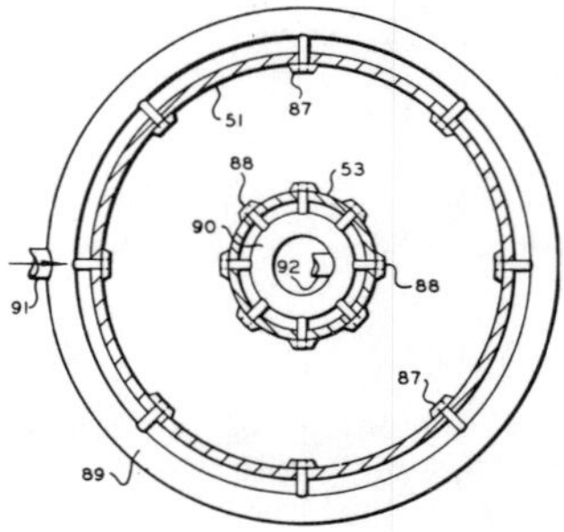

Horizontal Cross Section of Retorting Vessel

Source: R.B. Needham; U.S. Patent 3,487,001; December 30, 1969

In combustion region **24** fuel is introduced via line **25** and oxygen-containing gas, for example, air, is introduced via line **26**. Upon flowing downward through pyrolysis region **15**, diffuse region **17**, and combustion region **24**, residue solids are removed from retorting column **14** through line **27**.

A suitable product mist separation system is illustrated in conjunction with the retorting system in Figure 1.6a. Product mist, removed from pyrolysis region **15** via line **16**, flows through heat exchanger **28** to cooling zone **29** where steam and a portion of the hydrocarbon vapors are condensed. Water-hydrocarbon product mixture flows through line **30** to separator **31**. Hydrocarbon vapors are removed from the separator overhead through line **32**. If desired, a portion of the hydrocarbon vapors can be transferred via line **25** to the combustion region to be used as fuel.

Liquid hydrocarbons and water are removed from the separator **31** via line **33** to separator **34** wherein a portion of the water is separated from the mixture. Liquid hydrocarbon-water mixture is removed from separator **34** via line **35** and water is removed via line **36**. The mixture passes through the heat exchanger wherein the water is vaporized. Steam-liquid hydrocarbon mixture is transferred from the heat exchanger via line **37** to separator **38**. In this separator the steam is removed overhead via line **39** and liquid hydrocarbon products are removed via line **40**.

Steam flowing through line **39** and water flowing through line **36** are mixed and flow to steam generator **41**. Steam is removed from this steam generator through line **42** and flows through heat exchanger **43** wherein additional heat is transferred from the combustion gases to the steam. From heat exchanger **43** steam is transferred through line **18** to diffuse region **17**. Combustion gases are removed from the diffuse region and flow through line **19** to heat exchanger **43** to warm condensible gases. Combustion gases are removed from heat exchanger **43** through line **44**.

If desired, a portion of the combustion gases can be recycled through line **45** and admixed in line **25** with the fuel gas flowing to combustion region **24**. One advantageous use of excess combustion gas is to cycle the gases through line **46** to preheater **12**, thereby supplying a portion of the heat required to preheat the particulate solids. In the annular retort of Figure 1.6b, a plurality of ports **50** are positioned in the wall of outer retort shell **51** and a plurality of ports **52** are positioned in the wall of inner wall shell **53** directly across from ports **50**.

Ports **50** communicate with a manifold **54** and product mist is drawn into the manifold through the ports and removed via conduit **55**. Ports **52** communicate with manifold **56** providing for the removal of product mist through conduit **57**. Thus ports **50** and ports **52** define the upper limit of a pyrolysis region within the retorting apparatus.

A plurality of ports **58** in shell wall **51** and a plurality of ports **59** in shell wall **52** are used to introduce a superheated condensible gas into a compact bed of oil-yielding particles. Superheated condensible gas is introduced through conduits **60** and **61** into manifolds **62** and **63** which communicate with the ports. The introduction of steam at this location defines the lower limits of the pyrolysis region and the upper limits of a diffuse region with the retorting vessel.

Removal of fines-gas mixture is effected through a plurality of ports **64** and **65** communicating with manifolds **66** and **67**. A blower **68** is used to transfer the mixture to cyclone separator **69**. One separator is shown but if desired, a plurality of separators can be used. Conduit **70** communicating with manifolds **71** and **72** recycles the gas through a plurality of ports **73** and **74** to the retorting apparatus. A plurality of ports **75** in shell wall **51** and a plurality of ports **76** in shell wall **53** are utilized to remove noncondensible combustion gases.

Combustion gases are drawn through ports **75** and **76** into manifolds **77** and **78** and are removed through conduits **79** and **80** thus defining the lower limit of the diffuse region and the upper limit of a combustion region within the retorting apparatus. Oxygen-containing gas is supplied to the retorting apparatus through ports **81** and **82** in combination with manifolds **83** and **84** and conduits **85** and **86**. Fuel is supplied to the retorting apparatus through ports **87** and **88** in combination with manifolds **89** and **90** and conduits **91** and **92**.

Figure 1.6c is a plan view taken on line **3–3** in Figure 1.6b and illustrates the port and manifold system. Like numerals in Figures 1.6b and 1.6c denote like elements. A plurality of ports **87** is positioned in outer shell wall **51** and connected into manifold **89**. A communicating conduit **91** supplies manifold **89** with fuel to be fed to the retorting apparatus. Fuel is also introduced through a plurality of ports **88** in wall **53** in combination with manifold **90** and conduit **92**. Ports have been illustrated as being slightly raised from the shell walls but it is obvious that the ports can be set flush or recessed in the shell walls.

Example: Mahogany Ledge shale from the Green River formation, having oil content of 28 gal/ton by Fisher assay, is crushed to obtain particles with a major dimension of 3" or less. The crushed shale is screened to remove particles having a major dimension of ½" or less. The crushed screened shale is preheated to 250°F. Preheated particulate shale is introduced into retorting zone **14** at a rate of 600 tph. Spent solids are removed via line **27** at a rate of 474 tph. 810,000 lb/hr of superheated steam at 1000°F under a pressure of 13 psia is introduced into diffuse region **17** through line **18**. 13.4 M ft^3/hr of combustion gases are removed via line **19**.

Attrition during the downward movement of the compact particulate bed creates fines which, if not removed, cause gas channeling. 800 lb/hr of particles having less than 0.05" major dimension are removed in separator **21**. 10 M ft^3/hr of combustion gases are removed via line **22**. Product mist is removed through line **16**. Separation results in 188 M cubic feet per hour of gas recovered through line **32** and 360 barrels per hour of oil are recovered through line **40**. Thus, it can be seen that large quantities of oil shale can be retorted to obtain a high yield of hydrocarbons in the form of a mist from which the hydrocarbons can be easily separated.

Mixtures of Rich and Lean Oil Shale

W.S. Bergen; U.S. Patent 3,574,087; April 6, 1971; assigned to Mobil Oil Corporation has found that by mixing rich oil shale with relatively lean oil shale to obtain an average Fischer assay of between 15 and 36 gal/ton, that the gas combustion retorting of the mixture can be accomplished without encountering flooding and associated problems within the retort. Further, it has been found that significantly increased yields are obtained from the rich-lean shale mixture, having a Fischer assay within the range from 27 to 36 gal/ton. Therefore, oil product yields are significantly increased under stable retorting conditions.

Furthermore, considerable savings are realized in reducing mining and material handling costs. Thus, in the method, it is desirable to convert an oil shale mixture having as high an oil content as possible while maintaining at the same time stable retorting conditions so as to maintain and effect a thermally efficient utilization of a gas combustion retorting system. The process provides among other things, a means for retorting relatively lean as well as relatively rich oil shale materials in a manner where the amount of recovered oil is greater than that which would be recovered if separately retorted. The following example illustrates the process.

Example: Gas combustion retorting was carried out on shale of varying richness as shown in the table below. The shale was crushed to a particle size ranging from ¾" to 1½". The shale particles in runs 6, 7 and 8 were obtained by mixing shale having a Fischer assay of 45 gal/ton and shale having a Fischer assay of 28 gal/ton.

Run Number	1	2	3	4	5	6	7	8	9	10
Raw shale Fischer assay, gal./ton	18.8	19.4	19.6	28.8	27.5	36.0	35.4	35.4	38.9	39.8
Shale mass flow, rate lb./hr.-ft.^{2}	491	488	486	486	483	486	489	484	496	492
Air rate, s.c.f./ton shale	4,870	4,890	4,930	4,540	4,580	4,840	4,840	4,920	4,780	4,800
Recycle gas rate, s.c.f./ton shale	13,600	13,500	13,600	13,800	13,400	13,900	13,800	13,800	13,600	13,800
Yield—percent Fischer assay	86.3	83.7	83.1	84.9	89.8	88.6	89.9	91.7	86.5	85.2
Operation	Stable	Stable	Stable	Stable	Stable	Stable	Stable	Stable	(1)	(1)

[1] Unstable retort flooding.

From the table, it can be seen that retorting oil shale having a Fischer assay above 36 gal per ton results in unstable and undesirable operation. Further, by employing shale mixtures having a Fischer assay in the range from 15 to 36 gal/ton, it is possible to retort shales under stable operation conditions at high yields in a gas-combustion retort.

Feed Segregation and Shale Oil Recycle

G.E. Irish and R.F. Deering; U.S. Patent 3,133,010; May 12, 1964; assigned to Union Oil Company of California have found that in combustive retorting, when an oil shale feed mixture is separated into a high oil assay portion and a low oil assay portion, the high oil assay portion can be advantageously educted with substantial increased oil recovery (as a percent of Fischer assay) as compared to an unsegregated feed mixture, and with comparatively little change in solids retorting rate.

This finding can be combined in an integrated oil shale retorting process where the raw shale feed from the mine is separated into a high oil assay solids consist, which is fed to a combustion retort, and a low oil assay solids consist, which is fed to a noncombustive hot recycle gas retort. Thus, in this integrated oil shale retorting process, every particle of shale fed to the system is retorted under conditions best suited to obtain the highest possible yield of liquid shale oil in the most efficient manner.

More specifically, in processing a subdivided solids charge which varies in oil assay value, the solids charge is divided into two fractions which differ substantially in average oil assay value. The fraction having the higher oil assay is fed to a combustive retort while the fraction having the lower oil assay is fed to a noncombustive retort.

Any difference in the oil assay value of the two portions is advantageous, and a preferred solids feed has an assay value no lower than about 15 gal/ton at one extreme and about 60 gal/ton at the other. In a typical commercial oil shale operation, the shale is segregated into a lean fraction having an assay in the range of 20 to 35 gal/ton, and a rich fraction having an assay in the range of 35 to 50 gal/ton. A convenient manner of segregating the shale is to separate the rich shale from the lean shale at the mine and to arrange the solids transportation from the mine to the retort to maintain the segregation. Since low oil assay shale has a higher specific gravity than high oil assay shale, separation according to specific gravity can provide the required oil assay segregation. Further, it has been found that run of the mine shale, when crushed, usually provides an inherent segregation by size, in which case the solids fines consistently are leaner in oil assay value than the larger particles.

Thus, size segregation by conventional means, i.e., screening, can sometimes provide the desired assay segregation for this process. The preferred combustive retorting of this process utilizes an upflow of shale solids and a downflow of flue gas, e.g., the combustion retort of U.S. Patents 2,501,153; 2,640,014 and 2,640,019.

G.E. Irish; U.S. Patent 3,228,869; January 11, 1966; assigned to Union Oil Company of California describes a hot gas retorting process where at least a portion of the liquid product oil is recycled to the eduction zone of the retort as a component of the hot eduction fluid. The liquid oil recycle stream can be either the liquid oil phase directly from the retort, or it can be a portion of the retort liquid phase which has been passed through a fractionation column.

A heavy fraction obtained from the distillation of the retort oil can be recycled to the retort as the liquid oil recycle, or the oil recycle can be a middle-cut fraction from the aforementioned distillation. Also, any combination of the above streams derived from the retort liquid phase can comprise the liquid oil recycle stream to the retort eduction zone.

Although the recycle of any portion of the retort liquid phase is advantageous, it is most desirable to return to the retort eduction zone the heavier fractions of the hydrocarbons educted from the shale. No matter which stream is recycled to the retort, the essence of the process is to recycle a sufficient quantity of oil to effect a significant improvement in the quality of the net liquid oil product, particularly as evidenced by a reaction in the pour-point and viscosity.

No significant effect on pour-point reduction is apparent at recycle rates lower than about 5%, and it is usually uneconomical to recycle more than about 50% of the product oil. Therefore, although higher recycle reates can be employed, the preferable form of the process comprises recycle of at least 5% and not more than 50% of the volume of net shale oil product from the retorting process.

Furthermore, it is preferred that the recycled oil fraction have an initial normal (atmospheric pressure) boiling point in excess of 500°F, and preferably an initial over 600°F. Thus, the recycle stream is essentially comprised of at least a substantial portion of the heavy deleterious materials which contribute markedly to high viscosity and high pour-point of shale oils produced by existing retorting processes.

These materials, when recycled to the hot recycle gas retort eduction zone are converted to more desirable products which yield an oil of substantially lower viscosity and substantially lower pour-point with essentially no loss in overall yield. It is essential that the recycle oil be returned to the inlet of the retort with the hot eduction gases for maximum quality enhancement.

In related work, *R.F. Deering and G.E. Irish; U.S. Patent 3,058,904; October 16, 1962; assigned to Union Oil Company of California* describe an integrated oil shale retorting process where solids-upflow combustion retorting is integrated with hot gas eduction in a solids-upflow, fluid-downflow retort in which eduction is effected without combustion by recirculating an externally heated portion of the rich product gas to the retort.

R.F. Deering; U.S. Patent 3,361,644; January 2, 1968; assigned to Union Oil Company of California describes an oil shale retorting process where eduction is accomplished by recycling a preheated portion of the rich product shale gas to a solids-upflow, fluid-downflow retort, and where the eduction fluid is indirectly heated to retorting temperatures and is maintained essentially oxygen-free throughout the process.

Retorting Total Oil Shale

In a process described by *B.L. Schuman and H.P. Dengler; U.S. Patent 3,489,672; Jan. 13, 1970; assigned to Esso Research and Engineering Company* shale fines are retorted in a mixture with larger fines by slurrying the fines in a liquid vehicle, such as oil or water, and introducing the slurry into the retort at a selected point where the temperature is sufficient to decompose the kerogen in the fines to shale oil but insufficient to crack the shale oil.

Referring to Figure 1.7, **11** designates a feed line in which raw oil shale, having particle diameters larger than about ¼" and usually within the range from ⅜" to 3" in diameter, depending on the crushing procedures used, is introduced into a gas combustion retorting vessel **12**. This gas combustion retorting vessel comprises a preheating zone **13**, a retorting zone **14**, a combustion zone **15**, and a cooling zone **16**. The raw oil shale flows serially downward through the several zones and the spent shale, which comprises about 75 to 90% of the raw oil shale, is discharged by way of line **17** to a waste pile which may be located some distance from the gas retorting vessel.

Heat and gases for retorting and preheating the oil shale are provided by introducing into the combustion zone by way of line **18**, controlled by valve **19**, a sufficient amount of air to cause combustion of carbonaceous material on the descending shale. The amount of air employed may range from 4,000 to 6,000 scf/ton of raw oil shale introduced into the operation, including fines.

FIGURE 1.7: RETORTING PROCESS UTILIZING TOTAL RAW SHALE

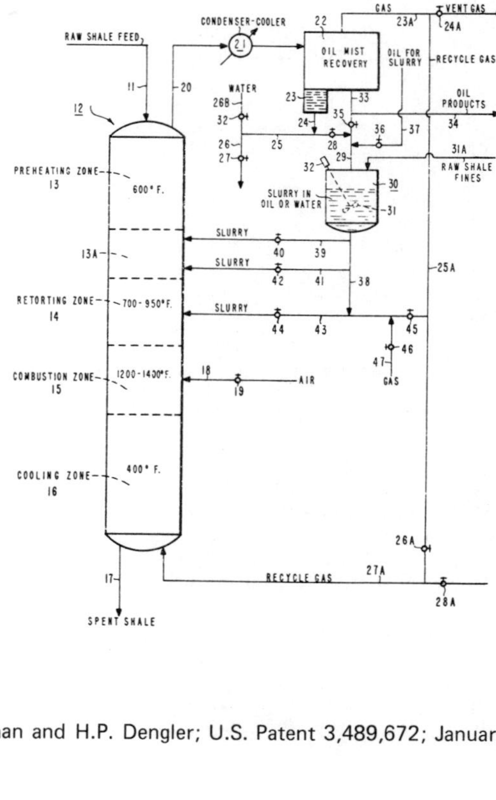

Source: B.L. Schuman and H.P. Dengler; U.S. Patent 3,489,672; January 13, 1970

This combustion operation provides a temperature in the combustion zone which may suitably range from 1000° to 1400°F, but usually within the range of 1200° to 1400°F. The resultant heat and the ascendant flue gases cause retorting and decomposition of the kerogen in the oil shale to shale oil in retorting zone **14**. The released shale oil flows upward into preheating zone **13** and the heat preheats the raw oil shale introduced by line **11**. Thus, the temperature in the preheating zone may be within the range from 400° to 800°F, usually about 600°F. The temperatures in the retorting zone may be within the range of 600° to 1000°F, usually from 700° to 950°F.

The temperatures in the cooling zone **16** will usually range from 1400° to 200°F, usually about 400°F. As stated, the temperature where the slurry is introduced may be within the range of 500° to 700°F. Pressures in the several zones may range from 0 to 100 psig. Preferably, a pressure within the range from 20 to 60 psig may be used.

As a result of the gas combustion retorting operation in vessel **12**, a vaporous product is withdrawn by way of line **20** and passed through a condenser-cooler **21**, and then into an oil mist recovery drum **22**, provided with a water leg **23**, where a separation is made among noncondensable gas, shale oil and water produced in the operation. The noncondensable gases may be discharged by line **23A** from the system by opening valve **24A**, but preferably a major part of the gas is recycled by line **25A** controlled by valve **26A** to the gas combustion retorting vessel **12** where the recycle gas serves to cool the spent shale in cooling zone **16**.

The amount of recycle gas may be within the range from 12,000 to 20,000 scf/ton of raw oil shale introduced into the operation, including fines. If desirable, the recycle gas may be augmented by an extraneously supplied gas through line **27A** which connects to line **25A** controlled by valve **28A**.

The water accumulating in water leg **23** may be discharged by line **24**, and then from the system by branch line **25**, and line **26** controlled by valve **27**. Preferably however, the water in water leg **23** is introduced by opening valve **28** in line **25** into line **29** and then into a slurry drum **30** into which raw shale fines, having particle diameters less than ¼" are introduced by line **31A** from the oil shale mining and crushing operation. The slurry is formed in slurry drum **30** by means of a mixing device **31**, which may be powdered with a suitable power means. The slurry in the slurry drum may then be introduced into the gas combustion retorting vessel.

It may be desirable to use an extraneous source of water as the liquid vehicle in making up the slurry, and under these instances the water from an extraneous source, which may be fresh water or brine, may be connected to line **26B** and introduced by opening valve **32**. The slurry may also be made up using a portion of the shale oil. In this instance, shale oil is discharged from oil mist recovery drum **22** by line **33**, and a portion thereof withdrawn as product by line **34**.

A smaller portion is discharged by line **33** controlled by valve **35** into the slurry drum. Of course, the oil may be from an extraneous source, and when it is desired to use oil produced outside the operation, or from any source, valve **36** in line **37** connecting to the outside source of oil may be opened.

When the liquid vehicle is oil, the slurry will preferably be introduced above the retorting zone **14** into a zone **13A**, which may be at a temperature within the range of 500° to 700°F. To this end, the slurry is withdrawn from the slurry drum by line **38** and introduced into zone **13A** by charge line **39** controlled by valve **40**.

When the liquid vehicle is water, however, the slurry may be introduced at a point nearer to, or just adjacent to, the retorting zone **13** and to this end, line **41** controlled by valve **42** is provided to allow the slurry comprising water and fine shale particles, to be introduced at least adjacent to the retorting zone **14** and sometimes in the retorting zone. Also, when the liquid vehicle is water, the slurry may be introduced into the retorting zone, particularly when a gas is also introduced with the slurry. To this end, line **43**, controlled by valve **44**, is provided for introduction of the water-raw shale fines slurry directly into the retorting zone **14** with gas which may be recycle gas obtained from line **25A**, to which line **43** connects, by opening valve **45**.

The process is quite useful in that numerous advantages occur by introduction of raw shale fines into the gas combustion retorting operation as a slurry in a liquid vehicle. Thus, when oil is employed, it has been found that if such oil is introduced above the retorting zone the oil would not be cracked since the oil is evaporated and recovered without loss, thus providing for introduction of fines from which shale oil is recovered.

The fines may usually comprise from 10 to 20% by weight of the oil shale which is mined and crushed, which ordinarily would have to be discarded. It is to be understood however, that the fines may comprise a lesser amount of the oil shale and may be within the range of 1 to 20% by weight and may be from 5 to 10% of the raw oil shale. Thus, the fines provide an additional source of oil shale in a very economical fashion, thus providing for recovery of a greater amount of shale oil than previously.

Recovery and Conversion of Shale Oil Mist

In a process described by *H.W. Parker; U.S. Patent 3,560,367; February 2, 1971; assigned to Phillips Petroleum Company* shale oil mist is recovered by collecting it in a solid subdivided mass, e.g., coke or catalyst, as in a fluidized bed and then converted in a conversion zone. In a specific example, cooled vapors from an oil shale retorting zone are treated to separate liquid oil and an oil mist. The oil mist is adsorbed on a solid, subdivided mass and the mass containing the adsorbed mist is passed to a conversion zone.

Referring to Figure 1.8, crushed shale is fed at **1** into retort **2** from which spent or retorted shale is removed at **3**. To create the heat necessary to pyrolyze or to decompose the kerogen, air is passed by **4** into retort **2**. In the retort, the kerogen will begin to decompose at a temperature in the range of 440° to 660°F. A substantial amount, if not substantially all of the oil or oily materials formed from the decomposition of the kerogen will have been removed from the shale at a temperature within the range of 840° to 1020°F.

Vaporous products are taken overhead from retort **2** by **5** and passed by coolers **6** and **7** to cyclone mist separator **8**. Continuous type crushed shale retort operation is shown. The temperature in this operation can be varied as indicated, but in any event, the cooler **6** is operated to cool the vaporous products to a temperature of 100°F, and accordingly, the interior of the cyclone **8** will be at a temperature of 98°F.

Approximately 3% of the oil will leave the cyclone as a fine mist by **9**, the remainder or bottoms liquid are passed by **10A** to fluid coker **11** in which the fluid coking operation is conducted according to known manner and conditions. The oil from **10A** is mixed with hot coke in **11** and as a result of the coking operations there are produced a stream of oil vapor coking products, taken overhead by **12** to oil recovery, and coke removed at **13** and passed into coke cooler **14**, then to fluid bed mist collector **15** by **14A**, where oil mist from **9** is adsorbed upon the cook coke. A blower **9A** moves the gas through the retorting, mist collecting and coke cooling sections.

The cool coke containing the oil mist is passed by **16** into fluid coker **11**. There is contained in the oil mist considerable gas and this shale gas is passed by **17** into counter flow with the coke in cooler **14**, cooling the coke while itself is heated and then recycled by **10** to retort **2**.

Excess shale gas is vented at **17A**. Retorted shale is passed by **3** to shale burner **18** from which spent shale is removed at **19**. Hot gases from **18** are passed by **20** to coke heater **21** into which coke from fluid coker **11** is passed by **13A**. Excess coke is removed at **13B**. Hot coke for the coking operation in **11** is passed from **21** by **22**. The products from fluid coker **11** are removed by **12** and treated in conventional manner.

FIGURE 1.8: RECOVERY AND CONVERSION OF SHALE OIL MIST

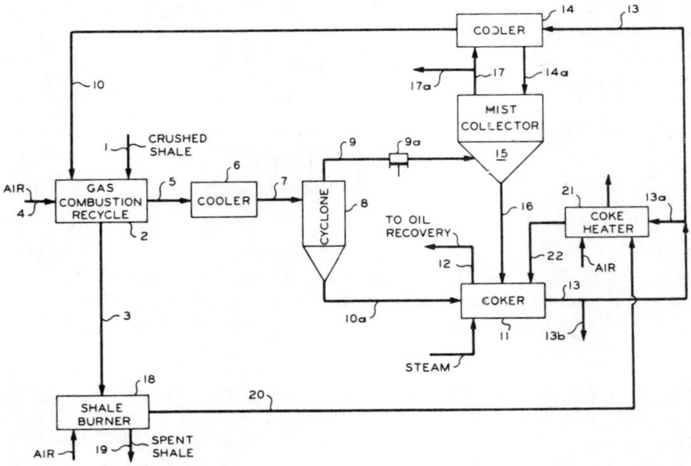

Source: H.W. Parker; U.S. Patent 3,560,367; February 2, 1971

The following tabulation is of conditions, etc., given by way of specific example of an operation in which 50,000 barrels per day of oil is produced.

Example:

	Temperatures (°F)
Retort **2** (maximum)	1200
Pipe **5**	125
Pipe **7**	100
Cyclone **8**	98
Pipe **10**	995
Pipe **14A**	105
Pipe **17**	100
Fluid coker **11**	1000
Pipe **20**	1200

	Pressures (psia)
Retort **2**	12
Pipe **5**	11.5
Fluid coker **11**	14
Fluid bed mist collector	16
Coke cooler	14

	Flow Rates
Crushed shale	80,000 tpd
Recycle gas, pipe **10**	4,400 cfm
Gas containing oil mist pipe **9**	4,400 cfm
Vapors from coking operation on liquid basis, pipe **12**	50,000 bpd

Rotatable Grate

D.A. Vorum; U.S. Patent 3,480,512; November 25, 1969; assigned to Phillips Petroleum

Company describes a retorting vessel which has rotatable grates comprising rotatable jacket tubes which can be provided with means for introducing coolants into the annulus of the jacketed tubes. The retorting vessel is provided with a first rotatable grate means for introducing an oxygen-containing gas into the vessel to support combustion of the hydrocarbons in the oil-bearing solids and a second rotatable grate means for removing the educed hydrocarbons and the products of combustion.

In one form of the process, the second rotatable grate comprises a pipe in the form of a tube positioned horizontally across the vessel and spaced beneath the first rotatable grate means. This pipe functions to collect and remove the hydrocarbon educt from within the vessel.

In another example of the process, the second rotatable grate comprises a horizontal pipe in the form of a tube positioned across the vessel and spaced below the first rotatable grate and a second pipe in the form of a tube positioned above and spaced from the first rotatable grate. The first and second pipes serve to remove the hydrocarbon educt from above and below the zone of combustion.

A third rotatable grate positioned horizontally within and near the bottom of the vessel serves to introduce a coolant into the vessel for the purpose of lowering the temperature of the residue particles. The coolant passes upward in the vessel to the second rotatable grate where it is removed with the hydrocarbon educt. This path of the coolant through the lower portion of the vessel will tend to scrub any remaining hydrocarbon educt from the residue particles, thereby increasing the overall efficiency of the method.

The third rotatable grate means can also be used to pass steam or air through the descending mass of hot residue particles. Since the residue particles will contain a certain quantity of carbonaceous material, a reaction between the steam and the carbon of the carbonaceous material will produce a combustible gas which will add to the total fuel gas produced. This reaction product will travel upward in the retort to the combustion zone.

The rotatable grates of this process can also be used to supply the combustion air in the so-called Bureau of Mines Gas Combustion Retort. In this environment, the rotatable grates provide an even descent of particles through the retort and a grinding up of the clinkers which form. This prevents channeling of the combustion zone thus providing a more even distribution of heat. When the rotatable grates are provided with cooling means, their life is much longer because they are better able to withstand the heat in the gas combustion retort.

Tunnel Oven with Separate Compartments

A process described by *J.D. Bennett; U.S. Patent 3,464,892; September 2, 1969; assigned to Sun Oil Company* employs a long tunnel oven containing a suitable flotation liquid which is separated by movable gas-tight partitions into separate compartments, through which barges, carrying the solid material to be treated and floating on the liquid, are moved step-by-step from one compartment to another.

Inlet and outlet gas-conveying ducts open into the tops and bottoms of the compartments, the gas flowing from each duct to the outlet duct of the same compartment through the material in the respective barge, the barges having foraminous load carriers which permit this gas flow through the material to take place. The flotation liquid is utilized to seal the bottom of the barge to the bottom duct in each respective compartment.

In each compartment a separate treatment step is carried out. A typical process comprises the following steps. In the first step (carried out in the first compartment), the fresh raw shale (after being loaded on the barge) is preheated to a temperature below the retorting temperature, by passing through gases at an elevated temperature derived from a burning compartment. In the second step (carried out in the second compartment), the preheated shale is retorted (distilled) to drive off volatilizable constituents (which are collected) by passing through hot lean gases; these hot lean gases comprise noncondensable gases (derived from the volatilizable constituents) which are recycled through the process and are heated by contact with the hot solids in a heat exchange zone. The retorting of the shale causes the same to become spent, which leaves a carbonaceous residue.

The third process step comprises the burning of the residue on the spent material, by passing through a hot combustion-supporting gas which has been heated by contact with the hot solids in a cooling zone. From this burning operation are derived the elevated temperature gases which are utilized for heating the raw shale in the preheating zone.

The fourth step takes place in a heat exchange zone, where the hot solids derived from the burning compartment give up heat to recycled lean gases passing through this hot material. The fifth and final step utilizes a cooling, where solids at an elevated temperature derived from the heat exchange zone give up heat to a combustion-supporting gas (e.g., fresh air) passing through this solid material. From the cooling compartment, the treated, decoked and relatively cool material is unloaded from the barge and piled on a dump.

The process steps set out above actually each comprise two phases, the action described previously taking place during the first phase. In the second phase, the combustion-supporting gas (e.g., air) passed through the burning compartment (or zone) is replaced by steam, thereby generating hydrogen by a water-gas reaction. This steam is developed in the cooling zone. During this hydrogen-generating second phase, the generated hydrogen is led off to a hydrogen collecting system.

Artificially Induced Condensation Nuclei

M.W. Putman; U.S. Patent 2,813,823; November 19, 1957; assigned to the U.S. Secretary of the Interior has found that by providing artificially induced condensation nuclei in the condensing zone, the size of the fog or mist droplets may be controlled so that none of the droplets are so large that they are disengaged from the gas stream by impingement in passing through the shale bed, nor none so small that they are difficult to recover from the relatively cool gas stream by conventional methods, irrespective of the gas cooling rate in that portion of the shale bed where vapor condensation is initiated. The formation of a fog or mist is accomplished at low degrees of supersaturation only in the presence of a large number of nuclei on which the saturated vapors may condense.

Sodium chloride is particularly useful as the condensation nuclei source in view of its low cost and ready availability. Other salts which have been found to be effective are potassium chloride and calcium chloride. In general, any substance which has an appreciable vapor pressure and is thermally stable at the retort temperature may be employed to supply the artificially induced condensation nuclei. Artificially induced condensation nuclei may be provided in the retort gas in several ways. In one method sodium chloride crystals are vaporized in the combustion zone of the retort; and in the second method sodium chloride crystals are vaporized outside the retort and injected into the recycle gases.

The first method is more convenient but harder to control, while the second method is more difficult to carry out but the results are more reproducible. The process involves continuously passing the oil shale, to which has been added a small amount of sodium chloride in water solution, downward as a bed of broken solids in a substantially vertical column. The solid residue is removed in a cool condition at the bottom of the column while the distillation and combustion products, including a noncondensable gas, are removed from the top of the column.

At least a portion of this noncondensable gas, in a cool condition, is recycled to the bottom of the column and passes upward through the downward moving residue, thus cooling the hot residue and itself becoming heated. At this point in the retort the gas stream is raised to a still higher temperature, preferably by passing through a combustion zone in the retort itself. As the gas stream passes on up through the column of shale, it delivers its heat to the cold incoming shale, thus gradually heating it to progressively higher temperatures.

The descending shale, consequently, passes successively through a preheating zone, where it meets the still hot vapor-gas mixture rising from the distillation zone, through a distillation zone where it reaches retorting temperature, and through a sublimation zone where it reaches a still higher temperature which is sufficient to cause a portion of the sodium chloride, which has been evaporated to a thin film on the shale particles in the preheating zone above, to sublime.

The sodium chloride vapors so formed, are swept upward by the gas stream, and as the gas stream is cooled by the descending cooler shale, a sodium chloride fume is formed. This fume consists of minute solid sodium chloride particles which are suitable condensation nuclei for the oil vapors to condense upon. In the distillation zone the organic content of the shale undergoes thermal decomposition producing condensable product vapors which are carried upward in the gas stream.

The gas-vapor mixture rising from the distillation zone encounters progressively cooler shale as it passes upward through the shale preheating zone, and of course, in this way, itself becomes progressively cooled. Eventually the gas-vapor mixture encounters shale below the initial dew point temperature of the mixture and condensation of the vapor begins. By withdrawing the gas stream from the top of the shale bed at a sufficiently low temperature, viz, between 100° and 200°F and preferably between 115° and 175°F, substantially the entire vapor content of the gas stream undergoes condensation in the shale bed on the sodium chloride nuclei in the gas stream.

In related work *M.W. Putman; U.S. Patent 2,901,402; August 25, 1959; assigned to the U.S. Secretary of the Interior* has found that the oxides of aluminum, tin and titanium are very effective condensation nuclei. The aluminum oxide nuclei may be prepared by vaporizing aluminum chloride, and then introducing the vapors into the recycle gas stream where the aluminum chloride reacts with a portion of the water vapor to form aluminum oxide nuclei.

Addition of Coke

In a process described by *C.E. Hemminger and R.I. Bergman; U.S. Patent 3,162,583; December 22, 1964; assigned to Esso Research and Engineering Company* coal and/or coke is added to the oil shale to be distilled. The coke is preferably coke made by subjecting bottoms from the distillation of shale oil to fluid coking as this coke is generally produced at the shale mines where it has no value other than as fuel and is burned up during retorting to supply part of the heat of distillation or retorting.

When using fluid coke and coal in a mixture with oil shale in distilling and retorting, higher yields of oil and better operability are obtained than when using a mixture of coal and shale alone. Thus, using the underfed type of retort, the efficiency of the retort is increased by adding coal such as bituminous coal, or coal and fluid coke to the shale being fed to the retort. In this type of retort the shale temperature increases from atmospheric to a maximum temperature of 2000° to 2400°F and then to 100° to 300°F as it passes down the retort with the result that in the high temperature zone a high degree of carbonate decomposition of the inorganic portion of the shale is produced.

With this type of retort a large amount of heat is required and if there is insufficient heat from the spent shale, some of the retorted oil is burned and this is definitely a loss in the process. By feeding coal with the oil shale not only is the liquid hydrocarbon from the coal made available but also the recovery of oil from the shale itself is increased from 75 to 95 to 105% of the Fischer assay of the shale.

Since the heat load in this type of retort is high and practically all of the carbonates are decomposed there is not sufficient residual carbon in the shale to provide the necessary heat for the shale retorting and carbonate decomposition. As a result in the countercurrent flow of air in the retort, air for which there is not sufficient carbon for combustion passes down through the shale bed and consumes a portion of the retorted hydrocarbon product.

By adding coal or coke to the shale feed, the coke from the coal distillation and the added coke are burned and this provides the necessary carbon to burn this excess air and oxygen-free hot gas as it enters the distillation or retorting zone at a lower level in the retort. Consequently there is a substantially complete recovery of oil from the oil shale as well as that from the coal. The fluid coke supplies heat by being burned.

The fluid coke will not fuse like coal being distilled and therefore better operability of the retort is obtained. Because of the fusing property of coals, which causes uneven distribution of the air by blocking air passages, the amount of coal is limited to about 25% by weight of the shale. Coke does not have any limitation, but it in itself, give little or no liquid hydrocarbon.

However, more coal can be distilled if coke is added with the shale; the 25% maximum content of coal can be increased to 40% on shale if 30% of coke on shale is added to the feed mixture of the retort. The amount of coal which is added should be at least sufficient to have 1 to 5% carbon in the ash or spent shale leaving the top of the retort. This is the minimum amount of coal and the amount will depend upon the richness of the shale, more coal being used for a 20 gal/ton than a 40 gal/ton shale.

For a 30 gal/ton shale, 25 weight percent coal is utilized in the coal-shale mixture. If too much coal is added with the mixture, channelling results in the retort due to the coking and fusion of the coke particles. When channelling occurs, there is inefficient contacting of the solids and the gases. The maximum amount of coal to be used is on the order of a 50-50 weight mixture of coal and shale.

Retorting and Coking in Single Vessel

F.E. Campion; U.S. Patent 3,112,255; November 26, 1963; assigned to Pan American Petroleum Corporation describes a process for the retorting of hydrocarbon yielding materials such as oil shale under conditions such that the retorting and coking operations are both effected within the same vessel or unit. In such an operation, oil shale and an oxidizing gas such as air, are concurrently introduced into a suitable retort. A zone of combustion is established within the retort at a level determined by the rates of injection of shale and air or oxygen and the temperature of the shale and oxidizing gas.

Oil produced from the shale passes through the combustion zone where it is thermally cracked into lower weight molecular hydrocarbons. In this manner the coking process, ordinarily effected in a separate unit, is accomplished in the retort thus eliminating the cost of constructing and operating individual coking units. The oil thus produced is of suitable viscosity for pipeline transmission or, if desired, may be further refined at the retorting site.

Externally Heated Flue Gas for Temperature Control

In a process described by *K.L. Berry; U.S. Patent 3,464,913; September 2, 1969; assigned to Pan American Petroleum Corporation* a portion of the heat for retorting oil shale is supplied by injecting externally generated hot flue gas at 800° to 1200°F; the remainder of the retorting heat is supplied by combustion in the spent shale bed. The purpose is to minimize overheating the shale, which causes carbonate decomposition and excessive cracking. In the proposed method, a portion of the shale is retorted before heat from the combustion zone reaches it.

The process is illustrated by reference to Figure 1.9, in which a vertical retort **2** is filled with crushed raw oil shale continuously introduced through hopper **4**. The process is initiated by first introducing a hot gaseous stream at a temperature from 800° to 1100°F into the retort via distributor **5** under conditions such that the major portion of the kerogen and/or hydrocarbons is removed from the shale particles, leaving primarily a carbon residue thereon. This hot gaseous stream may, if desired, be derived by burning a portion of the fuel gas recovered from the retorting operation, as will be explained in detail below.

The spent shale containing this residue is then passed into the combustion zone adjacent spray burner **6** where combustion of the residue in a deficiency of oxygen is accomplished by the introduction of air into the system via line **8**. The heat thus generated in large part passes up the bed and assists in the liberation of additional valuable products from the partially retorted crushed raw shale that has passed on downwardly from distributor **5**.

Alternately, the retorting process may be initiated by supplying an air-gas mixture to operate burner **6** via lines **8** and **10** respectively, and igniting the mixture to effect combustion of the oil shale in the vicinity of the burner. Once combustion is established, valve **12** is closed and the burning operation is sustained by the continued introduction of air through line **8**. Retorting in the upper portion of the bed may then proceed as described above.

After the process has lined out, the burner is likewise preferably operated at a temperature in a range of from 800° to 1100°F. At this temperature level, the retorting step is substantially completed with only a portion of the residual carbon on the shale approaching burner **6** generally being consumed. Thus, after the initial retorting step occurring in the vicinity of distributor **5**, the carbon residue on the shale amounts to about 3%.

At the burner from about 5 to 25% of the residual carbon is burned by combustion with air introduced through line **8**. Some of the heat is also generated by combustion in the zone of the cooled recycled gas introduced into the retort at distributor **34**. At all times, however, a deficiency of air is maintained in the system so that the temperature both at the burner and distributor **5** does not appreciably exceed 1100°F.

FIGURE 1.9: RETORT PROCESS EMPLOYING HOT FLUE GAS

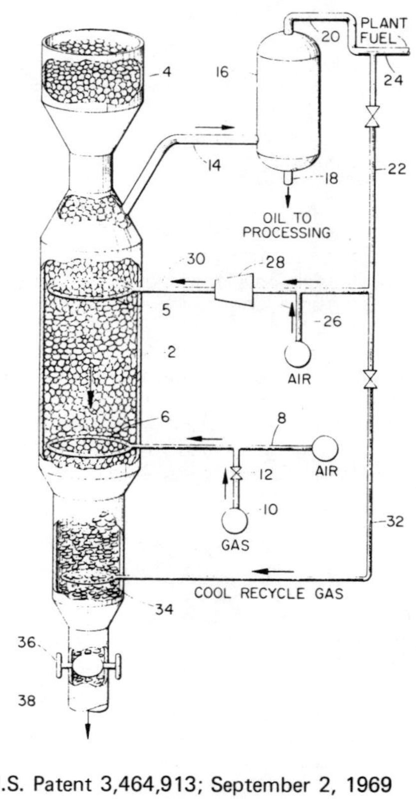

Source: K.L. Berry; U.S. Patent 3,464,913; September 2, 1969

Shale oil vapors rise through the descending bed of shale particles, preheating the latter, and at the same time lowering the temperature of the rising vapors and combustion products. At the top of retort **2** shale oil mist and products of combustion are removed by means of line **14**. This product stream is sent through separator **16** where the liquid portion is withdrawn through line **18** and sent to further refining while the uncondensed fraction is taken through line **20** to in-line burner **28**.

A portion of this gaseous stream is diverted through line **22** and the rest taken from the system via line **24**. The gaseous fraction in line **22** contains light hydrocarbons, hydrogen, carbon monoxide, and products of combustion and is mixed with air from line **26** after which the resulting mixture is fed to in-line burner **28**. Sufficient air is present in the mixture to form a stream of combustion products in line **30**, having a temperature of from 800° to 1100°F.

These hot gases then flow into the bed of raw shale via suitably shaped distributor **5** which is preferably spaced from about 2 to 5 ft above the burner **6**. At this level in the downward moving bed, the oil shale particles are heated to a temperature within the last mentioned range, resulting in the release of a substantial portion of the volatile carbon compounds from the particles. At these temperatures, however, essentially none of the carbonates or other inorganic materials present decompose. Cool fuel gas is returned to the system via line **32** and distributor **34** to take up heat from spent shale being removed from the retort through valve **36** and outlet **38**.

Once the supply of fuel gas is generated as described above, the amount of air supplied to the system through line **8** can be reduced to the volume just necessary to maintain the temperature at the burner in the range of 800° to 1100°F. The fuel required to produce temperatures in this range is derived partly from residual carbon on the spent shale and partly from the recycled flue gas from distributor **34**. Ordinarily it is considered that about 20,000 scf of gas, i.e., fuel gas plus air, is required to retort one ton of oil shale.

Under retorting conditions the shale moves through the retort at a rate of from 3 to 10 feet per hour. This volume of introduced gas may be somewhat less than 20,000 scf/ton because the heat in the combustion products resulting from burning of the residual carbon on the spent shale supplied an appreciable amount of heat to the downcoming raw shale. An additional heat source is the spent solids coming in contact with cool fuel gas introduced via line **32**.

In a typical case, the 20,000 scf of gas and air may be proportioned as follows: 1,000 scf of fuel gas in line **22** and 1,000 scf of air through line **26**, 1,000 scf of air via line **8**, and 17,000 scf of fuel gas through line **32**. In carrying out the process from 1,500 to 2,500 scf of air per ton of shale processed is used in order to maintain the temperature within the above range. In most cases, and as indicated above, the total volume of air employed is divided evenly between the supply going to line **30** and that introduced into the system via line **8**. This constitutes a substantial savings in air pumping costs over previous methods which require from 5,000 to 6,000 scf per ton of shale.

Hydrogen Produced as By-Product

P.H. Gifford II; U.S. Patent 3,577,338; May 4, 1971 describes a process for retorting oil shale and simultaneously producing hydrogen as a by-product with high yields of both oil and hydrogen. The process comprises continuously passing oil shale through a combustion zone in a retort while continuously introducing steam and oxygen into the combustion zone, removing spent shale below the combustion zone and recovering oil and hydrogen from the retort off gas from the combustion zone. Hydrogen is produced by the reaction of carbon-containing material with steam.

Example 1: A reactor was filled with about 1.6 lb of raw shale analyzing about 30 gal of oil per ton of rock (Fischer assay). Superheated steam at a temperature of 1000°F was fed into the reactor at a rate of 1 scfm to preheat the shale; this was also the temperature and rate during the run. After preheating, oxygen was introduced at a rate of 0.105 scfm until oxygen breakthrough occurred. The reactor was then allowed to cool, then was opened, and the spent shale inspected. During the run the off gas was tested every 5 to 10 min for the presence of hydrogen. Hydrogen appeared in every run immediately upon the introduction of oxygen.

The off gas burned with a violent, almost invisible, blue flame, as opposed to a yellow-orange smoky flame which was obtained during the preheat or from off gases from the prior art processes. The recovery of oil from the shale was from 80 to 90% FA (Fischer assay), that is, 80 to 90% of the oil present in the shale was recovered.

The examination of the spent shale revealed that it was very light in appearance; this is directly opposite from what was found in conventional oil shale processes in which the spent shale is almost black. It was evident that a carbon consuming reaction, after completely removing all external carbon, had penetrated by diffusion deep into the particles. The spent shale was completely retorted, and there was no evidence of sintering.

It was apparent that hydrogen had been produced in significant amounts by the reaction of carbon with the steam which was introduced. As all of the carbon on the outside of the particles and most of the carbon inside the particles had been consumed and since there was no sintering of the particles, it was apparent that clinkering could not have occurred.

There was no binder left to cohere the particles together to form the necessary bridge. The carbonate decomposition seemed to be about the same as experienced in the prior art process of gas combustion. This was determined by a comparison of the structural strength of the spent shale produced by either process.

Example 2: The purpose of this run was to determine the effectiveness of the process on rich shale. With all prior art processes it is difficult to process oil shale having a Fischer assay above 30 gal/ton, and none of them can process shale of a Fischer assay above 40 gallons per ton.

About 1.6 lb of raw shale having a Fischer assay of about 50 gal/ton of rock was introduced into the reactor. Steam and oxygen were introduced into the reactor in accordance with the procedure used in Example 1. The oxygen was fed until oxygen breakthrough occurred; the reactor was allowed to cool, and when opened, the spent shale was observed again to be very light in color.

The shale was completely retorted; there was no evidence of sintering, and all of the carbon had been chemically removed; in other words, the spent shale was very flowable. The tests for hydrogen indicated that about twice as much of this gas was produced as in Example 1, and it had a high purity. The recovery of oil from the shale was approximately 80 to 90% Fischer assay.

Examination of the spent shale showed that all of the carbon, external and internal, had been reacted, and there was no evidence of sintering of the particles even though, in this example, the reactor reached a higher temperature than normal. Most of the spent shale particles were friable as would be expected with rich spent shale because of the lower percentage of carbonates present in rich shale. Again, since all of the external and internal carbon on the particles had been completely reacted, and there was no evidence of sintering, it would be impossible for them to cohere, to bridge across large sections, and ultimately to clinker.

Inert Gas Used in Place of Hydrogen

In a process described by *J.R. Hopper, E.L. Wilson, Jr. and N.P. Peet; U.S. Patent 3,573,194; March 30, 1971; assigned to Esso Research and Engineering Company* shale oil of improved quality is recovered from oil shale retorted in a plural stage retorting system at an elevated temperature in the presence of an inert gas under an elevated pressure. The pressure on the retorting zone is maintained by injection of an inert gas which is preferably nitrogen, but may be a mixture of inert gases containing nitrogen and carbon dioxide. Such mixtures may be obtained by burning of residual carbonaceous material remaining on the oil shale after retorting and provide heat for the retorting operation.

The pressure on the retorting zone is suitably in excess of 250 psig and may be within the range of 250 to 2,500 psig. Below 250 psig the effect of pressure with an inert gas is not significant. Above 2,500 psig pressure, the added benefits accruing to higher pressures than 2,500 are marginal and are offset by the added operating problems and costs required. Therefore, the range of 250 to 2,500 psig is considered to be necessary and important in the process.

The pressure in the other zones of the retorting system may be substantially below that of the retorting zone and allows use of less expensive equipment. It is not necessary to use high pressure except in the retorting zone because pressure has no effect except in the retorting zone. Pressures in the other zones may range from 0 to 100 psig.

Continuous Fluidized Process

A process described by *G.E. Parr; U.S. Patent 3,663,421; May 16, 1972; assigned to Sun Oil Company* comprises the steps of feeding hydrocarbonaceous solids of an economically obtained particle size into a thermal treatment zone where the solids are maintained as a fluidized bed on shallow-bed concatenated, slanting trays. The solids are contacted with the hot fluidizing gas whereby volatile constituents from the solids are formed and these gaseous condensable hydrocarbon products are withdrawn from the thermal treating zone by a plurality of sidestream outlets where any solids present are separated and the product vapors condensed into liquid hydrocarbons.

The spent solids from the thermal treatment zone are passed through a seal of injected gas at the bottom of the thermal treatment zone into a spent solids combustion zone which is contiguous to the thermal treatment zone. The gas seal prevents passage of gas from the thermal treatment zone to the solids combustion zone or from the solids combustion zone into the thermal treatment zone, but at the same time the seal allows passage of the spent solids into the combustion zone.

The process further comprises burning the spent solids to produce hot combustion gas which is passed in noncontaminating heat exchange contact with the cooled fluidizing gas taken from the thermal treatment zone, and recycling the reheated gas into the thermal treatment zone as the hot fluidizing gas.

The particle size of the hydrocarbonaceous solids used in the process is readily and economically obtained. The hydrocarbonaceous solids need only be reduced in size to 0.25". Because of the poor heat conductivity of the material, particles which are much above 0.25" in size will result in a less efficient process where yields of hydrocarbon products are low. On the other hand, any particle size below one-quarter inch is operable, but there is no need to expend energy and money to obtain much smaller particles.

An important requirement for the process to operate efficiently is that the concatenated, slanting trays be shallow and that the fluidized solids on these trays be no greater in depth than about 12", preferably about 2" to 8". As the solids enter at the top of the thermal treatment retort zone they are maintained as fluidized solids by hot gases introduced into the retort at various levels as shown in Figure 1.10 to be described.

These fluidized solids, by virtue of the fact that they will act as a liquid and because of the slanting concatenated trays, will flow from top to bottom of the retort zone. As they flow downwardly, the volatile products generated by the elevated temperature within the retort, and which will be between about 50° and 500°C, will be removed by an overhead and sidestream outlets, equipped with separators for any solids present, and the volatile hydrocarbons will be condensed and subjected to further processing as is well known in the art. The apparatus of the above described process comprises (a) a retort vessel for the thermal treatment of hydrocarbonaceous solids to distill volatile constituents from the solids to produce spent solids and gaseous products; (b) a spent solids combustion vessel in contiguous relationship to the retort vessel for the burning of the spent solids to thereby produce hot gas; and (c) a gas injection inlet between the combustion vessel and the retort vessel which inlet serves to feed a gas to form a gas seal between the vessels to allow the passage of spent solids from one zone to the other while prohibiting the passage of gases between the zones.

Referring to Figure 1.10, the raw carbonaceous solids such as oil shale, coal, etc., are charged (as shown in the figure) from hopper **1** through flow control **2** and line **3** into the top tray **5** of vessel **4**. The flow control **2** can comprise any control means such as, for example, a star valve. Vessel **4** which contains the thermal treatment zone **26** of the process contains a plurality of trays slightly inclined to promote cross flow of the solids. The tray system is provided with dams **6** and downcomers **7** which control the bed height of fluidized solids and the tray-to-tray sealing. The trays can be perforated, slotted or can be provided with gas inlet heads.

FIGURE 1.10: CONTINUOUS, FLUIDIZED RETORTING PROCESS

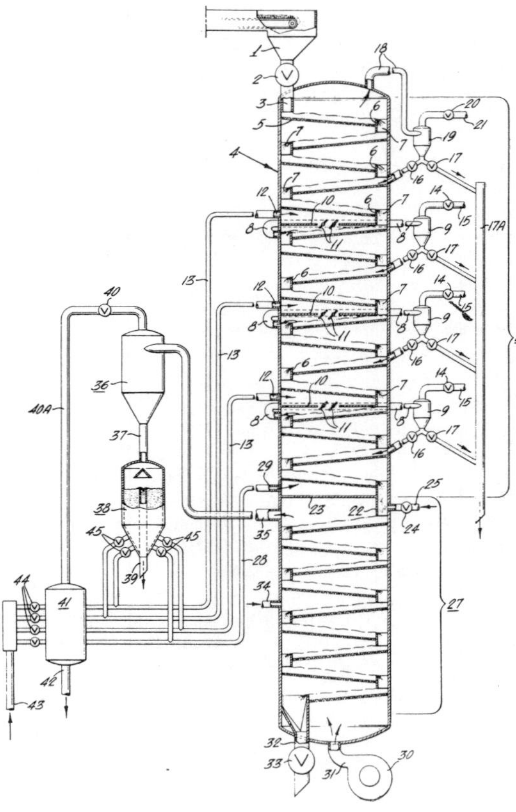

Source: G.E. Parr; U.S. Patent 3,663,421; May 16, 1972

The crushed solids, fluidized by upflowing hot gases as described below, flow across each tray and downwardly through vessel **4** from tray to tray. As the solids pass downwardly, heat transfer between the solids and upflowing hot gases, containing controlled amounts of hot solids fines, takes place in a combination of countercurrent and cross flow heat exchange. Since, in fluidized systems, temperature equalization occurs essentially within a few inches of depth, bed depths of one or two inches to about a foot are adequate for efficient heat exchange. This combination of cross flow of solids and low bed depths affords pressure drop advantages, particularly under high throughput conditions.

As the solids move downwardly in vessel **4**, they are progressively heated to temperatures at which pyrolytic action causes the hydrocarbonaceous materials to evolve as gases and vapors. Small fines resulting from decrepitation of the crushed solids are elutriated by the upflowing gases. This eliminates any adverse effects their accumulation might produce in the fluidized system.

At preselected locations in the vessel, determined by the processing requirements, sidestream outlets **8** connected to the vessel and to appropriate solids removal equipment **9** (e.g., single or multistage cyclones, mechanical separators, precipitators and/or others known in the art) permit the removal of hydrocarbonaceous cuts and undesirable small fines. Three such sidestream outlets are illustrated, however, more or fewer outlets can be used according to the needs of the process and/or the type of operation desired. Also, more than one outlet connection from the vessel to solids separator **9** can be made at alternate locations through appropriate manifolding and valving. Such details are obvious to those practiced in the art.

The sidestream cuts afforded by this system can be preheated individually before refining by more efficient and economical procedures than can a single product from other thermal recovery systems. Furthermore, the concomitant removal of small fines through the sidestream outlets prevents their adverse accumulation in the upper solids beds in the vessel. Located above each sidestream outlet is a disphragm partition **10** which, except for controllable opening **11** in the diaphragm, effectively seals off the vessel at that location from upflowing gases.

Above each diaphragm partition and above diaphragm partition **23**, hot inlets **12** and **29** respectively, connected respectively to external lines **13** and line **28** provide entry for hot gases, at controlled velocities and containing controlled amounts of hot solids, to the vessel to heat and/or react with the downflowing solids on the upper trays. The amount of sidestream material removed through outlets **8** can be varied and controlled by suitable adjustment of control valves **14** in outlet line **15**.

In turn, proper adjustment of variable openings **11** (e.g., damper, butterfly, covered slots, variable orifices or others known in the art) in diaphragm partitions **10** maintain the desired pressure gradient in the vessel. Thus, the vessel is divided into zones in each of which the velocity, temperature and hot solids content of the fluidizing gases can be set and controlled to independently provide various sidestream cuts. This feature allows an adjustment of operating conditions within each zone for optimum processing.

The solids removed in separators **9** are either returned to the vessel through control valves **16** into a higher temperature zone for further removal of any remaining hydrocarbonaceous material or a portion of the solids or all the solids can be passed through control valves **17** into fines manifold **17A** and thence to a waste heat exchanger before ultimate disposal.

Overhead gases and solid fines plus any light vapors evolved in preheating the crushed solids leave the vessel through line **18** into solids separator **19** which can be arranged similarly to the sidestream system described above. The essentially solids-free gases and vapors leave separator **19** through control valve **20** and line **21** to the downstream processing units. Operating conditions in the vessel are maintained so that all recoverable hydrocarbonaceous material is removed before the downflowing solids reach seal leg **22**. The solids after passing through the seal leg enter the fluidized combustion and cooling zone **27** of the vessel. This zone **27** is mechanically sealed from the upper sections by diaphragm partition **23** and vapor sealed by seal gas (e.g., flue gas) entering seal leg **22** through control valve **24** and line **25**. The solids in the lower section of the thermal treatment zone **26** above diaphragm partition **23** are fluidized on trays as shown and heated by hot gases and solid fines entering the vessel through line **28** which is connected to inlet **29**.

Ambient air mixed, if necessary, with enough cool flue gas to fluidize the downwardly flowing solids enters the combustion portion of the vessel through compressor **30** and line **31**. The air cools the solids by a combination of countercurrent and cross flow heat exchange. The cooled solids, in aggregate form, leave the vessel through line **32** and flow control **33** to spent aggregate disposal. The upflowing air, with or without flue gas as required, is heated by the downwardly flowing hot solids to temperatures at which combustion of any remaining carbonaceous material in the solids is effected.

The rate of combustion is controlled by introducing regulated amounts of partially cooled flue gas into the vessel through line **34**. Hot gases comprising air, introduced flue gas, and products of combustion plus entrained solids fines leave the vessel through line **35** and enter solids separator **36**. This separator can comprise suitable cyclones, mechanical separators, precipitators or any other separating equipment well known in the art.

The separated hot solid fines flow through line **37** into hopper **38**. The hopper is equipped with a requisite number of outlets with valves **45** to supply hot solids to fluidizing gas in lines **13** and **28**. Any excess hot solid fines leave the hopper through overflow line **39** and proceed to a waste heat exchanger before final spent fines disposal. Hot gases leave separator **36** through control valve **40** and line **40A** and pass through gas-to-gas heat exchanger **41** and then through line **42** to a waste heat exchanger and exhaust stack.

Fluidizing gas, part of which may be noncondensables from the product gases leaving the vessel, enter the system through line **43**, control valves **44**, and heat exchanger **41**. These fluidizing gas streams, partly or completely heated in exchanger **41**, pick up and carry into the vessel controlled quantities, if any, of hot solid fines, as desired or required, which fines are introduced into lines **13** and **28** through control valves **45** as mentioned above.

The process and apparatus described above provides an extremely efficient means to release valuable hydrocarbon products from hydrocarbonaceous materials and enables at least about 90% yield (Fisher assay) of organic products in the hydrocarbonaceous material to be recovered.

Catalytic Oxidation

K.L. Berry; U.S. Patent 3,487,004; December 30, 1969; assigned to Pan American Petroleum Corporation describes a retort for the recovery of oil from oil shale, in which hot gases for the retorting are prepared by controlled oxidation of recycled product gases. The oxidation is confined to burners within the retort by the placing of an oxidation catalyst within the burners and the limiting of air introduced into the burners. Retorting temperature is thus maintained between 800° and 1300°F.

Since the heat requirements are reduced, air requirements are less and there is a resulting decrease in nitrogen and carbon dioxide in product gases. Methane and ethane and lower molecular weight oil vapors production are also less due to the lower temperatures. Decomposition of inorganic carbonates is essentially eliminated by operating at temperatures below 1300°F.

Consequently, carbon dioxide production is further reduced. In addition, the heat of decomposition of inorganic carbonates need not be added, so even less air is required. The overall result is that organic materials are present in much higher concentrations in product gases leaving the retort. These higher concentrations, in turn, facilitate recovery of a larger percentage of the kerogen decomposition products as a liquid product. A smaller percentage of kerogen decomposition products appears in the vent gas from the separator.

Heat Recovery

A process described by *L.P. Evans; U.S. Patent 2,885,338; May 5, 1969; assigned to Socony Mobil Oil Company Inc.* is concerned with an improved apparatus and process where a hydrocarbonaceous solid is retorted economically and efficiently, with a substantially complete recovery of heat and with internal fractionation of the liquid and gaseous products.

Referring to Figure 1.11a, the basic apparatus of this process is provided. In general, the apparatus comprises an elongated cylindrical vessel **2**, provided with a seal leg **3**, a conical shaped internal cover disposed at the lower extremity of the seal leg and having a plurality of orifices **10**. In the upper, or condensing and heat exchange zone of the reactor proper, there is provided a trap tray **4** disposed around the inner wall of the reactor.

The trap tray is provided with a pipe **7** to permit draw-off of liquid products. The design of the trap tray provides an annular opening for constricted passage **35**. In the lower region of the reactor **2** there is provided a gas distributing device **23** and a series of baffle plates **26** and **27** having orifices **28** and **29**. Below the baffles there is an air inlet **30** and a draw-off tube **24**.

In operation, raw crushed hydrocarbonaceous material, such as oil shale, is carried to the top of the reactor by means of a suitable lift device **1**, e.g., a bucket elevator. The crushed shale thus fed into the top of the reactor gravitates downward in a compact mass through the seal leg and into the main portion of the reactor. Inert gases such as flue gas are introduced into the fed leg via pipe **9** to provide a seal which prevents the escape of gaseous products from the reactor.

FIGURE 1.11: FLUIDIZED BED RETORT PROCESS

a.

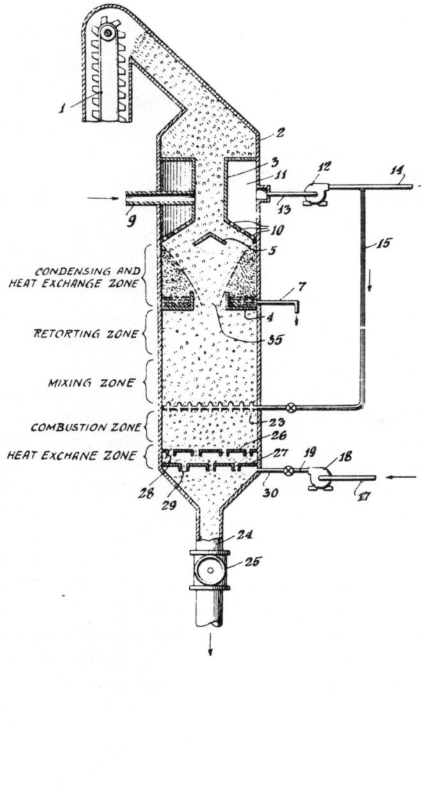

(continued)

FIGURE 1.11: (continued)

b.

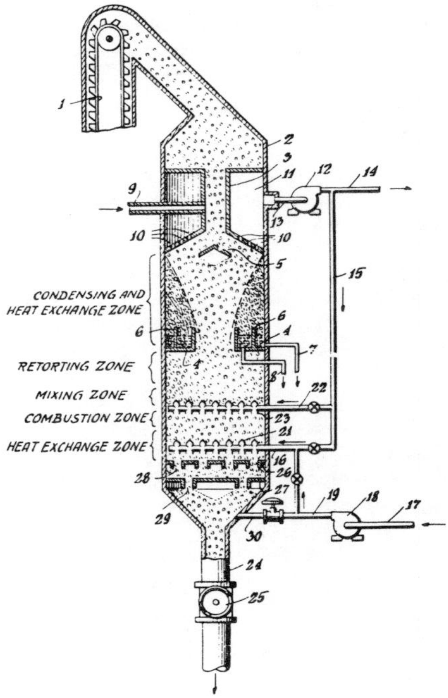

Source: L.P. Evans; U.S. Patent 2,885,338; May 5, 1969

As the cold shale gravitates through the condensing and heat exchange zone, it is contacted with the hot condensible and noncondensible vaporous products of the process. Upon contact with the hot vaporous products, the cold shale is heated and the condensible products are liquefied and gravitate downward through the bed of shale. Due to the restricted opening **35** a portion of the bed of shale, immediately above the trap tray **4** becomes substantially immobile.

The condensed liquid products gradually trickle through the zone of immobile shale and eventually reach the trap tray. The noncondensible hot gaseous products of the process are passed through openings **10** into an annular collecting zone **11**. These gases are then withdrawn via pipe **13** through pump **12**. A portion of the hot gases is sent to a gas recovery plant through pipe **14**. The remainder of the hot gases are recycled through pipe **15** into the gas distributing device **23**. By means of this operation, the heat requirements of the shale retorting are satisfied.

The shale gravitating downward passes through a combustion zone and then through the openings **28** and **29** in the baffles **26** and **27**. Below the baffles, a combustion-supporting gas, such as air or oxygen, is introduced into pipe **30**, pressured by means of pump **18**. As the combustion-supporting gas passes upward through the baffled region, it is heat exchanged with the hot spent shale so that the gas attains a temperature sufficient to support combustion. Thereafter, the spent shale is withdrawn through conduit **24**. In order to provide a more efficient distribution of the raw shale in the upper portion of the reactor, a conically shaped baffle **5** is positioned immediately below the opening of the seal leg **3**. Figure 1.11b represents another form of the apparatus in which the raw crushed shale is introduced into the upper portion of the reactor where it is heat exchanged with the noncondensible and condensible vaporous products of the reaction. In this example, however, the trap tray **4** is provided with one or more baffles **6** which divide the trap trays into two or more sections. Each section is provided with a separate draw-off tube **7** and **8**. By means of this arrangement, a rough fractionation of the liquid product is achieved in the reactor itself. Accordingly, the amount of fractionation required externally is greatly diminished.

Another improvement shown in Figure 1.11b relates to the utilization of the recycled gas via pipe **15**. The stream of recycle gas is split into two streams; one stream passes through pipe **22** into gas distributing device **23**. The remainder of the recycle gas passes through pipe **16**, where it is mixed with a portion of the combustion-supporting gas supplied via pipe **20**. The resulting mixture of combustion-supporting gas and recycle gas is then passed through a gas distributing device **21** into the bed of shale.

As was described above, the remainder of the combustion-supporting gas is passed into the lower portion of the reactor via pipe **30**. The remaining operations involved are substantially the same as those described in conjunction with the discussion of Figure 1.11a. The oxygen-containing gases absorbed heat from the spent shale as they passed through. The temperature of the gaseous mixture is thus raised to the ignition point at which the oxygen reacts with the combustibles present in the system, such as the combustible components of the gaseous mixture and the coke contained in the spent shale.

The combustion thus initiated generates the required amount of heat to maintain the entire process in thermal balance. This is an important feature of the process which serves to minimize heat loss. Since the heat recovery from the incoming and outgoing process is substantially complete, only a small amount of net combustion heat must be supplied to make up heat loss and heat reaction.

The temperature required to achieve combustion is ordinarily within the range between 1000° and 1500°F. Accordingly, in order to utilize more effectively the heat evolving from the combustion zone, it is preferred to split the recycle gas stream into two or more streams as was illustrated in Figure 1.11b.

Generally, the amount of recycle gas conducted to the retorting zone is between 1 and 3 unit weights per unit weight of shale charge. The temperature of the gas issuing from the combustion zone is adjusted by means of the flow of residual cold recycle gas lead to the pipe **22** and gas distributor **23**. Contact between the residual recycle gas and the gas from the combustion zone takes place in the mixing zone of the reaction vessel. The resulting gaseous mixture rises to the retorting zone.

In general, the temperature of the gas entering the retorting zone varies between 800° and 1000°F. The temperatures in excess of the foregoing range are usually to be avoided in the retorting zone, because excessively high temperatures promote undesired secondary distillation and/or cracking of the shale oil released in the retorting zone. In a typical operation, the volume and temperature of the gas conducted to the retorting zone varies between 120 and 160 lb of gas per 100 lb of shale charge at a temperature of 1000°F.

A number of other processes are described which involve increasing the efficiency of the retorting process. For example, *J.C. Todd; U.S. Patent 3,617,466; November 2, 1971; assigned to Atlantic Richfield Company* describes a method for obtaining oil from oil shale by the gas combustion retort process. The off-gas from the retort containing oil, fuel and other combustion products is treated first to remove the oil and second, to burn the fuel components. The resulting gas is recycled into the retort to pass countercurrently to the shale travel. Temperatures in the combustion zone of the retort are moderated and clinker formation is eliminated. The heat from the fuel burning can be used as a power source.

A.L. Saxton; U.S. Patent 3,526,586; September 1, 1970 describes the use of a single vessel oil shale retorting system which comprises a preheating zone, a retorting zone, a combustion zone, and a cooling zone. The crushed oil shale is introduced into the preheating zone and flows serially through the zones and is subjected to preheating, retorting, combustion and cooling conditions as it flows through the zones. During the flow of the oil shale through the zones and subjecting of the oil shale to the several conditions, the zones are maintained under superatmospheric pressure within the range of 5 to 100 psig and thereafter the spent shale is discharged from the cooling zone whereby increased capacity of the system is obtained.

J.H.V. Finney; U.S. Patent 3,006,816; October 31, 1961; assigned to Union Oil Company of California describes a retort which includes essentially three zones, that is, a combustion zone, a distillation or retort zone, and a preheating zone for combustion air which also cools the ash from the combustion zone. The retort includes the desirable feature of having an upward flow of shale and downward flow of gas and oil in the retort zone, but in the burning and ash cooling zone there is the desirable downward flow of shale and ash and an upward flow of air.

The retort includes a lower combustion zone, and an upper retort zone subject to the gases of combustion from the lower combustion zone to perform the pyrolysis and the distillation of shale oil. The retort includes a simplified design where the interior walls of the retort may be made of fire brick in those parts which are subjected to high heat, and which may be constructed so that the depth of the beds of shale in the previous parts are not controlled by the size of the device but may be determined by the optimum operation of the retort. The simplified design provides means for distributing the weight of contained shale so that it is easily supported, and so that there is no substantial increase of friction of the shale on the supporting walls of the vessel or an increase of size of the retort.

In a process described by *D.W. Peacock; U.S. Patent 3,476,650; November 4, 1969; assigned to Phillips Petroleum Company* oil shale is combusted under a partial pressure of oxygen of 15 to 105 psia at 500° to 700°F thereby producing a valuable metallurgical coke suitable for smelting iron, a heavy hydrocarbon fraction and a gaseous fraction.

A process described by *R.W. Whiteacre 3rd; U.S. Patent 3,546,092; December 8, 1970; assigned to Koppers Company, Inc.* comprises treating shale oil ore particles that have been retorted in a first stage to recover residual energy, by comminuting the particles while hot; collecting the comminuted particles in a receptacle; flowing a gaseous fluid through the comminuted particles; and removing the residual energy in the form of product gas and oil from the receptacle. Thereafter, the product gas and oil flow through a separator to remove ore fines, and the product gas and oil may be conducted into a product removal conduit, or burned at a fuel distributor in order to retort the ore particles in the first stage.

In a process described by *L.D. Friedman and L.P. Gaucher; U.S. Patent 3,051,644; Aug. 28, 1962; and by L.D. Friedman; U.S. Patent 3,074,877; January 22, 1963; both assigned to Texaco Inc.* oil shale particles are subjected to treatment with steam at a pressure in the range of 1,000 to 3,000 psig and a temperature in the range of 700° to 900°F, for a period of time within the range of 20 minutes to 6 hours.

In a preferred operation, shale particles are mixed with sufficient water to form a pumpable mixture and the mixture passed under pressure through an elongated heating zone of restricted cross-sectional area in which the water is vaporized, entraining the shale particles in steam, and the mixture is discharged into a soaking zone of large volume where the oil shale is subjected to the stripping and heating action of steam for the required period and from which recovered oil from the oil shale is withdrawn in vapor form admixed with the steam. Oil yields of more than 100% in comparison with the standard Fischer assay, are obtained from commercial grade oil shales. The quality of the recovered oil is comparable with that obtained by other methods of retorting.

K.I. Jagel, Jr., D. Liederman and L.J. Skowronek; U.S. Patent 3,562,143; February 9, 1971; assigned to Mobil Oil Corporation have found that the operation of a countercurrent retort for oil shale can be improved by reducing the refluxing and percolation in the bed.

Localized liquid accumulations in the bed form carbonaceous agglomerates which wedge in the air distribution assembly and burn in this hot oxidizing environment to form clinkers that interfere with retort operability. Removal of these liquid accumulations from the shale bed reduces the formation of carbonaceous agglomerates and vitreous clinkers. Localized liquid accumulations in the bed may be initiated by pieces of rich oil shale or by locally high concentrations of fine shale in the shale bed. These fines are more heavily wet with oil by the impaction of mist or by the surface condensation of oil than are large shale pieces.

As liquid is vaporized and cracked from these local accumulations, a tarry, carbonaceous residue is left behind which binds pieces of shale and dust together to form cohesive masses. This occurs at a temperature of about 700°F. The cohesive masses of shale, dust, and tarry binder are converted to dry hard agglomerates at temperatures between 800° and 1200°F.

In this process, flooding is caused to occur in a controlled fashion at a section of the retort where liquid can be disengaged. This reduces the likelihood of uncontrolled localized liquid accumulations and thus breaks the chain leading to clinkering and retort inoperability. Complete details of the liquid disengaging system are provided.

C.R. Garbett; U.S. Patent 3,634,225; January 11, 1972; assigned to Shell Oil Company describes a process for retorting oil-bearing shale including passing shale through a preheating zone and a retorting zone. Preheated shale is introduced into the retorting zone where it is contacted countercurrently with a hot gas stream which releases an oil vapor product at high temperature from the shale.

The oil vapor product is rapidly quenched by heat exchange with water and the resulting heated water is passed to a steam generating plant where a fuel fired furnace is employed to produce high temperature steam. The steam is employed to generate electric power and in expanding steam through power recovery means, it is cooled enough to be useful to quench the oil vapor product. The exhaust gas from the fuel fired furnace is passed to the preheating zone to preheat the shale.

Controlled Retort Atmosphere

A.K. Reyburn and R.T. Ellington, Jr.; U.S. Patents 3,617,467; November 2, 1971; and 3,617,468; November 2, 1971; both assigned to Atlantic Richfield Company have found that the composition of the atmosphere surrounding the oil shale particles at the time of pyrolysis has a definite effect on the oil yield obtained from the retorting pyrolysis. The atmosphere corresponding to combustion within the bed with air or air partially denuded of oxygen, as frequently used in gas combustion retorting, gives the lowest yields.

For example, oil shale was retorted in the laboratory in a 4" diameter batch fluidized bed at 915°F. A fluidizing gas was used containing 5% oxygen, as might be used in a typical conventional, gas combustion retort, and the oil yield was 85 to 88 weight percent Fischer assay oil. Natural gas in the same equipment and at the same temperature gave a higher yield of 102 weight percent Fischer assay oil. A mixture of 90 volume percent natural gas and 10 volume percent carbon dioxide gave the highest of all, 107 weight percent of Fischer assay oil.

Example 1: A batch-type oil shale retorting process was undertaken utilizing natural gas atmosphere that contained various percentages of carbon dioxide. The retorting temperature, pressure and residence time were approximately constant. The results are given in Table 1.

This data shows that the maximum oil yield is obtained at 10% carbon dioxide, balance natural gas with yields greater than that obtainable with natural gas alone being obtained when using up to 20 volume percent carbon dioxide diluent.

Example 2: A retorting process and retorting environment were used where the bed composition was kept essentially constant throughout the run with a variety of residence times utilized, none of which were sufficient for complete retorting of all the feed. Regardless of the retorting temperature, retorting pressure and residence time employed for a specific series of runs, the effect of increasing yields by adding carbon dioxide to a natural gas fluidizing gas as the process retort fluidizing gas is exhibited. The data are given in Table 2.

TABLE 1

Vol. percent CO_2 (balance natural gas)	Retort temperature (° F.)	Residue time (min.)	Oil yield (wt. percent F.A.)
0	915	16	102.12
10.52	920	16	107.41
25.00	915	16	99.30
49.92	915	16	89.72
100	905	16	85.27

TABLE 2

Fluidizing gas composition	Retort temperature (° F.)	Average residence time (min.)	Oil yield (wt. percent F.A.)
Natural gas	940	15.44	96.09
Do	940	9.12	88.67
Do	960	15.44	99.28
Do	960	9.12	91.42
90% N.G., 10% CO_2	940	15.44	97.87
90% N.G , 10% CO_2	940	9.12	90.52
90% N.G., 10% CO_2	960	15.44	106.32
90% N.G., 10% CO_2	960	9.12	100.38

The data of the examples thus shows that the use of a natural gas atmosphere and natural gas containing up to 20 volume percent carbon dioxide either in batch or continuous processes and under varying conditions greatly improves oil yields.

In general, the process for the recovery of volatile material from solid carbonaceous material, such as oil shale, comprises introducing the solid carbonaceous material into a retorting zone and fluidizing the solid carbonaceous material within the retorting zone by introducing a gaseous material. The fluidized solid carbonaceous material in the retorting zone is heated and volatile material is retorted and withdrawn from the retorting zone.

The spent solid carbonaceous material is withdrawn from the retorting zone and transferred to a burn-off zone. The spent solid carbonaceous material is fluidized in the burn-off zone by introducing oxygen-containing gas into the burn-off zone and carbon is burned off the fluidized spent solid carbonaceous material therein. The burned off spent solid carbonaceous material and combustion products are then withdrawn from the burn-off zone. A part of the stream of gaseous combustion products from the burn-off zone can be recycled and mixed with air to provide the oxygen-containing gas which is introduced into the burn-off zone. In this manner, the partial pressure of oxygen in the oxygen-containing gas can be controlled in such a way as to limit the peak temperature in the burn-off zone to a desired value.

A portion of the gaseous material introduced into the retorting zone can be a portion of the gaseous material withdrawn from the retorting zone. In such a form, the gaseous material is noncombustion-supporting and the chemical composition and gas velocity of the fluidizing gaseous material can be controlled to maximize product yield. This yield would be maximized by these methods because the gas composition to be used, as shown in the examples, gives greater oil yields than gases used conventionally in such a process.

When the fluidizing gaseous material is a noncombustion-supporting material, there is no loss of product through combustion in the retorting zone. The gaseous material can be preheated prior to introduction to the retorting zone.

SOLID HEAT TRANSFER MEDIA

Special Pellets and Steam Stripping

D.K. Wunderlich and J.L. Skinner; U.S. Patents 3,844,930; October 29, 1974; 3,844,929; October 29, 1974; 3,841,994; October 15, 1974; 3,841,993; October 15, 1974; all assigned to Atlantic Richfield Company describe a process in which hot special pellets are cycled to a retort zone to retort crushed oil shale thereby producing gas and oil products, particulate spent shale, and a combustible deposition on the pellets and spent shale. The main source of heat for retorting is derived from burning in a pellet deposition burning zone the combustible deposition formed on the pellets as they are cycled in the process. The amount of deposition on at least a portion of the pellets at the time that the pellets are passed to the burning zone is partially controlled by contacting after retorting at least a portion of the pellets with steam at a temperature hot enough to strip some of the deposition previously formed on the pellets.

Steam stripping may be carried out on the pellets and the spent shale, or on just the pellets, or on both. Preferably, steam stripping will be conducted after the pellets have been separated from substantially all of the spent shale smaller than the pellets. Postretort steam stripping allows more leeway in operation of the retort zone, recovers product materials that would otherwise be burned, and aids in more efficient and controlled burning the pellet deposition without excessively damaging the surface area of the pellets. The special pellets are characterized primarily by their effective surface area, size, and quantity relative to the oil shale.

Referring to Figure 1.12, raw or fresh oil shale which has been mined and pulverized, crushed or ground for the most part to a predetermined maximum size for handling in a retorting system by any suitable particle diminution process is fed directly from a crusher or from a hopper or accumulator by way of shale inlet line **11** into retort zone **13**. At the same time, special heat-carrying pellets substantially hotter than the shale feed are fed by gravity or other mechanical means to the retort zone by way of pellet inlet pipe **15**. The pellets and shale feedstock could be fed to the retort zone by way of a common retort zone inlet.

Crushing of the raw mined shale expedites more uniform contact and heat transfer between the shale feedstock and hot pellets and more uniform deposition of a carbon-containing deposition on the pellets. In normal practice, the degree of crushing is simply dictated by an economic balance between the cost of crushing and the advantages to be gained by crushing when retorting the kerogen from the shale. Generally the shale feedstock is crushed to about minus three-fourths inch and no particular care is taken to produce or restrict production of finer material.

In this process, crushing has a special purpose and aids in a preburn separation step. In one embodiment, for reasons which will be hereinafter shown and despite the added costs and standard practice, the mined shale is crushed to a substantially finer size wherein at least 95% by weight of the crushed oil shale will pass through a U.S. Sieve Series size 6 screen. The crushed oil shale may or may not be preheated by direct or indirect heat from any source including indirect heat exchange with pellets or flue gases generated during this retorting process. If the shale feedstock is preheated, the temperature of the feedstock will not exceed 600°F. The shale feedstock will usually be fed by way of a metered weight controller system. The preferred system for preheating the raw shale is to lift the shale in lift pipes with the hot flue gases generated in the combustion phase of the process.

The hot special heat-carrying pellets are especially characterized by having a principal size during use of between approximately about 0.055 and 0.5 inch and preferably between about 0.055 and 0.375 inch, and a surface area during use of between 10 and 150 square meters per gram. The surface area is the average effective surface area of the pellets as they enter the retort zone. The heat-carrying pellets are at temperatures ranging between 1000° and 1400°F, which is about 100° to 500°F higher than the designed retorting temperature within the retort zone. The most favorable practical temperature range depends on the process variables and more particularly on the specific advantages and characteristics of this process. The quantity of pellet heat carriers is controlled to cooperate with other variables so that the pellet-to-shale feedstock ratio on a weight basis is between one and three with a ratio between 1.5 and 2.5 being preferred. This ratio is, moreover, such that the sensible heat in the pellets is sufficient to provide at least 50% of the heat required to heat the shale feedstock from its retort zone feed temperature to the designed pyrolysis temperature. The retort zone feed temperature is the temperature of the oil shale after preheating, that is, the temperature of the shale upon entry into the retort.

The average retort temperature ranges between about 850° and 1200°F depending on the nature of the shale feedstock, the pellet-to-shale ratio, the type of product distribution desired, heat losses, and the like. The relative mass and size of the pellets are selected in a manner which facilitates separation of the pellets from spent shale, controls the amount of combustible carbon-containing deposition deposited on the pellets, optimizes other facets of the retorting process, and makes allowance for wear or size reduction of the pellets as they are cycled and recycled through the retorting process.

The term pellets refers to subdivided or particulate bodies a majority of which have the characteristics and properties herein required and which are composed of the same or dissimilar materials having the specified surface area and strength and of irregular shape, cylindrical shape, approximately oval or spherical shape, or purely spherical shape. The preferred pellets have a sphericity factor of at least 0.9 which, in addition to the usual advantages of facilitating movement of the pellets through the retorting process and of providing optimum solid-to-solid heat transfer and contact between the pellets and oil shale feedstock, has an advantage particularly useful in separating the pellets from other solids produced in the process as hereinafter set forth. The sphericity factor is the external or geometric surface area of a sphere having the same volume as the pellet divided by the external surface area of the pellet.

The pellets are made up of materials, such as, alumina or silica alumina, which are not consumed in the process and which are subdivided or particulate matter having significantly high internal surface area, but not excessively high. The pellets are sufficiently wear or breakage resistant and heat resistant to maintain enough of their physical characteristics under the conditions employed in the process to satisfy the requirements herein set forth, to effect retorting of the oil shale, and to permit controlled burning of a carbon-containing deposition formed on the pellets during the process. More specifically, the pellets do not disintegrate or decompose, melt or fuse, or undergo excessive surface area reduction at the temperatures encountered during such burning and the thermal stresses inherent in the process. The pellets will, of course, undergo some gradual wear or size reduction.

The size of the pellets is related to the other variables and to the preburn spent shale separation step of this process. The original or fresh pellets are generally comprised of particulate sensible heat carriers in a size range between about 0.1 and 0.5 inch and preferably between 0.1 and 0.375 inch, and are for the most part maintained during use at a plus 14 U.S. Sieve Series screen size, that is, approximately about 0.055 inch or greater. Finer pellet grain sizes are undesirable in the process. Suitable pellet materials are also found in cracking catalysts; however, the retorting process is not to be considered as relying on active catalytic sites. Many catalysts have surface areas far in excess of the maximum surface area of 150 square meters per gram provided in this process.

For example, some silica alumina catalyst may have a surface area ranging between 180 and 700 square meters per gram. As will hereinafter be discussed and as indicated by the trend shown in Table 1, high surface areas tend to cause too much combustion carbon-containing deposition or coke-like deposition being deposited in the retort zone. As herein pointed out, a post-retorting steam stripping stage or stages overcome some of the undesirable effects of too much deposition; however, it is preferred that these undesirable effects be avoided by selecting pellets with a surface area of 150 square meters per gram or less, there thereby maintaining the flexibility provided by the steam stripping stage or stages. Active catalytic sites tend to have effects similar to excessively high surface areas. As a result, in this process, although cracking catalysts may be used, it is preferred that the pellets bear no added active acid cracking catalytic catalyst sites or the like when the pellets are added to the retort zone.

The pellet surface area is considered one of the most important variables. The effect of pellet surface area is illustrated by the test results set forth in Tables 1, 2, and 3. The effect of pellet surface area on the amount of combustible carbon-containing deposition formed on pellets in the retort zone and on the distribution of carbon deposition between the pellets and spent shale is illustrated in Table 1. The effect of pellet surface area on liquid product distribution when a modified Fischer retort was used is illustrated in Table 2. The effect of pellet-to-shale ratio and, therefore, the total surface area of the pellets is illustrated in Table 3.

The total surface area is determined by the surface area per gram of pellets and the total pellet weight which in turn is controlled by the pellet-to-shale ratio and shale throughput rate. The results illustrated in these tables lead to several conclusions. First, if the surface area exceeds 150 square meters per gram, too much coke-like deposition may be produced on the pellets during the retorting stage of the process when the pellet-to-shale ratios specified herein are used. This in turn indicates an undesirable or excessive shift toward gaseous products in the retort zone. This also affects the flexibility of the process to adjust to expected fluctuations in the other process variations and objectives.

If the surface area of the pellets is less than 10 square meters per gram, either too little total coke-like deposition will be formed in the retort zone or the burning of the deposition will not be sufficient to provide a major portion of the heat required to heat the pellets to the desired temperature and to carry out the retorting phase of this process.

TABLE 1: EFFECT OF PELLET SURFACE AREA ON CARBON DEPOSITION

PELLET AREA m²/g	WT. PERCENTAGE OF CARBON ON PELLETS	WT. PERCENTAGE OF CARBON ON RESIDUAL SHALE
28.0	0.725	3.90
28.5	0.89	3.65
29.0	0.93	3.95
32.5	0.94	3.71
46	1.24	3.49
No Pellets	—	4.30

TABLE 2: EFFECT OF PELLET SURFACE AREA ON LIQUID PRODUCT DISTRIBUTION

PRODUCT BOILING RANGE	NO PELLETS	PELLET AREA*	
		47m²/g	96m²/g
150° - 400°F.	12%	27%	34%
400° - 700°F.	37%	46%	48%
700° - 900°F.	32%	22%	14%
900°F.+	19%	5%	4%

*Pellet : Shale Ratio = 2:1

TABLE 3: EFFECT OF PELLET:SHALE RATIO ON LIQUID PRODUCT DISTRIBUTION

PRODUCT BOILING RANGE	NO PELLETS	PELLET : SHALE RATIO		
		1:1	1.5:1	2:1
150° - 400°F.	14%	25%	30%	34%
400° - 700°F.	38%	45%	47%	48%
700° - 900°F.	31%	22%	17%	14%
900°F.+	17%	8%	6%	4%

As illustrated in Tables 2 and 3, the effective surface area of the pellets and the pellet-to-shale ratio in the retort zone affect liquid product distribution of the products from the retort zone. Increasing the surface area and the pellet-to-shale ratio tends to decrease the yield of condensable product vapors and increase production of gases. This process seeks to optimize the balance between gas, oil products and combustible deposition on the pellets. Any combustible deposition exceeding fuel requirements is wasteful. This balance, therefore, places restrictions on the retorting stage which restrictions are partially alleviated in this process by a postretorting steam stripping stage which recovers some of the deposition.

As a result, the operator has additional leeway when selecting the pellet-to-shale ratio and the original surface area of the pellets. All variables considered, it has been concluded that an original pellet surface area between 10 and 150 square meters per gram is acceptable with the range of 10 to 100 being preferred and that operating with a pellet-to-shale ratio between one and three is preferred with a ratio between 1.5 and 2.5 being more preferred.

As illustrated in Table 1, of particular additional significance is the fact that a substantial portion of the carbon-containing deposition on the pellets does not come from a loss in products, but instead is derived from the residual carbonaceous material that would normally be left on the spent shale. In other words, the residual carbonaceous material that would have otherwise been left on the spent shale divides itself partially between the pellets and the spent shale. The postretorting steam stripping stage can also be used to recover some of this deposition from the pellets.

The mixture of pellets and shale moves through the retort zone toward retort exit **17** and the gaseous and vaporous effluents containing the desired hydrocarbon values separate from the mixture. Since there is no need to use carrier, fluidizing, or retorting gases in the retort zone, the gaseous and vaporous effluents are able to leave the retort essentially undiluted by extraneous fluids except for any water or steam vapor added to prevent or retard carbonization, or to sweep product vapors from the solids, or for other reasons to the retort or effluent collection chamber. In a rotating retort system, the mixture movement is continuous and is aided by the action or design of the retort and by continuous withdrawing of pellets and spent shale from the exit end of the retort zone.

If a rotating retort zone is used, caking or coking together of the heat-carrying pellets or spent shale will be kept low. Moreover, a rotating-type of retort zone is especially suited to varying the residence time, that is, the length of time that the shale and pellets remain in the retort zone by allowing variations in pellet-to-shale ratio and volume of shale throughput. As previously indicated, greater than normal leeway in control over these variables is especially advantageous to regulation of the amount of deposition deposited on the pellets during the retort stage of the process.

The residence time for the pellets required to effect retorting and deposition of the pellet deposition is on the order of about 3 to 20 minutes with residence times of less than 12 minutes for the pellets being preferred. The shale residence time depends on its flow or movement characteristics and since the shale is not uniform in size and shape, the shale residence time varies.

The mixture of pellets and spent shale exits from retort zone **13** at a temperature between 800° and 1150°F by way of retort exit **17** into separation zone **19** for separation of the vapor, pellets, and spent shale. The separation zone may be any sort of exiting and separation system accomplishing the functions hereinafter mentioned and may be comprised of any number of units of equipment for separating and recovering one or more of these three classes of retort zone effluents either simultaneously, or in combination, or individually. In the process, it is critical that at least 75% of the total spent shale be separated from the pellets in the separation zone to eventually be collected in separation zone exit line **21**. In addition, at least 95% of the spent shale smaller than the pellets is separated. As shown in Figure 1.12a, the retort zone mixture is first passed through revolving screen or trommel **23** which has openings or apertures sized to pass the pellets and spent shale of the same or smaller size than the pellets. The trommel extends into product recovery chamber **25**. In the trommel, the gaseous and vaporous products separate from the mixture of pellets and spent shale and, at the same time, at least a portion of the larger spent shale particles or agglomerates are separated from the pellets and spent shale. Spent shale and pellets flow through the openings in trommel **23** and drop to the bottom of recovery chamber **25** to exit via retort exit line **27**. Any spent shale too large to pass through the openings in the trommel pass outward through exit **29**. The product vapors and gases resulting from retorting the oil shale collect overhead in recovery chamber **25** and rapidly pass to overhead retort products line **31** at an exit temperature between about 750° and 1050°F where the product vapors are usually subjected to hot dust separation and/or fractionation or partial fractionation, and/or other stages of the overall operation.

The hot dust separation may be interior or exterior, or both, of recovery chamber **25** and the dust thus collected may be combined and handled with other spent shale. Hot dust or fines separation may be accomplished by hot gas cyclones, quenching and washing, agglomeration with sludge or a separately condensed heavy product fraction, centrifuging, filtration, or the like. Partial fractionation may be accomplished by condensing only a high boiling fraction of the vapors, e.g., above 900°F.

The gases are not diluted by other gases and are, therefore, readily used in the overall shale operation. Some gas may be needed for supplementary fuel and some for production in the usual manner of hydrogen if hydrogenation is used in the overall shale operation. The optimum amount of gas production is just enough to satisfy these requirements as this process stresses the liquid oil products produced in the overall shale operation.

As shown in Figure 1.12a, the spent shale and pellets in recovery chamber **25** are discharged via exit line **27** at a temperature between about 750° and 1050°F, where these particulate solids are passed or conducted by gravity or other means of conveyance to gas elutriation system **33** which is a part of separation zone **19**. In the elutriation system, a major portion, and more preferably substantially all, of the remaining spent shale is separated from the pellets. It is essential that elutriation be accomplished in a way which retains the desired amount of carbon-containing deposition on the pellets; consequently, the elutriating gas fed by line **35** is a noncombustion supporting gas.

By conducting the process with pellets in the size range between about 0.055 and 0.5 inch and preferably between 0.055 and 0.375 inch, at least 75% of the total spent shale may be separated by action of the trommel and subsequent gas elutriation at a velocity of between 18 and 25 feet per second if most of the raw shale feedstock was crushed to about minus three-fourths inch. Based on an average of six sieve analyses of spent shale produced by retorting of minus ¾ inch shale feedstock in a rotating retort using ceramic ½ inch balls, about 16% by weight (analyses range 8 to 27%) of the spent shale is retained on a U.S. Sieve Series size 14 screen which is in a size range similar to the pellets.

Gas elutriation with irregular or cylindrical shaped pellets only separates about 2.0 to 4.0% of this portion of the spent shale from the pellets. Therefore, on an average between 12 and 13% of the spent shale is difficult to separate by screening and elutriation depending on whether the pellets cover the entire size range of this part of the spent shale. As mentioned previously, retention of more than 25% of the spent shale interferes with proper operation of the pellet deposition burning zone even if most of the spent shale entering the burning zone is originally in the same size range as the pellets. Upon combusiton, this spent shale would disintegrate further to fine ash and cause erratic operation of the combustion zone.

Separation of the spent shale prior to the supplemental deposition stage of this process is also desirable. Otherwise, some of the combustible deposition would be deposited on the spent shale. In addition, some allowance is made for spent shale and ash build-up as the pellets are cycled and recycled through the process. Since the spent shale having a size similar to the pellets is difficult to elutriate while the spent shale smaller than the pellets is readily separated by elutriation, and practically complete, it is desirable to alter the character of the spent shale or of the pellets to accomplish a greater degree of separation while holding heat losses in the pellets to a resonable level. One way to accomplish this objective is to crush at least 95 percent by weight of the shale feedstock to a minus 6 screen size. This results in a separation of at least 95 percent of the total spent shale from the pellets and the trommel may also be eliminated. Crushing to this size is costly and normally not done; however, in view of the fact that in this process it is essential that the bulk of the spent shale be separated from the pellets prior to reheating of the pellets, the cost of additional crushing may be justified. Another way to accomplish the objective of this separation prior to reheating the pellets has been discovered.

It has been found that if the pellets are essentially spherical, that is, have a sphericity factor of at least 0.9, the efficiency of separation by gas elutriation is greatly increased when the raw shale is crushed to a minus three-fourths size. Spherical pellets have improved flow properties over the spent shale and for a given screen size particle exhibit greater weight per particle. Gas elutriation with spherical pellets will separate about 97 percent or more of the spent shale retained on a U.S. Sieve Series size 14 screen and complete separation of the smaller spent shale. Thus, if spherical pellets are used, gas elutriation will separate at least 95 percent of the spent shale in the separation zone. As mentioned previously, therefore, the preferred shape of the pellets is spherical, that is, the preferred pellet should have a sphericity factor of at least 0.9.

The separated spent shale is carried out of the elutriating chamber overhead through line **37**. The spent shale is collected and may be combined and handled with other spent shale for eventual compaction and waste disposal or sale for use in manufacturing other products. At least a portion of the pellets with the combustible carbon deposition previously deposited on the pellets are subjected after the retorting stage to steam stripping to remove or recover some of the combustible deposition. The deposition is removed by contacting the pellets with steam at a temperature sufficient to strip some of the combustible material from the pellets. The combustible material is carried with the steam and is recovered as additional product or yield from the process. Steam stripping may be conducted in one or more stages.

The steam stripping stage or stages of the process may be conducted on the mixture of pellets and spent shale, or on at least a portion of the pellets after separation from the spent shale, or on both by use of two stripping stages. Treatment of the mixture has the advantage of recovering more of the combustible material, but has the disadvantages of increasing fine solids in the products, of increasing steam requirements, of providing less positive control over the amount of deposition left on the pellets, and of creating other operating difficulties. It is, therefore, much preferred that the steam stripping of the pellets be conducted on the pellets after the spent shale separation stage of the process. In addition, the increased yield advantage obtained by steam stripping the entire solids mixture can be partially offset by selecting pellets with a higher surface area.

As previously illustrated and noted in relation to Table 1, the combustible material normally left or deposited on the spent shale tends to divide itself between the spent shale and pellets depending on the surface area of the pellets. Less combustible material is left on the spent shale if higher surface area pellets are used. Steam stripping permits the use of higher surface area pellets since stripping can be used to remove excess combustible deposition prior to burning the deposition, provided however, that the increased recovery advantages are not offset by an excessive shift in the balance between gas and oil production and product quality.

The percentage of pellets treated by steam stripping and the amount of combustible deposition removed by steam stripping, therefore, provide flexibility and adaptability in controlling the total amount of combustible deposition on the pellets passed to the pellet deposition burning stage of the process and permit greater leeway in the amount of the combustible deposition formed on the pellets during earlier stages of the process, especially the retorting stage. Accordingly, as illustrated, steam from line **39** is passed either by way of steam line **41** to the bottom of collection chamber **25** which acts as a steam stripping zone to contact the mixture of pellets and spent shale, or by way of steam line **43** into steam stripping zone **45** to contact at least a portion of the separated pellets. The pellets separated in elutriation system **33** pass by gravity or other means from the separation zone via exit line **47** either by way of stripper feed line **49** to steam stripping zone **45** or by way of bypass line **51** to bypass the steam stripper and flow directly to pellet return line **53**. The steam stripping stage or stages should be carried out at a temperature between 800° and 1150°F, and preferably at a temperature between 850° and 1050°F. At temperatures higher than 1150°F, the probability of adversely reducing surface area is increased. Whether or not a surface area reduction occurs, depends primarily on temperature, steam pressure, and the surface area and nature of the pellets.

With higher surface areas and/or temperatures, a surface area reduction is more likely to occur. At temperatures below 800°F, the steam stripping action is inefficient. The primary factors controlling the amount of organic deposition stripped are the amount of pellets or total solids treated, the steam-to-pellet or solids ratio, the amount and nature of the combustible deposition, the surface area of the pellets, the stripping zone temperature, and residence time. The residence time of the pellets or pellets and spent shale during stripping will preferably be less than 10 minutes. Preferably, the stripppng will be conducted in a manner such that the final amount of deposition left on the pellets is less than 1.5 percent by weight of the pellets subjected to steam stripping, and will on an average range between 0.3 and 1.5 percent by weight depending primarily on the percentage of pellets subjected to steam stripping.

When the steam stripping stage is carried out on some or all of the pellets after the spent shale has been separated, steam stripping zone **45** acts as first stripping zone and the treated pellets collect in stripper exit line **55** and pass to pellet return line **53**. The steam and stripped deposition products collect in the top of the steam stripper and exit the stripper by way of overhead line **57**. The steam plus stripped products is then either passed in the usual manner for eventual product recovery operations, or to a second steam stripping zone. As illustrated in Figure 1.12a, at least a portion of the steam plus products may be simply passed to overhead retort products line **31**.

When the stripping stage is carried out on the mixture of pellets and spent shale in collection chamber **25**, the steam and stripped products collect with the retort products in chamber **25** and are thereby at least partially recovered. One way of steam stripping both separated pellets and the mixture of pellets and spent shale with some steam is illustrated in the drawings. The separated pellets in steam stripping zone **45** are first contacted with steam to remove a portion of the combustible deposit on the pellets in zone **45**. At least a portion of the steam plus stripped or removed deposition are then passed through overhead line **57** to bypass line **59** and then to steam line **41** to flow into collection chamber **25** which acts as a second steam stripping zone.

In collection chamber **25** the steam removes additional combustible material from the pellets and residue in chamber **25**. The steam plus the depositions removed in both zone **45** and chamber **25** collect with the retort products and the removed depositions are thereby eventually recovered. The pellets in pellet return line **53** pass to pellet lifting system **61** where the pellets are lifted preferably to an elevation which allows gravity feed to retort zone **13** by way of lift line **63** to pellet deposition burning zone **65**, which as shown in Figure 1.12a has surge hopper **67** for collecting the lifted pellets and leveling out fluctuations and from which the pellets fall into pellet deposition burning zone **65**. While any conveying and lifting system holding heat losses to a reasonable value may be used, it is preferred as shown in Figure 1.12a that the pellet lifting system be a pneumatic conveying system which will operate in the conventional manner to lift the pellets to the pellet deposition burning zone.

The lift gas enters the lift system via line **69** at a velocity between 25 and 70 feet per second and the lift time is, therefore, very short. As a result, air may be used as the lift gas without causing uncontrolled combustion of the deposition on the pellets and the detrimental effects attendant to such uncontrolled burning. As mentioned previously, the pellets bear a combustible carbon-containing deposition which was absorbed or deposited during the process and which was not removed in the steam stripping zone. This combustible deposition is burned in combustion or pellet deposition burning zone **65** to provide at least 50 percent or more of the heat required to reheat the pellets to the temperature required to effect retorting of the shale. The combustible deposition is burned in a manner similar to the way that catalytic cracking catalysts particles are regenerated and which is controlled to avoid excessive heating of the pellets which would excessively reduce the effective surface area of the pellets to less than 10 square meters per gram. A progressive bed burner with a gas flow of about 1 to 2 feet per second is preferred.

A combustion supporting gas, for example air, a mixture of air and fuel gas generated in the process, flue gas with the desired amount of free oxygen, is blown into the pellet deposition burning zone at a temperature at which the combustible deposition on the pellets is ignited by way of combustion gas inlet **71** which in Figure 1.12a includes a blower. Steam may also be used to control burning provided that the steam does not excessively reduce the surface area of the pellets. The combustion supporting gas may be preheated in heaters **73** by burning some of the gases produced in the process to reheat the pellets to the minimum ignition temperature.

The quantity of combustion supporting gas, for example, about 10 to 15 pounds of air per pound of carbon deposit, affects the total amount of carbon-containing deposition burned and the heat generated by such burning and in turn the temperature of the pellets. The bulk density of the pellets is about 40 to 50 pounds per cubic foot and the specific heat of the pellets varies between about 0.2 and 0.3 Btu per pound per degree F. The gross heating value of the carbon-containing deposition is estimated to be about 15,000 to 18,000 Btu per pound.

The amounts of carbon dioxide and carbon monoxide produced in the fuel gases created by burning the pellet deposition indicate the amount of combustion supporting gas required of used and the amount of carbon-containing deposition not burned. Generally, it is desirable to attempt to free the pellets of combustible deposition. In any case, at least 50 percent of the combustible deposition is burned.

The unburned deposition stays with the pellets and affects to some degree the total amount of combustible carbon-containing deposition deposited in the next cycle. It should be noted that this type of controlled burning does not selectively burn the same amount of deposit from every pellet. Other factors taken into consideration during burning of the pellet deposition are the pellet porosity, density, and size, the burner chamber size and pellet bed size, residence burning time, the desired temperature for the pellets, heat losses and inputs, the pellet and shale feed rates to the retort zone and the like.

The residence burning time will usually be rather long and up to about 30 to 40 minutes. Combustion of the deposition should be controlled in a manner which does not heat the pellets to above 1400°F. The hot flue gases generated in the pellet deposition burning zone may be removed by burner overhead exit line **75** and used to preheat cool raw shale feedstock or for heat transfer to any other phase or part of the shale operation.

For example, this stream could be fed to a carbon monoxide boiler and the heat available from the boiler could be used for processing product vapors or to drive turbines. Of course, additional fuel material or gases may be used to supplement burning of the pellet deposition if this is necessary, but it is to be understood that the pellet deposition supplies the major portion of the sensible heat required for retorting the oil shale and that the variables are set to accomplish this objective along with the other advantages and objectives of this process.

A continuous stream of hot pellets having a temperature above 950°F and not exceeding 1400°F is thereby produced for return and introduction via burner bottom exit line **77** back through pellet inlet line **15** either by gravity and/or mechanical means into retort zone **13**. The rate of passage of the pellets from the combustion zone will be metered or controlled in conventional manners to eventually provide the optimum pellet-to-oil shale feedstock ratio to the retort zone. The optimum ratio is governed by the pellet properties, the surface area of the pellets as they enter the retort zone, the organic content of the raw oil shale, and the other process variables.

Although the retorting process is carried out in a manner to hold loss of pellets to a minimum, some pellets will be lost in the process and a relatively small quantity of pellets may be added continuously to maintain the pellet quantity.

FIGURE 1.12: RETORTING WITH SPECIAL PELLETS AND STEAM STRIPPING

a.

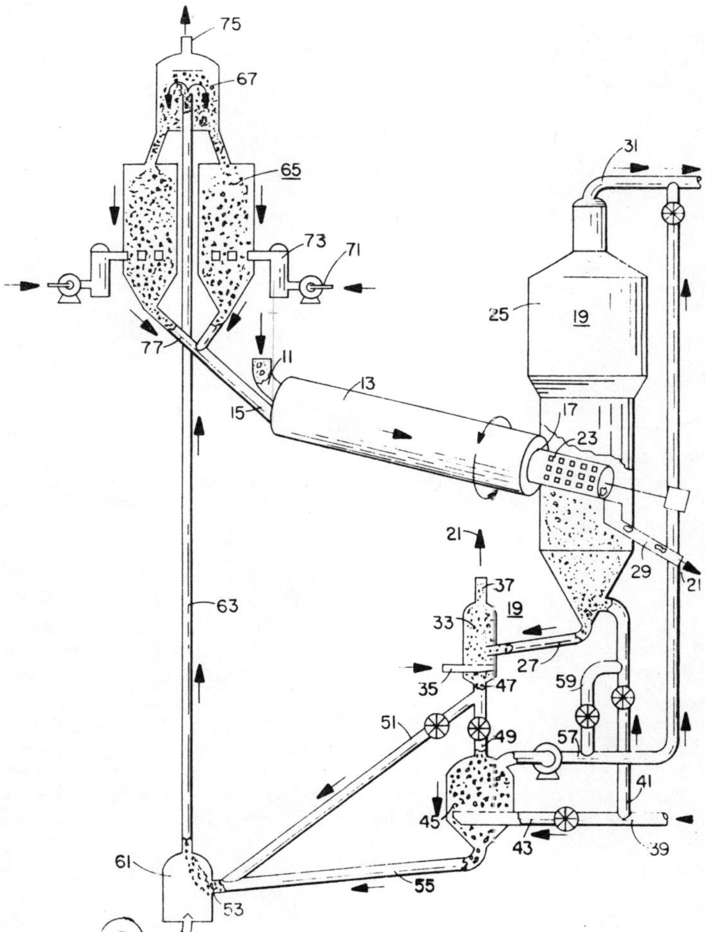

Process Flow

(continued)

FIGURE 1.12: (continued)

b.

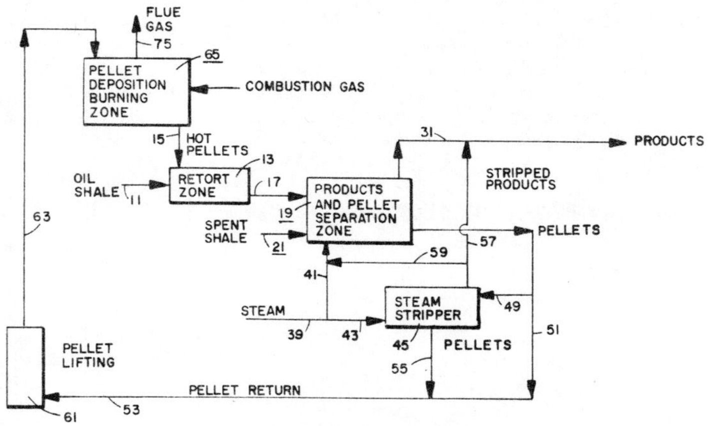

Flowsheet of Process

Source: D.K. Wunderlich and J.L. Skinner; U.S. Patent 3,844,930; October 19, 1974

Special Pellets and Precoking Stage

According to a process described by *D.K. Wunderlich and J.L. Skinner; U.S. Patent 3,850,739; November 26, 1974; assigned to Atlantic Richfield Company* hot special heat-carrying pellets are continuously cycled to a retort zone to mix with and continuously retort fresh crushed oil shale, thereby producing carbonaceous products and spent shale. The main source of heat for retorting is derived from controlled burning of a coke-like carbon-containing deposition which is deposited on the pellets as they are cycled in the process. At least a portion of the pellets are coked to some extent by cracking or stabilizing oil products in a thermal precoking zone. Some of the pellets from the precoking zone may be cycled to the retort zone.

The amount of preretort coking provides flexible regulation over the total deposition formed on the pellets during the process especially that portion of the deposition formed during the retorting of oil shale and over product quality. After retorting, the pellets passed to the retort zone are separated from substantially all of the spent shale smaller than the pellets prior to combustion of the carbon deposition on the pellets.

The pellets from the precoking zone and from the retorting zone are passed or lifted to a pellet deposition burning zone where the deposition is burned. The special pellets are characterized primarily by their effective surface area, size, and quantity relative to the oil shale. The process stresses improvement in the quality of the liquid oil products and at the same time greater useful recovery of the carbonaceous matter in oil shale.

Heated Spent Oil Shale

G.W. Frick; U.S. Patent 3,784,462; January 8, 1974; assigned to Cities Service Research & Development Company describes a process and apparatus for the continuous steady state retorting of ground oil shale in the absence of air. Retorting is accomplished by countercurrently contacting heated spent oil shale with fresh ground oil shale in a vessel from which air is excluded. The spent oil shale is heated by combustion of its carbonaceous residue to form a hot heat transfer medium which, when contacted with fresh oil shale in the retorting process, provides the energy for the recovery of hydrocarbons. Referring to Figure 1.13, fresh ground oil shale **11** is introduced continuously into an insert vessel **14** contained within a retorting vessel **13** by means of a screw feeder **12**, although other possibilities for fresh oil shale introduction exist. For example, use of a lock hopper or introduction by pumping a slurry consisting of ground shale mixed with a heavy oil into the retorting vessel **13** would accomplish the same end. The retorting vessel **13** can operate at or near atmospheric pressure, thus facilitating continuous fresh oil shale input. Heated spent shale flows in through a downward tangential entry point by means of an inlet pipe **22**. Fresh oil shale is introduced into the incoming heated spent oil shale countercurrently to promote turbulent mixing and favorable heat transfer of energy from the heated spent oil shale to the fresh oil shale.

FIGURE 1.13: RETORTING PROCESS EMPLOYING HEATED SPENT OIL SHALE

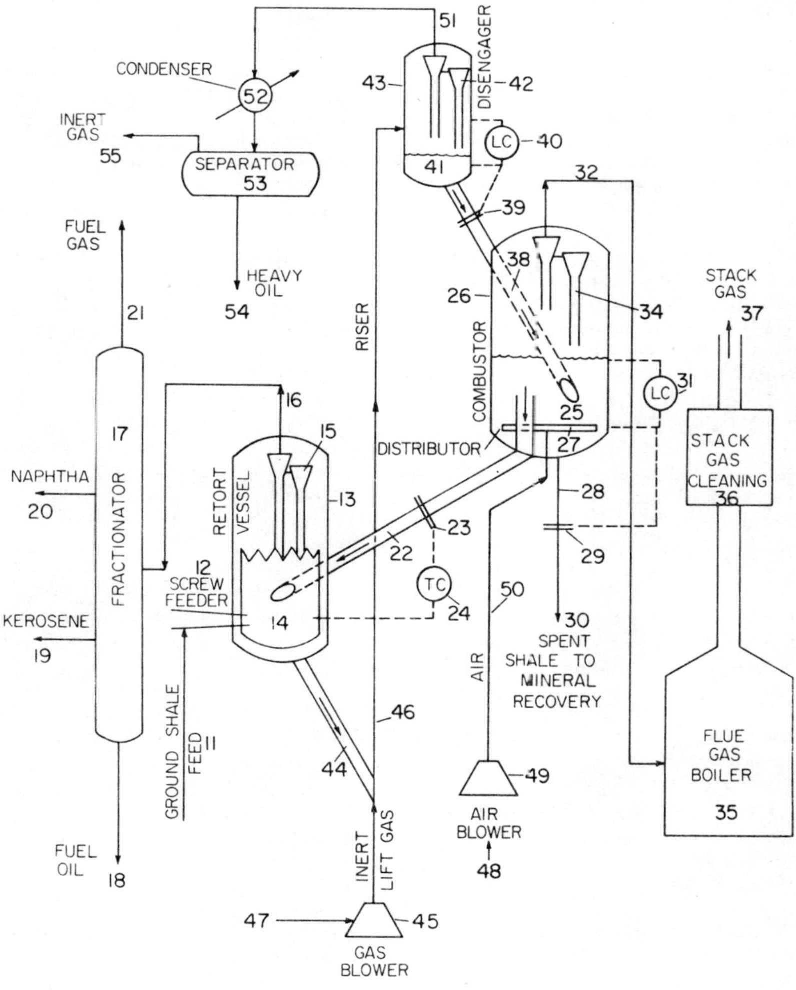

Source: G.W. Frick; U.S. Patent 3,784,462; January 8, 1974

The retorting bed level maintained in the insert vessel **14** by means of a notched circular overflow weir forming the top of the insert vessel **14**. Bed temperature is automatically controlled by regulation of the inlet valve **23**. Inlet valve **23** is controlled by a temperature controller **24**, temperature responsive to the temperature of the oil shale within insert vessel **14**. In general, retorting temperature will be controlled between 500° and 1500°F, and preferably between 800° and 1200°F, depending on yield, structure of the apparatus, and the oil products desired.

Higher temperatures will promote cracking of the shale oil and increasing yields of gas and light ends, while lower temperatures will increase the fuel oil and kerosene yields. The hydrocarbon vapors leave the retorting vessel **13** through stacks **15**, having cyclone inlets to retain the oil shale fines wihin retorting vessel **13**, and pass through a pipe **16** to a fractionator **17**. The fractionator **17** consists of a multiplate fractionation column which separates the retorted products into the desired boiling range materials, for example, fuel oil **18**, representing the bottoms of the fractionator **17**, kerosene **19**, naphtha **20**, and fuel gas **21**.

In order to maintain smooth circulation of spent oil shale, it is desirable to lift the spent oil shale particles leaving the retorting vessel **13**. It is also economically desirable to remove as much heavy oil from the spent oil shale as possible before combusting the carbonaceous residue or hydrocarbon material remaining therein by the further heating of the spent oil shale. Both these requirements may be accomplished by blowing the spent oil shale upward with a fast moving inert lift gas, preferably heated, through a vertical pipe. This is depicted in the figure by a discharge line **44**, discharging spent oil shale from retorting vessel **13** into an inert lift gas **47**, provided by a gas blower **45**.

The spent oil shale and inert lift gas **47** pass through a riser pipe **46**. Preferred lift gases include flue gas and steam, but inherently an inert gas should be utilized. The riser pipe **46** terminates at a disengager vessel **43** which serves to separate the lift gas, through stacks **42**, also having cyclones therein to retain fines. The lift gas passes through line **51** and is introduced into a condenser **52**, which may be water or air cooled to condense the normally liquid components of the gas, for example, any heavy oils.

The gas-liquid mixture is then passed to a separator **53** in which heavy oil **54** is separated therefrom. Spent oil shale forms a level **41** in the bottom of the disengager vessel **43**. This material is then fed to a combustor **26** to form a level of spent oil shale **25** therein. The level of spent oil shale **41** in the disengager **43** is regulated by a level controller **40** connected to a responsive valve **39** connected within the discharge tube **38**, from the disengager vessel **43**, which feeds spent oil shale to the combustor **26**.

The hot bed of shale particles is maintained in the combustor **26** by continuously burning the coke contained within the spent oil shale with air **48** blown into the bottom of the combustor **26**. The air **48** is introduced through an air blower **49** and air line **50**, and passed passed through a distributor plate **27** to uniformly heat and combust the carbonaceous residue contained within the spent oil shale bed **25**.

The distributor **27** may consist of a perforated grid or a network of pipes having holes at close intervals. Spent oil shale flows in from the disengager **43** through the discharge pipe **38** in a downward tangential entry which promotes turbulent mixing with air **48** introduced into the combustor **26**, resulting in complete combustion of the carbonaceous residue of the spent oil shale. Heated spent oil shale is introduced into the retorting vessel **13** through an overflow well comprising the inlet pipe **22** extending within the retorting vessel **13** and is controlled by temperature control **24**. In addition, excess heated spent shale oil is continually withdrawn from the combustor **26** by the use of a level control **31** having a responsive valve **29** connected to a discharge line **28** such that spent oil shale **30** is provided for mineral recovery and/or disposal as it is continually purged from the combustor **26**.

As it would be possible to produce a totally carbon free spent oil shale by contacting the excess spent oil shale withdrawal stream with oxygen, a totally clean spent oil shale may be produced from the apparatus of this process for total environmental control and no adverse pollution effects. The process, therefore, is not inhibited by pollution criteria. Flue gas which is vented through stack **34** and line **32** from the combustor **26** is passed through the cyclones contained within stacks **34**, to remove fines, and through line **32** into a waste heat boiler **35**. The waste heat boiler **35** recovers combustion heat value not used in the retorting process. This further allows for minimum waste of oil shale energy potential.

Cool flue gas, leaving the waste heat boiler **35**, may be subjected to various gas treating operations for removal of particulates and pollutants such as SO_2 by passing the flue gas through stack gas cleaning stage **36**. The clean flue gas produced is then sent through stack **37** into the atmosphere. Therefore, through this process, production of clean flue gas and clean spent oil shale insures minimum environmental debasement. As the combustor **26** may be operated in the range of 800° to 2000°F, the assurance of complete combustion of carbonaceous materials and with the absence of conventionally produced carbonaceous residues normally associated with spent oil shale removed, environmental control and reclamation of any heat content or energy value contained within the oil shale is provided.

Accordingly, by use of the apparatus, oil shale is heated by the combustion of its own carbonaceous residue through a process whereby steady state conditions are maintained in supplying retort heat by means of a recirculating stream of heated spent oil shale particles. Through the operating conditions and system utilized, one is capable of producing high-quality oil products with the production of clean flue gas and oil shale completely free of carbonaceous materials such that environmental control and pollution inhibition are afforded conjunctively with an economically desirable process for oil shale recovery.

R.T. Ellington, Jr.; U.S. Patent 3,597,347; August 3, 1971; assigned to The Oil Shale Corp. describes a process for retorting carbonaceous material, such as oil shale, where the heat of pyrolysis is supplied by mixing hot spent shale ash with raw shale feed. Shale particles lose their strength progressively above about 500°F so crushing of the feed to minus ⅛" size does not take place until after mixing or retorting. Cold shale feed fines are added to the mixed hot shale ash and larger shale feed before retorting.

One of the processes of interest is the sandcracking process. In operation, spent shale ash preheated to about 1200°F is mixed with cold raw shale feed that has been previously crushed to minus ⅛" size. The spent shale ash thus provides the heat of pyrolysis. After mixing, the shale ash and feed are discharged into a retorting vessel. The size of the vessel and the rate of withdrawal of spent shale ash are in such relationship that the shale feed has adequate time to absorb heat from the shale ash and to be fully retorted before it is discharged from the retort vessel.

Air is introduced into the solid effluent discharged from the retorting vessel and the air-entrained effluent is passed into a lift-tube combustor where the carbon content of the spent shale is burned off to heat the air and the shale ash to the desired temperature. At the top of the lift pipe, part of the entrained shale ash is knocked out and dropped in the retention bin to provide heat for retorting. The remainder is carried off with the hot combustion products, through a waste heat boiler and/or any other heat recovery means and then through a cyclone where the ash is separated from the cooled combustion products and sent to disposal. The gaseous retort products are withdrawn from the top of the retort vessel and transferred to a hot-dust separator where the entrapped waste dust is removed. The thus purified gaseous retort products are then conventionally treated to remove oil and other products.

While this process, as outlined above, offers oil yields of at least 90% Fischer Assay, it also presents problems. The raw oil shale feed must be crushed from a size of 1" or more to the minus ⅛" size to give optimum operation of the lift-tube combustor. Also, there is the possibility that the burned off shale ash will absorb shale oil vapors in the retorting drum and thus carry part of the yield from new shale out to the combustor, thus reducing the yield in the recovery system. It has been found that these problems may be overcome by mixing shale ash of a temperature of 1000° to 1500°F, preferably 1200° to 1400°F with cold raw oil shale feed fines and larger feed of a size of minus 1.5" to minus 1" down to ⅛", the larger feed preferably being preheated to 500° to 550°F, transferring the mix to a retort vessel where the atmosphere is maintained noncombustion supporting (an atmosphere of natural gas or processed retort gas containing 10 volume percent CO_2 is preferred), retorting off volatile material at a preferred retort temperature of 800° to 950°F, such as oil, with the heat of pyrolysis supplied by the hot spent shale ash, crushing the shale mixture at temperature above about 500°F to essentially minus ¼" size, preferably minus ⅛" size, passing the spent shale through a lift-tube combustor by the addition of air where the carbon content of the spent shale is burned off to heat the air and shale ash to the desired temperatures, and recycling part of the heated spent shale back to the mixer for mixing with raw feed oil shale.

It has been found that oil shale commences to lose crushing strength above about 500°F and is very friable by the time it reaches retort temperature and is retorted. As a consequence, crushing of the oil shale above about 500°F is much easier and requires less extensive crushing apparatus to achieve the fineness normally sought in the low temperature crushing. Retorting for the oil can be done with the large pieces nearly as easily as with the finer particles normally used.

However, finer particles are desirable for flow through the lift-tube combustor under the action of air. The crusher can thus be placed in the process anywhere the shale is at a temperature of over 500°F and before the lift-tube combustor. It may thus be placed at the outlet of the mixer, or, if residence time in the mixer is not sufficient to raise the raw shale temperature to above 500°F and larger particles in the retort vessel do not affect yield, at the outlet of the retort vessel.

Recycle of Pellets

According to a process described by *Y.A.K. Abdul-Rahman; U.S. Patent 3,803,022; April 9, 1974; assigned to Atlantic Richfield Company* hot porous pellets are cycled to a retort zone to mix with and retort preheated crushed oil shale, thereby producing gas and oil products, a combustible deposition on the pellets, and a mixture of pellets and spent shale. The main source of heat for retorting is derived from controlled burning of the deposition on the pellets in a pellet deposition burning zone. The hot flue gas thus generated is passed to a secondary preretort oil shale heating zone to additionally preheat already preheated oil shale.

Thereafter, the hot flue gas is passed to a gas elutriation system where the mixture of spent shale and pellets is processed to separate and recover the pellets for return to the retorting process. From the gas elutriation system, the hot flue gas is used to lift the recovered pellets to the pellet deposition burning zone. Thereafter, the hot flue gas is passed to an initial preretort oil shale heating zone to preheat raw crushed oil shale before it is fed to the secondary preretort heating zone. The hot flue gas from the burning zone usually contains carbon monoxide and may, at any desirable point in the system, be passed through a carbon monoxide shift reaction zone.

Referring to Figure 1.14 pellets bearing a combustible deposition which was previously absorbed or deposited on the pellets in an earlier stage of the retorting system are passed through pellet return line **11** which acts as an inlet line to pneumatic pellet lift system **12** where the pellets are lifted by way of lift line **13** to pellet deposition burning zone **14**. Preferably, the pellets will be lifted to an elevation which allows gravity feed of the pellets to a retort zone. The pneumatic pellet conveying system operates in the conventional manner to lift and convey the pellets except that the lift gas which enters the lift system via line **15** and moves through lift system **12** at a velocity sufficient to elutriate or lift the pellets, for example, 25 to 60 feet per second, is at least in part comprised of a hot flue gas which originated in pellet deposition burning zone **14**.

As shown, pellet burning deposition burning zone **14** is comprised of surge hopper **16** for collecting the lifted pellets and leveling out fluctuations and from which the pellets fall into burning units **17**. The hot lift gas passes overhead of the surge hopper through exit line **18** where the hot lift gas is passed to oil shale first preretort heating zone **19** which may be comprised of one or more heating units. Raw or fresh oil shale which has been mined and pulverized, crushed or ground for the most part to a predetermined maximum size for handling in a retort system by any suitable particle diminution process and which is at a temperature lower than the hot lift gas is fed directly from a crusher or from a hopper or accumulator by way of shale inlet line **20** into first preretort heating zone **19**.

FIGURE 1.14: RETORTING SYSTEM

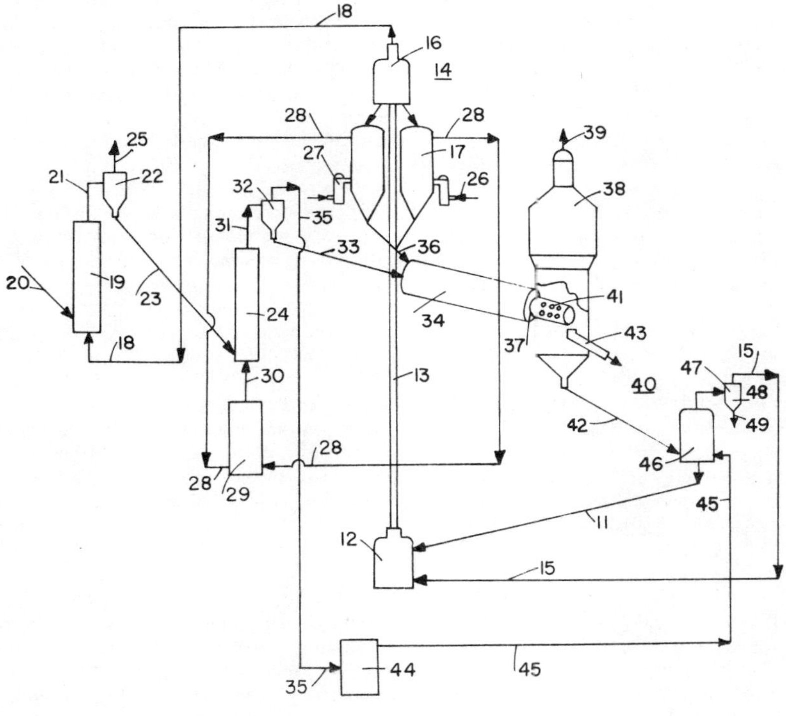

Source: Y.A.K. Abdul-Rahman; U.S. Patent 3,803,022; April 9, 1974

In the zone, the hot lift gas lifts, intimately contacts, and heats the cooler crushed oil shale to a temperature not to exceed 600°F in order to avoid retorting or loss of oil values in the raw shale. The hot lift gas and preheated oil shale pass overhead through exit **21** to collection chamber **22** where the oil shale settles from the lift gas and is gravity or mechanically fed through line **23** to second preretort heating zone **24**. The separated hot lift gas passes overhead of collection chamber **22** through exit line **25** where the gas may be passed to additional upstream preretort oil shale heating units.

Crushing of the raw mined shale expedites more uniform contact and heat transfer between the shale feedstock and hot lift gas and in a retort zone between the shale and pellets. In normal practice, the degree of crushing is simply dictated by an economic balance between the cost of crushing and the advantages to be gained by crushing when retorting the kerogen from the shale. Generally the shale feedstock is crushed to about ½ inch and no particular care is taken to produce or restrict production of finer materials. In burning units **17**, the combustible deposition on the pellets is burned to provide at least 50 percent or more of the heat required to reheat the pellets to the temperature required to effect retorting of the shale. The combustible deposition is burned in a manner similar to the way that catalytic cracking catalyst particles are regenerated and which is controlled to avoid excessive heating of the pellets which would excessively reduce the effective surface area of the pellets to less than 10 square meters per gram. A progressive bed burner is preferred. A combustion supporting gas, for example, air, a mixture of air and fuel gas generated in the process, and flue gas with the desired amount of free oxygen, is compressed or blown into the pellet deposition burning zone at a temperature at which the deposition on the pellets is ignited by way of combustion gas inlet **26**.

Steam may also be used to control burning provided that the steam does not excessively reduce the surface area of the pellets. The combustion supporting gas may be preheated in heaters **27** by burning some of the gases produced in the process to reheat the pellets to the minimum ignition temperature. The quantity of combustion supporting gas, e.g., about 2 to 3 pounds and higher depending on compression pressure of oxygen per pound of carbon deposit, affects the total amount of deposition burned and the heat generated by such burning and in turn the temperature of the pellets.

The bulk density of the pellets is about 40 to 50 pounds per cubic foot and the specific heat of the pellets varies between about 0.2 and 0.3 Btu per pound per degree F. The gross heating value of the carbon-containing deposition is estimated to be about 15,000 to 18,000 Btu per pound. The amounts of carbon dioxide and carbon monoxide produced in the flue gases created by burning the pellet deposition indicate the amount of combustion supporting gas required or used and the amount of carbon-containing deposit not burned. Generally, it is desirable to attempt to free the pellets of all of the combustible deposition.

Other factors taken into consideration during burning of the pellet deposition are the pellet porosity, density, and size, the burner chamber size and pellet bed size, residence burning time, the desired temperature for the pellets, heat losses and inputs, the pellet and shale feed rates to the retort zone and the like. The residence burning time will usually be rather long and up to about 30 to 40 minutes. Combustion of the deposition should be controlled in a manner which does not heat the pellets to above 1500°F. Of course, additional fuel material or gases may be used to supplement burning of the pellet deposition if this is necessary.

The hot flue gases generated in the pellet deposition burning zone are passed through lines **28** to second preretort heating zone **24**, which may be comprised of one or more heating units. Generally, the hot flue gases will contain carbon monoxide which can be used in a conventional manner in a carbon monoxide shift reactor wherein the carbon monoxide is changed to carbon dioxide thereby exothermically producing additional heat. Therefore, whenever additional heat is required or if at some point an excess of heat is present, the hot flue gas may be passed through an optional carbon monoxide shift reaction zone. If an excess of heat is present, the heat can be used for boiler operation or the like. The carbon monoxide shift reaction zone may, as indicated, be located at any appropriate point along the path of the flue gas from the pellet deposition burning zone to the last stage of process wherein the flue gas is used.

As shown, the hot flue gas containing carbon monoxide is first passed through shift reaction zone **29** before being passed to secondary preretort heating zone **24**. If the hot flue gas contained an excess of oxygen, supplemental fuel could be added to consume the oxygen or the oxygen converted by other means. The hot flue gases via line **30** entering secondary preretort heating zone **24** are hotter than the preheated crushed oil shale entering the zone from the first preretort heating zone. The hot flue gas lifts, intimately contacts, and heats the oil shale to a higher temperature which does not exceed about 600°F. The hot flue gas and heated oil shale pass overhead through exit **31** to collection chamber **32** where the oil shale settles from the hot flue gas and is gravity or mechanically fed through retort inlet line **33** to retort zone **34**. The separated hot flue gas passes overhead of collection chamber **32** through exit line **35** where the gas may be passed to additional preretort oil shale heating units or passed to a separation zone. In the pellet deposition burning zone, there is produced a continuous stream of hot pellets having a temperature above 1000°F and not exceeding 1500°F. As used herein, the term pellets refers to subdivided or particulate heat-carrying bodies having, when they leave the combustion zone, a minimum relatively high effective surface area of 10 square meters per gram and higher, a size range between about 0.06 and 0.3 inch, including a narrower range therewithin, and a shape and density or particle weight such that for a particular size the elutriation velocity of a pellet of that size is significantly, that is, measurably and usefully, higher than the elutriation velocity of most of the particles of spent shale of that size. Preferably, the average surface of the pellets will be between 10 and 150 square meters per gram.

The surface area is the average effective of the pellets as they enter the pyrolysis zone. The surface area may be determined by the conventional nitrogen absorption method. The pellets may be cylindrical shape, approximately oval or spherical shape, or purely spherical shape. The much preferred pellets have a sphericity factor of at least 0.9 which shape gives the highest particle weight and is particularly useful in separating the pellets by elutriation and screening from spent shale produced in the retort zone. The sphericity factor is the external or geometric surface area of a sphere having the same volume as the pellet divided by the external surface area of the pellet.

The pellets are made up of materials such as alumina or silica alumina, which are not consumed in the process and which are subdivided or particulate matter having significantly high internal surface area. The pellets are sufficiently wear or breakage resistant and heat resistant to maintain enough of their physical characteristics under the conditions employed in the process to satisfy the requirements of the process, to affect retorting of the oil shale, and to permit controlled burning of a carbon-containing deposition formed on the pellets during the process. More specifically, the pellets do not disintegrate or decompose, melt or fuse, or undergo excessive surface area reduction at the temperatures encountered during such burning and the thermal stresses inherent in the process.

The stream of hot pellets is fed by gravity or other mehcanical means to retort zone **34** by way of pellet inlet pipe **36**. The pellets and shale feedstock could be fed to the retort zone by way of a common retort zone inlet. The hot pellets are at a temperature ranging between 1000° and 1500°F, which is about 100° to 500°F higher than the designed retort temperature within the retort zone. The most favorable practical temperature range depends on other process variables. The quantity of pellet heat carriers is controlled so that the pellet-to-shale feedstock ratio on a weight basis is between 1 and 3. This ratio is such that the sensible heat in the pellets is sufficient to provide at least 50 percent of the heat required to heat the shale feedstock from its retort zone feed temperature to the designated retort temperature.

The feestock feed temperature is the temperature of the oil shale after preheating, that is, the temperature of the shale upon entry into the retort. The average retort temperature ranges between about 850° to 1200°F, depending on the nature of the shale feedstock, the pellet-to-shale ratio, the product distribution desired, heat losses, and the like.

The mixture of pellets and shale moves through the retort zone toward retort exit **37**, and the gaseous and vaporous effluents containing the desired hydrocarbon values separate from the mixture. The residence time for the pellets required to effect retorting and deposition of the pellet deposition is on the order of about 3 to 20 minutes with residence times of less than 12 minutes for the pellets being preferred. The shale residence time depends on its flow or movement characteristics and since the shale is not uniform in size and shape, the shale residence time varies. The mixture of pellets and spent shale exits from retort zone **34** at a temperature between 800° and 1050°F by way of retort exit **37** into recovery chamber **38** where the gaseous and vaporous products, resulting from retorting the oil shale collect overhead and rapidly pass to overhead retort products line **39** at an exit temperature between about 750° and 1050°F. The product vapors are usually subjected to hot dust separation, and the dust thus collected may be combined and handled with other spent shale. Hot dust or fines separation may be accomplished by hot gas cyclones, quenching and washing, agglomeration with sludge or a separately condensed heavy product fraction, centrifuging, filtration, or the like.

The retort zone particulate mixture of pellets and spent shale are passed through separation zone **40** for separation of the spent shale from the pellets and for recovery and return of the pellets to the pellet deposition burning zone. The separation zone may be any sort of exiting and separation system accomplishing at least **75** percent by weight separation of the total spent shale from the pellets and at least **95** percent by weight separation of the spent shale that is smaller than the pellets, provided that, at least one stage of the separation zone involves gas elutriation of at least part of the spent shale.

The separation zone may be comprised of any number and types of units of equipment. As shown, the mixture of pellets and spent shale first passes through optional revolving screen or trommel **41** extending into product recovery chamber **38** and which has openings or apertures sized to pass the pellets and spent shale of the same or smaller size than the pellets and screen out any spent shale and agglomerates larger than the pellets. Most of the spent shale and the pellets flow through the openings in trommel **41** and drop to the bottom of recovery chamber **38** to exit via exit line **42**. Any spent shale, mineral matter, and agglomerates too large to pass through the openings in the trommel, pass outward through exit **43** to spent shale disposal. This prescreening or initial separation of spent shale larger than the pellets is optional and may also be delayed until a later of final stage of the separation system. The rotating retort makes initial screening easier.

The spent shale and pellets in the bottom of recovery chamber **38** are discharged via exit line **42** at a temperature between about 750° and 1050°F where these particulate solids are passed or conducted by gravity or other means of conveyance to a subsequent phase of the separation and recovery zone involving at least one stage wherein the mixture of pellets and spent shale is subjected to elutriation gas, preferably preheated or already hot, to elutriate and separate a part of the spent shale from the pellets. The elutriation gas is, at least in part, comprised of flue gas which originated in pellet deposition burning zone **14** and was used in seondary preretort oil shale heating zone **24**.

As shown, the flue gas in line **35** is first passed through optional heater **44**, wherein, if desired, the flue gas can be reheated. Heater **44** could be a carbon monoxide shift reaction zone as previously discussed or could use other means of direct or indirect heating including additional fuel burning. It should be noted, however, that the flue gas will already be hot and the need for additional fuel is, therefore, reduced. The hot flue gas then passes through line **45** to elutriation zone **46**. The elutriation zone may be in one or more stages conducted in batchwise fashion or in a combination of batch and separate units, or in two or separate units or stages.

The mixtures of pellets and spent shale in line **42** passes to elutriation zone **46** where the mixture is subjected to gas elutriation with gas entering the elutriation zone by way of line **45**. As stated previously, at least a part of this elutriation gas is hot flue gas from the pellet deposition burning zone. In the elutriation zone, a major portion of the spent shale is elutriated and separated from the pellets. For simplicity, only one elutriation stage is shown and the spent shale and elutriation gas pass overhead through line **47** to overhead separator **48**, for example, a cyclone, scrubber, or the like, to remove the spent shale to disposal by way of bottom line **49** and to clean and allow the elutriation gas to exit overhead through line **15** and pass from the separation zone through line **15** to pellet lift system **12** as previously described.

If additional elutriation stages are used, the elutriation gas may first be returned to such stages, if desired. The mixture of remaining pellets and spent shale in elutriation zone **46** is added to any other pellet return streams and passed through line **11** to pellet lift system **12**, where, as previously described, the pellets are relifted to the pellet deposition burning zone by the hot lift gas from line **15** at least part of which is comprised of hot flue gas originating in the pellet deposition burning zone. In a retorting system of the nature just described, it would be highly advantageous to initially produce the hot flue gas at a pressure which would eliminate or at least reduce the need for subsequent pressure increasing or compression stages especially at points where the gas contains entrained solids. This may be accomplished in the illustrated process by raising the inlet pressure of the combustion supporting gas in line **26** when the combustion supporting is more or less solids free.

It has previously been pointed out that the pressure in the retort zone may be the autogenous pressure. This retort zone pressure may be below the pressure to which the combustion supporting gas is compressed. If so, an appropriate conventional gas seal system will be placed in the pellet feed system between the pellet deposition burning zone and the retort zone. Preferably, the combustion supporting gas will be compressed to a pressure high enough to cause passage of the flue gas throughout the system as herein described. For example, if the retort system involves one lift heating unit in the secondary preretort heating zone, three stages of elutriation, and two lift heating units in the first preretort heating zone, the entry pressure of the combustion supporting gas could be increased to about 6 to 19 pounds per square inch above the atmospheric pressure.

In related work *Y.A.K. Abdul-Rahman; U.S. Patent 3,803,021; April 9, 1974; assigned to Atlantic Richfield Company* hot pellets in a size range between about 0.06 and 0.3 inch characterized primarily by their effective surface area, quantity, size, density, and elutriation velocity are cycled to a retort zone to mix with and retort crushed oil shale, thereby producing gas and oil products, a combustible deposition on the pellets, and spent shale. The main source of heat for retorting is derived from controlled burning of the deposition on the pellets.

After retorting and prior to burning the deposition, at least 95 percent by weight of the spent shale is separated from the pellets in a way which recovers at least 99 percent by weight of the pellets for reuse in the process. Separation and recovery of the pellets is accomplished by a multiple stage combination of gas elutriation at increasing elutriation velocities and appropriate screening of the elutriated particles when pellets are present.

Ball Heater-Equilibrator System

A process described by *J.A. Whitcombe; U.S. Patent 3,595,540; July 27, 1971; assigned to The Oil Shale Corporation* relates to a ball heater and ball temperature equilibration system. Solid heat exchange bodies, such as metallic or ceramic balls or pebbles, are used in a variety of processes for heating material to be processed.

Exemplary of such processes are those where oil shale or other solid hydrocarbonaceous materials are pyrolyzed by means of hotter heat exchange bodies to extract hydrocarbonaceous material therefrom. After the heat exchange bodies have exchanged their heat by contact with the material to be heated thereby, the heat exchange bodies must have their temperature brought back to an elevated temperature so that they can be economically reused or recycled for the processing of further material. A variety of heaters have been proposed previously for reheating and/or regenerating solid heat-carrying bodies such as balls, pebbles, catalyst and the like. The typical of such devices are those described in U.S. Patents 2,534,625; 2,609,332; 2,626,794; 2,735,880; and 2,851,403.

This process involves a method and apparatus for equalizing the temperature of heat-carrying bodies previously heated to temperatures which vary significantly. Broadly, the method involves introducing solid heat-carrying bodies of varying or nonuniform temperatures into a temperature equalization chamber, introducing flue gas having the average mean temperature of the heated heat-carrying bodies into the temperature equalization chamber, intimately intermixing the heat-carrying bodies and the temperature equalizing gas while gravitating the heat-carrying bodies through the temperature equalization chamber to provide flue gas and heat-carrying bodies at a substantially uniform temperature, and thereafter withdrawing the flue gas and heat-carrying bodies from the equilibration chamber. The uniformly heated bodies may thereafter be passed to a retort or other reaction chamber for use therein. The solid heat-carrying body heater and solid heat-carrying body temperature equilibrator system of the process comprises means for the cocurrent, rather than countercurrent, flow of hot flue gas and solid heat-carrying bodies through the heater and means for flowing, preferably cocurrently, the heated solid heat-carrying bodies and discharge flue gas from the heater through the solid heat-carrying body temperature equilibrator.

Referring to Figure 1.15, the ball heater-ball temperature equilibrator system comprises a gravity fed, cocurrently fired, ball heater **1** and a ball temperature equilibration chamber **2**. The equilibration chamber units are four in number spaced horizontally from one another at approximately 90° (only two of each of these units are shown).

The ball heater **1** comprises a combustion chamber **3** having at the top a fuel line **5** and an air line **6** connected to a burner **7** for the generation of the hot flue gases. Four hot flue gas outlets **8** are at the base thereof and at right angles to one another (only two are shown). Each of the four flue gas outlets **8** is connected respectively to the ball-heating chamber **9**.

FIGURE 1.15: BALL HEAT-EQUILIBRATOR SYSTEM

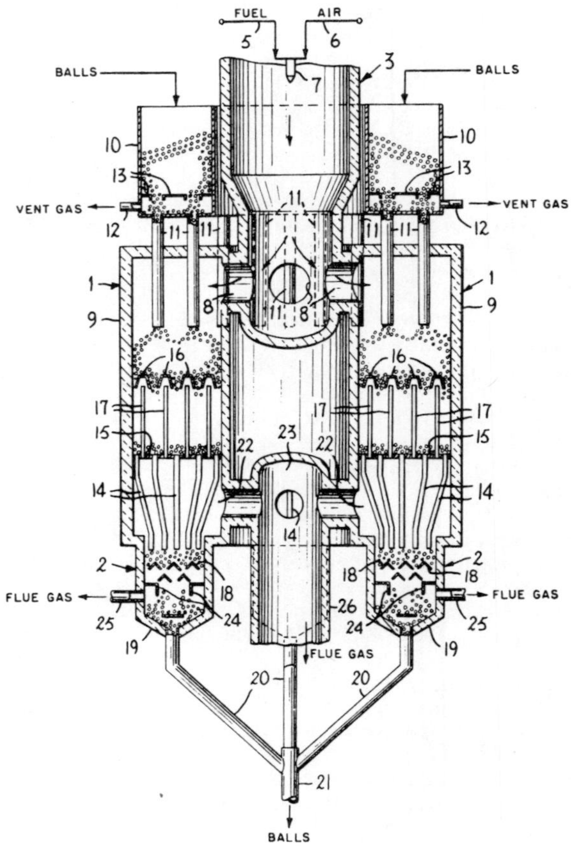

Source: J.A. Whitcombe; U.S. Patent 3,595,540; July 27, 1971

The ball-heating chamber **9** is provided with a ball surge vessel **10** of annular design for the receipt of balls to be reheated. The ball surge vessel **10** is connected at the base thereof to the top of the ball heating chamber **9** by a plurality of ball conduits **11**. The lower portion of the ball surge vessel **10** has vent gas outlets beneath vent gas collector channels **13**. The ball-heating chamber has a plurality of ball outlet conduits **14** at the base which extend downwardly from ball bed support plates **15** having orifices corresponding with the ball outlet conduits **14**. Within the ball-heating chamber **9** is a plurality of circumferential flue gas collector channels **16** for withdrawing the hot flue gas from the ball-heating chamber **9** via a plurality of flue gas conduits **17** which extend from beneath the channels **16** to orifices in the ball bed support plates **15**.

The four ball temperature equilibration chambers **2** are arranged horizontally at right angles to one another. Each ball temperature equilibration chambers **2** has beneath the ball inlet conduits **14** a plurality of baffle plates **18** for commingling the downwardly flowing heated balls. The tapered base **19** of each ball temperature equilibration chamber **2** is provided with a reheated ball outlet conduit **20** connected to a common ball transport conduit **21** for passage of the reheated balls to a heat exchanger, such as an oil shale pyrolyzer or retort.

The flue gas flowing downwardly through flue gas conduits **17** flows out of the heating apparatus through flue gas outlet zone **22** and then to the flue gas collection zone **23**, and by flowing downward through the equilibration chambers **2** for ball temperature equilibration purposes therein and then under downwardly extending circumferential right angle channel members **24** mounted on the walls of each of the ball temperature equilibration chambers **2** for disengaging the flue gas therefrom, and then to flue gas outlet conduits **25**. The relative amount of the gases that leave the heating apparatus through conduit **26** compared to that leaving through the equilibration chamber exit pipes **25** is controlled by control of the pressures at these two points. The flue gas collection zone **23** is connected to a major flue gas outlet conduit **26** for withdrawing most of the flue gas for other use, such as in a raw oil shale preheat lift pipe.

The balls or pebbles flow through the apparatus in the following manner. Balls to be reheated are fed into the top of the ball surge vessel **10**. The balls then pass by gravity from the base thereof through ball outlet conduits **11** and into ball-heating chamber **9**. The balls gradually pass downwardly through the ball-heating chamber **9** and exit therefrom through the plurality of ball outlet conduits **14** and then into ball temperature equilibration chambers **2**. There the balls are intermixed by baffle plates **18** before passing from the tapered base **19** of the ball temperature equilibrators **2** through outlet conduits **20** and common ball transport conduit **21**.

The passage of the flue gas through the apparatus occurs in the following manner. Fuel and air fed from lines **5** and **6** respectively are burned in burner **7** to produce hot flue gas. The hot flue gas passes from the combustion chamber **3** via the hot flue gas outlet **8** into the top of ball-heating chamber **9**. The hot flue gas descends in contact with the balls within the ball-heating chamber **9** and then passes via flue gas collector channels **16**, flue gas outlet conduits **17** and flue gas outlet zone **22** partly into the ball temperature equilibration chambers **2** and partly into the flue gas collection zone **23**. A small portion of the flue gas escapes from the ball-heating chamber **9** by ascending through the ball conduits **11** and out of the ball surge vessel **10** via vent gas collector channels **13** and vent gas outlets **12**.

The flue gas is withdrawn from the flue gas collection zone **23** via major flue gas outlet conduit **26**. After passage through the balls within the ball temperature equilibration chambers **2**, the flue gas exits therefrom via the channel members **24** and the flue gas outlet conduits **25**. In operation of the apparatus, balls or pebbles having a temperature of from about 850° to 950°F are fed into the ball surge vessel **10**. The balls pass downwardly therefrom via conduits **11** and into the ball-heating chamber **9**. Fuel and air are fed via lines **5** and **6** respectively into the burner **7** within combustion chamber **3** where the fuel is burned to produce hot flue gas having a temperature of from about 2300° to 3000°F. The hot flue gas is passed from the combustion chamber **3** via outlets **8** and into the ball-heating chamber **9**. The hot flue gas in contact with the balls in the ball-heating chamber **9** raises the temperature of the balls to varying temperatures such that the coldest balls generally have a temperature of about 950°F and the hottest balls generally have a temperature of about 1800°F.

The nonuniformly heated balls are withdrawn from the ball-heating chamber **9** via ball outlet conduits **14** and fed into the ball temperature equilibration chambers **2**. The flue gas withdrawn from the ball-heating chamber **9** via flue gas collector channels **16** and flue gas outlet conduits **17** and into the flue gas collection zone **23** has a temperature generally of about 1275°F in view of the heat exchange between the hot flue gas and the balls in the ball-heating chamber **9**.

The major portion of the flue gas is withdrawn from the flue gas collection zone **23** via flue gas outlet conduit **26** for further use, such as in a raw oil shale preheat lift pipe. The remainder of the flue gas having a temperature of about 1275°F passes through the balls in the ball temperature equilibration chambers **2** in a cocurrent manner to effect ball temperature equilibration. The flue gas is withdrawn from the ball temperature equilibration chambers **2** via channel members **24** and flue gas outlet conduits **25** wherein the flue gas has a temperature of about 1275°F.

The balls within the ball temperature equilibration chambers **2** have their temperature substantially equalized so that the hottest balls have a temperature of approximately 1400°F with the average temperature of the balls being about 1275°F. The substantially uniformly heated balls are withdrawn from the ball temperature equilibrators **2** via conduits **20** and **21** and used as heat exchange media in any desired process, such as in the pyrolysis of oil shale.

Inert Ball Heater

T.D. Nevens and W.J. Culbertson, Jr.; U.S. Patent 3,020,227; February 6, 1972; assigned to The Oil Shale Corporation describe a process for heating solid bodies, spherical or nonspherical, where the requisite amount of heat transfer to the solid bodies is obtained in a relatively short path length while avoiding the great pressure drop usually encountered in the packed tower type of heater.

The process for heating thermospheres, or balls, consists in fluidizing the balls by means of hot combustion gases, the combustion gases preferably containing finely divided solid material. This finely divided material is, generally speaking, combusted residue. For example, in the case of the combustion of shale coke, shale ash comprises the finely divided entrained material.

The gas velocity required for fluidization of the balls is necessarily sufficiently high so that any of the finely divided material will not be retained within the balls. At the same time, the rate of heat transferred to the balls, in their fluidized state, both by the finely divided solids and gases, is considerably greater than is the case in a packed bed, and further, the problem of retention of finely divided solids by the packed bed is completed obviated. In addition, balls are readily removed by such a process without the resulting channeling of gases or of balls, as would be liable to occur in packed beds.

H.E. Linden and T.D. Nevens; U.S. Patent 3,164,541; January 5, 1965; assigned to The Oil Shale Corporation describe a process which utilizes an oil for conveying solid bodies from point to point in a process. Thus, the balls used for pyrolyzing oil shale are generally quite hot having a temperature somewhat between 500° to 1050°F, depending upon the point in the process at which the temperature measurement is taken. These hot balls are fed, usually by gravity, into an elevator conduit in which an appropriate oil such as, for example, shale oil, or other high boiling point oil flows upwardly. The oil flows upwardly through this conduit due to the provision of an oil pump therein. As the oil moves upwardly, it carries the solid bodies with it. After the balls have been conveyed to a desired height, the balls and oil are separated by any suitable means, such as by screening means. The oil may then be recycled to the pipeline through the oil pump, or discharged from the system. If oil obtained from the pyrolysis of the oil shale is used as the transporting hydraulic medium, this oil may be coked to varying degrees.

Heat Supplied by Pyrolysis Products

A process described by *C.L. Crawford; U.S. Patent 3,265,608; August 9, 1966; assigned to Technikoil, Inc.* provides a method for producing an effluent vapor from a solid carbonaceous material leaving, upon pyrolysis, a spent solid residue. The process comprises the steps of: pyrolyzing the solid material in a pyrolysis zone by solid-to-solid milling contact with hotter heat-carrying bodies to obtain effluent vapor and hot spent solids; recovering the effluent vapor from the pyrolysis zone; separating the hot spent solids from the heat-carrying bodies; transferring the heat-carrying bodies to a pebble heater; heating the heat-carrying bodies in the pebble heater by means of heat derived from the hot spent solids under noncombustion conditions and heat derived by combusting fuel and recirculating the heated heat-carrying bodies to the pyrolysis zone to effect the pyrolysis of fresh solid carbonaceous material.

FIGURE 1.16: PROCESS EMPLOYING HEATED SOLID BODIES FOR THERMAL TREATMENT OF OIL SHALE

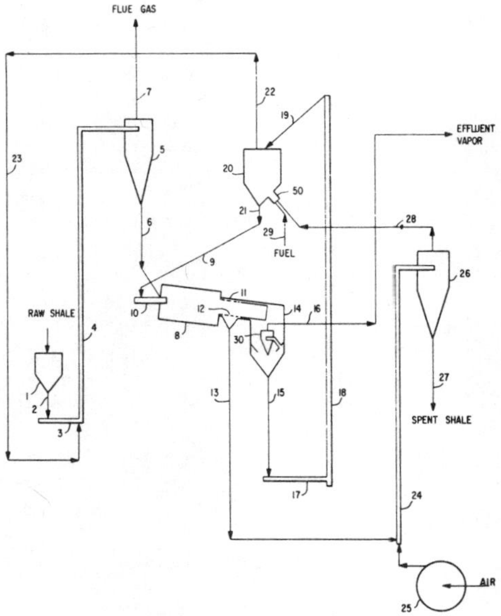

Source: C.L. Crawford; U.S. Patent 3,265,608; August 9, 1966

Referring to Figure 1.16, Colorado oil shale at about 50°F crushed to an average particle size of about ½" diameter is fed from a bin **1** via line **2** and feeder **3** to a raw shale lift heater **4** where it is contacted and entrained by flue gas having a temperature of about 1100°F. In the raw shale lift heater, the raw shale feed is heated to a temperature of about 300°F. The preheated raw shale is then removed from the gas-solids suspension in a separator **5** and passed to a horizontal rotating pyrolysis drum **8** via line **6**. The cooled flue gas from which the heated raw shale has been removed passes from gas solids separation zone **5** through line **7** and is vented to the atmosphere. In pyrolysis zone **8**, the preheated raw shale is contacted with hot alumina balls having a temperature of about 1570°F and entering pyrolysis drum **8** via line **9** and feeder **10**. The balls and shale pass concurrently through pyrolysis drum **8** whereby the heat of the balls is imparted to the shale with the production of an effluent vapor and spent shale solids. Effluent vapor and spent solids at about 870°F and cooled balls at about 900°F exit from pyrolysis drum **8** through line **11** which is adapted with a screen **12** having openings such that spent solids pass through while the passage of balls is precluded.

The effluent vapor and cooled balls are then passed by means of line **11** to a bin **14** where the vapor and balls are allowed to remain in contact for a longer period whereby the removal of dust from the effluent vapor is achieved. Substantially dust free vapor is removed from bin **14** through line **16** and passed to a recovery section. Cooled balls are removed from bin **14** by means of line **15** and passed to a ball elevator **18** via feeder **17**. In ball elevator **18** the cooled balls are lifted to the top of a pebble heater **20** into which they are introduced from elevator **18** by means of line **19**.

The spent shale solids removed from pyrolysis drum **8** and separated from the cooled balls by means of screen **12** are passed via line **13** to a gas lift line **24**. In gas lift line **24** the hot spent shale is contacted and entrained by means of air at about 95°F introduced into lift line **24** through a blower compressor **25**. The entrained spent shale imparts its sensible heat to the air raising the temperature of the solids lifting medium to about 700°F. The air spent solids suspension is then passed through a separator **26** where the entrained solids are removed from the air stream.

The separated solids pass from solid separator **26** through line **27** and are discarded. The heated air from which the spent solids have been removed passes from separator **26** by means of line **28** and is introduced into the bottom of pebble heater **20** via a fuel combustion zone **50** located at the inlet of line **28** to ball heater **20**. A gas oil fraction recovered from the effluent vapor produced in the process is introduced by means of line **29** into fuel combustion zone **50** where it is ignited and burned by the preheated air.

The flue gas produced by burning the gas oil fraction and having a temperature of about 3000°F is then passed upwardly through ball heater **20** to contact and heat balls. Heated balls leave ball heater **20** through line **21** and are recirculated to pyrolysis drum **8** via line **9** and feeder **10** to contact and effect the pyrolysis of additional raw shale. Flue gas having a temperature of about 1100°F is removed from the top of ball heater **20** by means of line **22** and passed to lift heater **4** where it contacts and preheats a stream of additional raw shale feed.

M.U. Zimmerman, Jr.; U.S. Patent 3,442,789; May 6, 1969; assigned to Technikoil, Inc. describes a process for producing oil from shale in which the shale is pyrolyzed in a rotating mill by rotating and milling with hot solid heat carrying material, e.g., grinding balls, in which the overhead hydrocarbons and residue are recovered and part of the product, e.g., a butanes fraction of the overhead, is heated with hot flue gas and the hot product is returned to the mill to directly heat the pyrolysis zone.

A process described by *C.L. Crawford; U.S. Patent 3,167,494; January 26, 1965; assigned to The Oil Shale Corporation* is based upon a finding that heat required for pyrolyzing solid carbonaceous materials efficiently and economically can be transferred to the carbonaceous solids from hotter heat-carrying bodies in a vertical pyrolysis zone without the necessity of establishing a completely fluidized solids bed containing the carbonaceous and heat-carrying solids throughout the pyrolysis zone.

In accordance with the process, the carbonaceous solids and heat-carrying bodies are mixed and contacted intimately immediately upon being charged to the top of the pyrolysis zone only by the action of upwardly rising effluent vapor produced in the pyrolysis zone. Thus, the method does not contemplate the utilization of an extraneous gasiform material and/or the formation of a completely fluidized solids bed throughout the pyrolysis zone. Accordingly, the process provides a method for pyrolyzing carbonaceous solids where an effluent vapor is produced from which the desired components easily and economically can be recovered in smaller and simpler equipment than that required in processes using an extraneous fluidizing gas. In addition, in not requiring that the solids in the pyrolysis zone be completely fluidized, the method suitably is adapted for handling raw carbonaceous solids feed streams containing appreciable amounts of the previously unusable finely divided material thereby enabling increased total product yields to be realized from available raw materials.

Heated Solids for Attrition and Heat Transfer

In a process described by *H.F. West; U.S. Patent 3,350,280, October 31, 1967; assigned to The Standard Oil Company* externally heated attrition resistant solid bodies such as steel or ceramic balls are used for transferring heat to the oil shale. The process includes the use of a stationary drum retort equipped with an internal rotating screw which contains throughout its length perforated or slotted screw flights together with baffles designed in such a way that the solid bodies or balls are retained on the screw and do not fall through the slots or perforations but the oil shale falls through. The oil shale which is crushed by action of the heated balls upon it is fed into the top of the retort and by gravity falls down through the perforations or slots in the screw in the retort to the bottom where the carbonaceous residue or shale coke is removed and may be burned to supply heat to the external heater which in turn supplies heat to the recycled balls.

The externally heated balls flow by gravity from the heater to the bottom of the retort where the rotating screw picks them up and works them to the top and out of the retort to a point where they flow by gravity back to the external heater where they are reheated and recycled to the bottom of the retort. Proper baffles are present on the screw to retain the balls in their upward movement and to prevent the balls from rolling backwards or toward the bottom of the retort. The screw can be driven by any means such as variable-speed motor and the degree of heat transfer or contact time between the balls and the oil or shale can be varied by adjustment of the speed of rotation of the screw.

Control of Entrained Solids

C.L. Crawford; U.S. Patent 3,252,886; May 24, 1966 describes a method for treating effluent vapor derived by contacting solid carbonaceous materials with heat-carrying bodies where an effluent vapor is provided which is characterized by a relatively low entrained-solids content.

In the pyrolysis process the raw solid carbonaceous material initially is crushed by any suitable method to be of a suitable particle size. Preferably the raw carbonaceous material is reduced in particle size to be of such an average diameter that it is of a smaller value than that of the heat-carrying bodies employed, so that separation of the spent material produced in the pyrolysis and the heat-carrying bodies readily can be accomplished.

Usually the average particle size of the raw carbonaceous solid feed stream is in the range of from ⅜" to ¾" and preferably is about ½". The crushed raw carbonaceous material at ambient temperatures then may be passed directly to the pyrolysis zone, although it is preferred that the raw solids feed stream be preheated.

Preheating of the crushed material suitably may be carried out either indirectly or directly with heat from any source available. One preferred method for preheating the crushed carbonaceous feed stream utilizes the residual sensible heat of flue gases previously employed in the process to supply heat to the heat-body heating zone. In such cases the technique preferred for utilization is to effect the desired heat transfer between the crushed raw material and the hot flue gases by entraining the carbonaceous solids in a gas lift line with the flue gas stream. The preheated raw material is then separated from the entraining gas in a gas-solids separation zone and passed to the pyrolysis zone. The cool or preheated carbonaceous material is then passed into a pyrolysis zone which suitably may be in any form for carrying out the pyrolysis of carbonaceous solids such as a fluidized bed, a horizontal or slightly inclined rotating drum or a vertical stacked solids bed where it is contacted in solid-to-solid heat exchange contact with heat-carrying bodies having sufficient available heat to effect the pyrolysis of the carbonaceous material and produce effluent vapor and hot spent solid residues. Preferably the pyrolysis zone is in the form of a rotating drum where solid-to-solid milling contact between the heat bodies and carbonaceous solids is achieved.

The available heat of the heat-carrying bodies is such that essentially complete pyrolysis of the carbonaceous solids is effected with a maximum yield of hydrocarbon values being thereby achieved. For example, when oil shale is subjected to pyrolytic treatment, the heat supplied by the heat-carrying bodies to the oil shale is such as to raise the temperature of the shale in the range of from 750° to 950°F, preferably from 800° to 900°F. Temperatures in the pyrolysis zone greater than about 950°F usually are avoided due to the fact that some slight cracking of valuable hydrocarbon product may occur under such conditions.

Although countercurrent flow of solids through the pyrolysis zone may be employed, it is preferred to use a rotating drum as the zone where pyrolysis is effected, the carbonaceous materials and heat bodies are passed through the pyrolysis drum in concurrent fashion. By the utilization of concurrent flow in lieu of countercurrent flow, the power requirements for moving the solids through the system are greatly reduced and the chance of coking occurring at the heat-body inlet end of the pyrolysis drum is essentially eliminated.

The actual temperature and amount of the heat bodies introduced into the pyrolysis zone in a particular embodiment of this method, will depend upon, inter alia, the type of carbonaceous material being treated, the degree of pyrolysis desired, the inlet temperature of the carbonaceous feed, and the heat transfer characteristics of the heat-carrying bodies. Where oil shale is treated and alumina balls are utilized as the heat-carrying bodies, the ratio of heat-carrying bodies to raw shale feed to the pyrolysis zone preferably is in the range of from 0.6:1 to 10:1, preferably from 0.8:1 to 3:1. In such oil shale treatments the alumina balls introduced into the pyrolysis zone usually are at a temperature in the range of from 1200° to 1800°F, preferably from 1350° to 1650°F.

The heat-carrying bodies, cooled by giving up a portion of their heat to the carbonaceous material during pyrolysis, the effluent vapor and spent materials produced in the pyrolysis of the carbonaceous solids are then removed from the pyrolysis zone. The condensible effluent vapor containing fine solid materials and the heat-carrying bodies are then sent to a dust removal treatment where the solids suspended in the vapor are removed preliminary to the vapor being passed to a recovery section.

In the preferred dust removal procedure, the vapor is subjected to at least a two stage treatment to effect the removal of suspended solids. In a first dust removal treatment, the solids-containing vapors are introduced into a first solids removal zone where the still hot heat-carrying bodies recovered from the pyrolysis zone are caused to collect in sufficient volume to form a bed of appreciable depth and are maintained in contact with the heat-carrying bodies from the pyrolysis zone. Part of the solids are removed from the vapor by contact with the heat-carrying bodies.

It is believed that by contact electrification, effected by previous particle-to-particle contact in the process, the dust and heat-carrying bodies become differentially charged, and that a portion of the dust particles entrained in the effluent vapor are attracted to and thereby removed from the vapor by the heat-carrying bodies. The vapor containing a reduced amount of suspended solids, is then passed through a vapor-solids separator, preferably of the cyclone type, constituting a second solids removal zone, where the remainder of the suspended solids in the vapor are substantially completely removed. When according to the preferred embodiment of the process the vapor solids separator is located within the first solids removal zone, heat from the heat-carrying bodies prevents condensation of the high boiling constitutents of the effluent vapor. To obtain maximum efficiency in removing entrained solids from the effluent vapor, the second vapor-solids separation zone preferably is located within the first solids removal zone in such a manner that the outlet of the second vapor-solids separation zone is in juxtaposition with the heat-carrying body outlet from the first zone.

If the hot spent solids and cooled heat-carrying bodies produced in the pyrolysis zone have not already been separated inside the pyrolysis zone, as is described in U.S. Patent 2,592,738, separation of the spent materials and heat-carrying bodies recovered from the pyrolysis zone is then carried out by any suitable means such as by a screening operation. Separation of the spent materials and heat-carrying bodies easily can be accomplished when a significant difference in the average particle size between the two solid streams is maintained. The heat-carrying bodies recovered from the pyrolysis zone and separated from the spent material are passed through the dust removal zone, as above described; then are reheated and recycled to the pyrolysis zone to effect the pyrolysis of additional raw carbonaceous material.

Finely Divided Heat Carrier

A process described by *P. Schmalfed, H. Sommers and H. Janssen; U.S. Patent 3,655,518; April 11, 1972; assigned to Metallgesellschaft AG and Ruhrgas AS, Germany* is directed to an apparatus for the dry distillation of bituminous or petroleum-containing materials such as coal, lignite, oil shale or oil sand in a finely granular state. The material to be distilled is heated by means of being thoroughly mechanically mixed with a circulating, finely divided heat carrier which is separated from the distillation vapors along with the solid distillation residue, heated in a pneumatic conveyor, and returned to the distillation apparatus.

The apparatus consists of a vertical pneumatic conveying and heating unit for the particulate heat carrier to the bottom end of which a free-oxygen containing gaseous propellant is fed axially as through a venturi tube and finely granular thermal carrier is introduced from a concentric annular chamber thereabout through slits by means of a flow of a control gas. A separator at the upper end of the conveyor is divided by a wall extending downward from the top into a separating chamber and a secondary chamber containing the discharge for the propelling gas and products of combustion, and a bottom solids collecting chamber that is common to both.

A distillation chamber is connected at its input end to the heat carrier discharged from the separator. It also receives the raw material or feed that is to be distilled. At its discharge end, one conduit carries away the hot mixture of thermal carrier and fresh solid distillation residue and another conduit carries away the distillate vapors. The distillation chamber has a screw conveyor to force and mix the solids. An intermediate reservoir or hopper is connected at one end to the distillation chamber and at the other end to the annular chamber at the bottom of the conveyor.

A dust gas separator is used to remove fines from the distillation vapors after which they are condensed in a product recovery system in which the vapors are condensed by spraying them with previously separated and cooled condensate. Another dust separator cleans the products of combustion discharged from the conveyor-separator and the gas is then utilized in heat exchange to heat the incoming propellant gas and/or in a waste heat boiler. The particulate solids removed from the process are cooled, preferably with water, in a solids cooler.

The finely divided solid distillation residue which serves as the heat carrier is carried and simultaneously heated in a vertical penumatic conveying and heating unit and then is separated from the combustion gases by free fall and inertial ejection in a separating chamber, and delivered to a mechanical mixer in the distillation chamber. In the mixer the thermal carrier, heated to 700°C, for example, is combined with the finely granulated starting material that is to be distilled, e.g., coal, oil shale, oil sand or the like, and blended intensely within seconds. A very rapid transfer of heat from the thermal carrier to the finely granulated starting material takes place. This heat quickly brings the starting material to the desired distillation temperature, usually between 450° and 650°C, breaks down the bitumen, and/or drives out in vapor form the oils contained in or developing from the raw material, together with water vapor formed from the moisture and from chemically bound water. The formation of light cracked gases while this is taking place is extremely slight.

The mixture of thermal carrier and fresh distillation residue flows from the distillation chamber into an intermediate reservoir or hopper for the after distillation of the feed material and for the driving of hydrocarbon vapors from the interstices and pores of the solids as by the injection of water vapor returned light distillation gases into the mixture. The solids are then fed back into the bottom part of the pneumatic conveying and heating unit, thereby completing the cycle of the thermal carrier through the heating and distillation steps.

The apparatus permits an effective and simple distillation of finely granulated oil-bearing materials, the condensation of the distillation vapors, and a good recovery of heat from the waste gas from the heating of the thermal carrier. The distillation vapors from the mixer-distilling chamber are processed in a cooling system providing for several stages of contact of the distillate vapors with their own cooled condensate, while the cooling of gases from the heater-conveyor is performed in an air preheater and/or a waste-heat boiler.

Finely Granular Carrier

R. Rammler and P. Schmalfed; U.S. Patent 3,703,442; November 21, 1972; assigned to Metallgesellschaft AG, Germany have found that the low-temperature distillation process that operates on the basis of finely granular heat carriers can also be applied to a starting material that leaves a pulverulent distillation residue if a finely granular solid foreign substance is used as the heat carrier, which is of sufficiently stable shape to withstand repeated circulation through the heating and distillation without substantial grain disintegration, and which is heavy and large enough to permit the pneumatic separation of the pulverulent residue.

The pulverulent distillation residue does not participate in the circulation of the heat carrier, but is separated from the latter pneumatically before the heat carrier returns to the distillation zone. This is done particularly for the purpose of keeping harmful amounts of dust out of the distillation zone. Besides, considerable amounts of dust are entrained by the flowing distillation gases and vapors from the residue that forms in the distillation. The feedback of dust as a heat carrier would greatly increase these amounts of dust and would only add to the difficulty of recovering a low-dust distillate. The separation of the heat carrier from the distillation residue is therefore to be as complete as possible so that little or no dust will be fed back to the distillation zone with the heat carrier. The process comprises:

(a) circulating a heat carrier between a distillation zone and a heating zone;
(b) introducing the raw material into the distillation zone and heating it for the distillation with the heat carrier, and for formation therefrom of distillation gases and the pulverulent residue;
(c) withdrawing distillation gases from the distillation zone;
(d) withdrawing the heat carrier and pulverulent distillation residue in a mixture from the distillation zone;
(e) pneumatically conveying the mixture through the heating zone with a propellant gas and reheating the heat carrier;
(f) returning the heated carrier to the distillation zone; the improvement which comprises
(g) separating the effluent of the heating zone, comprising the heated carrier, pulverulent distillation residue and propellant gas, into propellant gas and a mixture of heated carrier particles and pulverulent distillation residue, and
(h) sifting pulverulent distillation residue from the mixture of heated carrier and pulverulent material to provide the heated carrier returned to the distillation zone in step (f).

Example 1: In Figure 1.17, **1** designates a feed hopper from which oil shale of a grain size smaller than 4 mm is continuously fed into the distillation reactor **2** which is in the form of a mechanical mixing mechanism. Make-up heat carrier river sand having a grain size smaller than 1.5 mm is mixed continuously or periodically with the shale in a quantity of 0.8% by weight. Substantially dust-free heat carrier is fed continuously, via line **5a**, at a temperature of 630°C into the distillation reactor **2**. The weight ratio of heat carrier to starting material is about **6**. On account of the large surface areas of the finely granular heat carrier and of the finely granular shale starting material, and on account of the positive mixing action, a very intense heat exchange takes place in the distillation reactor, so that by the time the two components have reached the end of the mixer, both have substantially reached the equilibrium temperature of 530°C. Oil vapors and gases are thereupon released from the shale and the solid distillation residue disintegrates largely to a floury dust, a portion of which is carried out of the reactor with the distillation vapors and gases.

Most of the pulverulent distilltion residue leaves the mixer together with the finely granular heat carrier and, after flowing through the after-degassing hopper **3**, it passes into the lower part of the pneumatic updraft conveyor **4**, into which air, preheated to 450°C is injected from the bottom. This air drives the mixture of pulverulent distillation residue and finely granular heat carriers upward through the conveyor and causes a burning away of residues of organic substances which are left in the residue after distillation.

In this manner the solids are heated while they are being conveyed upward. The feed of air to the conveyor is controlled so that the temperature of the propellant gas and of the material being propelled is 590°C at the upper end of the conveyor. At this temperature the solids in the separator-collector hopper **5** are largely separated from the propellant gases and collected in the bottom of this hopper.

By a sufficiently high velocity of the propellant gases in the exit side of the separator-collector hopper **5** it can be brought about that the pulverulent residue will leave the separator with the propellant gas via line **5b** as will be described in connection with Figure 1.17b. The pulverulent distillation residue that is separated with the granular heat carrier is substantially sifted out of the heat carrier deposit by the introduction and uniform distribution of preheated air in the lower portion of the hopper. At the same time the burning off of the organic residues in the distillation residue is continued, so that the temperature of the heat carriers in the sifter increases from 590° to 630°C.

The finely granular heat carrier, substantially freed of the pulverulent distillation residue runs at this temperature into the distillation reactor, thus completing its circuit. On the other hand, the distillation residue does not participate in this circuit, but leaves the separator **5** together with the propellant and sifting gases and flows with them through a waste-heat recovery system consisting of an air preheater **6** and a waste-heat boiler **7**. To combat abrasion by the dust-laden gas the turns in the ductwork between the separator **5** and the air preheater **6** can be lined with wear-resistant material and the top level of the air preheater **6** can be protected by laying wear-resistant plates on it.

After cooling to 270°C, the exhaust gas is substantially freed of dust in the cyclone **8**. Then it is cooled to 200°C by the injection and evaporation of water. At this temperature it enters the electrostatic filter **9** where the final filtration is performed. The evaporation of water ahead of the electrostatic filter increases the electrical breakdown strength of the gas and at the same time reduces the electrical resistance of the dust, which is dependent upon temperature.

Although the sifted heat carrier from hopper **5** brings only a very small amount of pulverulent distillation residue into the distillation reactor **2**, an appreciable amount of dust is carried out of the mixer reactor **2** by the distillation gases and vapors as a result of grain disintegration, the temperature of these gases being 530°C. This flying dust is substantially precipitated in two cyclones **10** and **11** connected in series, and is added to the material entering into the pneumatic lift conveyor **4**.

FIGURE 1.17: LOW-TEMPERATURE DISTILLATION PROCESS

a.

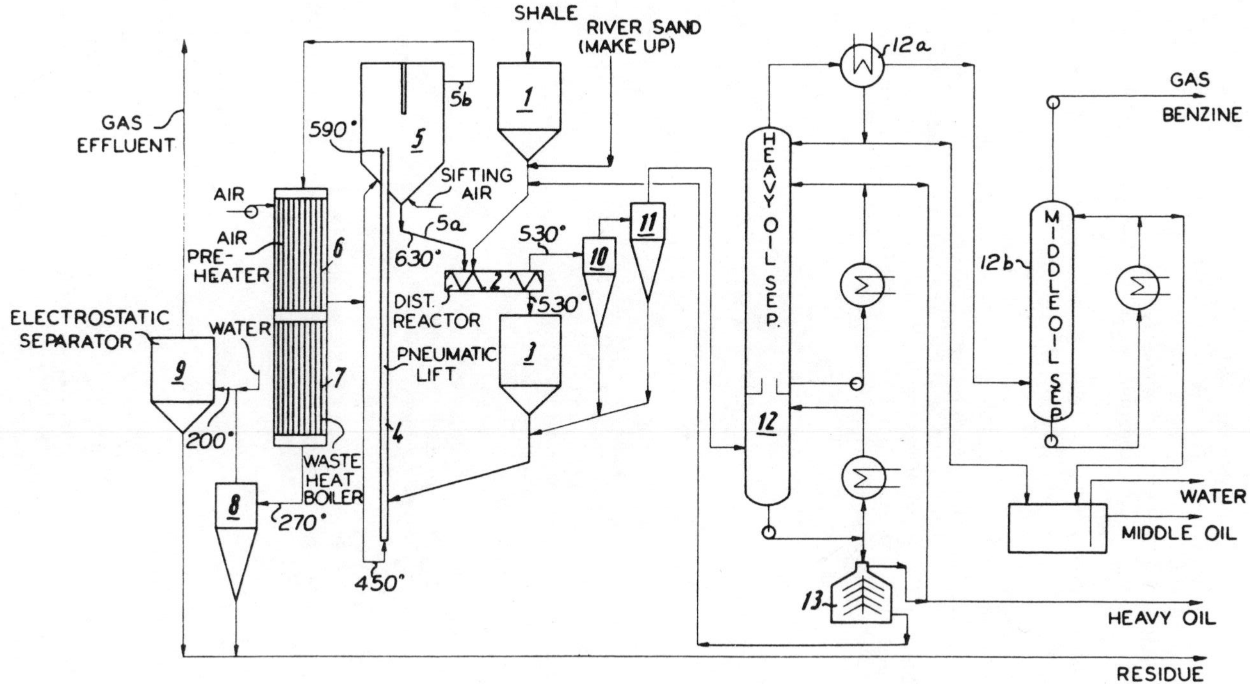

Flow Diagram of the Process with Two-Stage Sifting

(continued)

FIGURE 1.17: (continued)

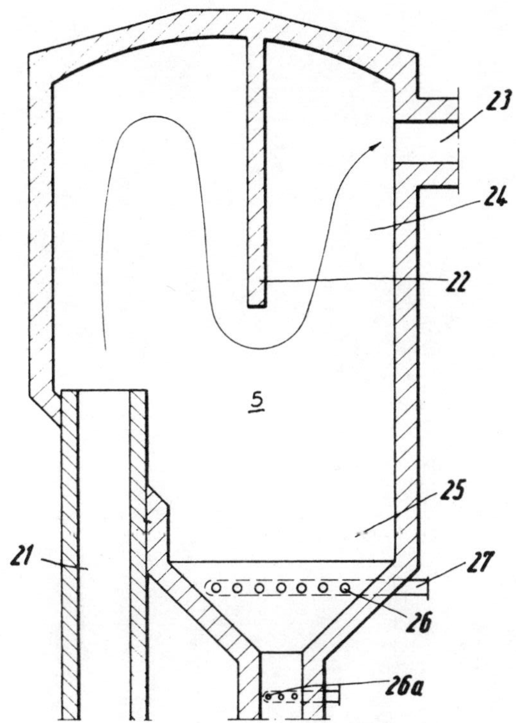

b. Vertical Section Through a Two-Stage Sifting Separator

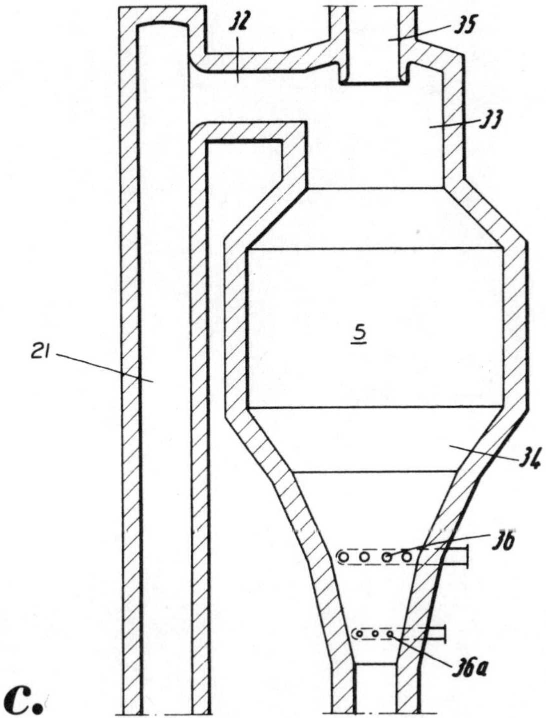

c. Vertical Section Through a Single Stage Cyclone Separator

(continued)

FIGURE 1.17: (continued)

d.

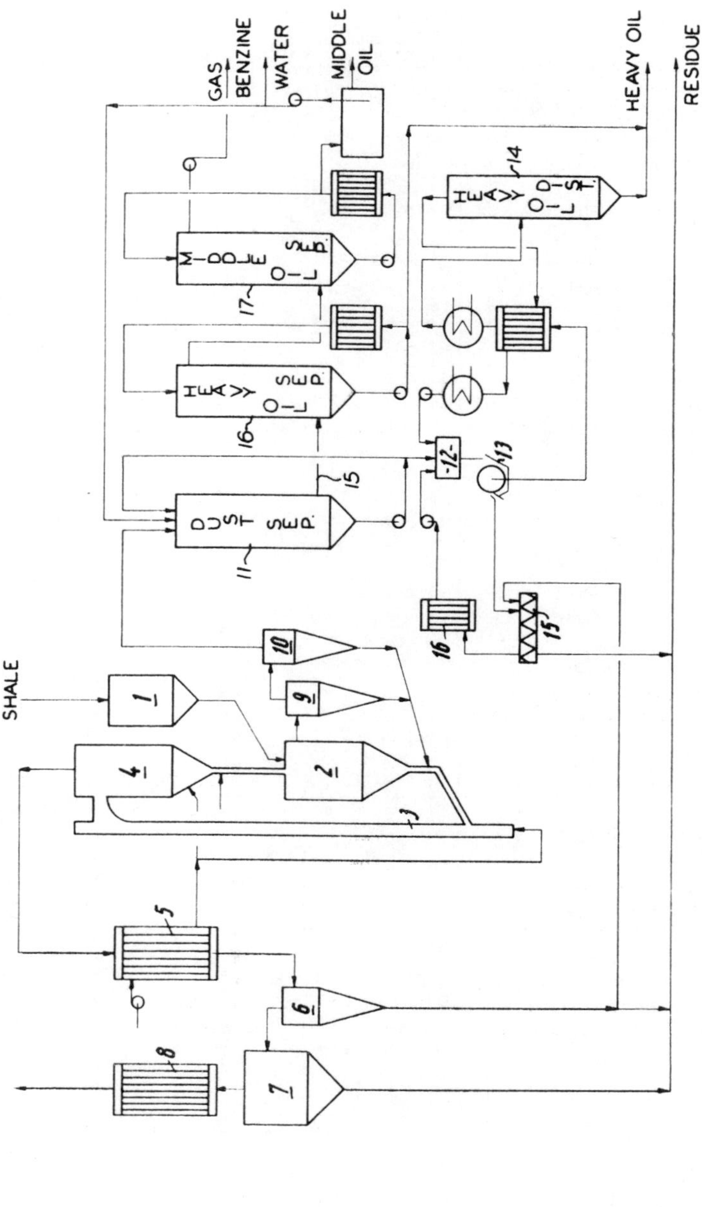

Flow Diagram of the Process with Single-Stage Sifting

Source: R. Rammler and P. Schmalfeld; U.S. Patent 3,703,442; November 21, 1972

The dust still contained in the gas-vapor mixture after it leaves the cyclones is completely freed of dust in the lower part of the heavy oil separator **12** by circulated heavy oil, the temperature in this stage being adjusted according to the amount of dust to be removed, so that the heavy oil that is withdrawn from the bottom of the separator **12** will have the highest possible dust content without losing its ability to be pumped.

Consequently the dust content of the heavy oil fraction is kept slightly below 50% by weight. The dust-laden heavy oil is cleaned in centrifugal separators **13**. The dust-laden centrifugation residue is returned to the mechanical mixer **2** for redistillation. The gases and vapors then flow through additional condensation stages in which oil fractions free of dust are produced. Heavy oil is also stripped from the distillation gases in the upper part of the separator **12**. The overhead distillation gas from separator **12** is treated in cooler **12a** and middle oil separator **12b** to separate the overhead into a middle oil fraction and a light gas fraction as is indicated in Figure 1.17a.

The construction of the separator and sifter at the upper end of the pneumatic lift, as chosen for use in Example 1, will be further explained with reference to Figure 1.17b. Broadly considered, the separator is for separating from a gas stream entraining solid particles of relatively large size, and solid particles of relatively small size, the solid particles of relatively large size. The separator includes a separating chamber having an inlet for the gas stream, an outlet for the gas stream, and an outlet for the separated large particles. The outlet for large particles is disposed in the bottom portion of the separator. The separator further includes means for routing the gas stream through the separator chamber so that the large solid particles are thrown out of the gas stream to the bottom of the separator for discharge through the outlet for large particles.

The process provides the improvement which comprises sifter means disposed in the bottom portion of the separator chamber adjacent the outlet for large particles for introducing a sifting gas for travel upwardly through the separated large particles to sift small particles from the separated large particles.

The mixture of solids and propellant gas passes up through the lift **21** into the separating chamber on the left side of the masonry partition **22**. The partition forces the gas, which is at first flowing freely upward, to perform two changes of direction in order to reach the outlet **23**. In the second reversal under the partition, the solids are hurled into the lower part of the hopper. The space **24** behind the partition is dimensioned to produce higher gas velocities, preferably of more than 1 m/sec, e.g., from 2 to 4 m/sec, in it. Thereby the separator can be made to perform a preliminary sifting in that certain amounts of pulverulent distillation residue will escape being separated from the heat carrier. This is particularly the case when the chamber **24** is omitted and the diameter of the separator is made smaller, the gas discharge aperture is placed lower down, and a baffle is placed ahead of the discharge aperture instead of the partition wall **22**.

The further sifting out of the pulverulent distillation residue takes place in the lower part **25** of the hopper, where a system of horizontal, parallel sparger tubes **26** are installed, through which a sifting medium is introduced into the material and is uniformly distributed over the sifter cross-section. This medium is fed to the sparger tubes through the manifold tube **27**. Underneath the hopper, in the pipe that carries the cleaned heat carrier to the distillation reactor, there is provided an additional system of spargers through which steam, for example, is introduced into the material, for the purpose of sweeping or displacing the sifting medium (e.g., air) from the interstitial areas.

Another possible mode of construction of the separator and sifter is represented schematically in Figure 1.17c. In this case a cyclone separator is used. The mixture of solids and gases coming from the pneumatic lift enters the cyclone **33** tangentially through the inlet passage **32**. The solids are substantially separated and collect in the lower portion **34**. The propellant gas leaves the cyclone through the pipe **35**. In the tapering bottom portion a sifting medium (e.g., air) is injected through the sparger system **36**. At a distance of at least 250 mm below that, sweeping or stripping steam or a stripping gas in introduced through the spargers **36a**. The dimensions of the bottom portion **34** are such that the accumulation of heat carrier and distillation residue that forms in it is put into a fluid state by the upwardly flowing sifting medium. In order that the fluidization may be as uniform as possible, the bottom portion **34** flares from the bottom up in the manner of a diffuser. The pulverulent distillation residue carried up out of the fluidized bed by the sifting medium is carried out together with the propellant gas through the discharge pipe **35**. This design of the separator-shifter is the basis of the flow diagram in Figure 1.17d, which is the subject of Example 2.

Example 2: **1** again designates the shale feeding hopper, provision being made for the admixture, with shale as is used in Example 1, of 1.5% of an oil-poor shale which has less of a tendency to grain disintegration in the low-temperature distillation process. In this embodiment, the heat carrier is distilled oil-poor shale and the 1.5% is make-up heat carrier. In this variant, no mechanical mixer is used, i.e., the starting material and heat carrier are fed into the low-temperature distillation reactor **2** at the top in such a manner that they are well mixed together. This method of construction is preferred in installations of medium and small throughput.

The solids elevated and heated in the pneumatic lift **3** are separated from the propellant gas in a cyclone **4** as is shown in Figure 1.17c. In the lower part of cyclone **4**, a grateless fluidized bed is maintained by the introduction of preheated air, the heating of the heat carrier being continued and the pulverulent distillation residue being separated from the heat carrier in the fluidized bed. The dust-laden sifting gases leave the separator together with the propellant gases, pass through the air preheater **5** and are then cleaned of dust in two stages **6** and **7**, separator **6** being a cyclone separator, and separator **7** being an electrostatic separator.

The electrical resistance of the dust, which varies with temperature, is in a desirable range at this relatively high temperature, and at very low temperatures as well, whereas at temperatures in between it passes through a maximum which is unfavorable to electrostatic dust removal. The cleaned gases are finally further cooled in the waste-heat boiler **8** before they are fed to the smokestack.

The distillation gases and vapors, after dust is removed in the cyclones **9** and **10**, are subjected to a treatment for removal of further dust in the dust separator **11**, where some condensation of heavy oil occurs, and the heavy oil scrubs the gas removing the dust. Then the distillation gases pass via line **15** to heavy oil separator **16**, and from the latter to middle oil separator **17**, these separations being carried out as is indicated in the flow diagram.

The dust-laden heavy oil that is produced in the dust separator **11** is mixed in mixer **12**, in a ratio of 1:1 by weight, with light oil to thin it. The suspension is then filtered on the continuous revolving suction filter **13**. The filtrate, after heating and the partial evaporation of the easily boiling fractions, is separated in heavy oil distillation column **14** into heavy oil and light oil.

The heavy oil is taken out as a product, and the light oil is pumped, after condensation, back into the mixer **12** to serve as the light oil diluent. The filter cake is mixed with hot distillation residue from cyclone separator **6** in the mixer **15** and heated to dry it. The benzene that evaporates is liquefied in the condenser **16** and is passed to mixer **12** to serve as diluent. Desirably a high performance separator can be installed between the mixer **15** and condenser **16** to separate solids entrained by the vapor. The dried filter cake, composed of distillation residue and rock sand purge is discharged from the process.

Horizontal Retort

R.N. Hall, R.H. Beaver, R.G. Vawter and C.J. Mains; U.S. Patent 3,616,266; October 26, 1971; assigned to The Oil Shale Corporation describe a pyrolyzer using a solid-to-solid heat transfer technique which provides a high degree of mixing of raw material and heat carrier required for an efficient and economical pyrolysis. The pyrolyzer is a substantially horizontal pyrolysis drum comprising a first inlet means for introducing the solid heat carriers, a second inlet means for introducing the raw material onto the surface of the heat carrier, an internal rotatable mechanism having a first conveyor section and a second mixing section. The conveyor section comprises a standard screw flight positioned so as to receive the heat carriers and raw material. The mixing section is positioned downsteam from the conveyor section and comprises a series of longitudinally spaced pyrolysis zones or chambers, each zone terminated at the opposite ends thereof by suitable flow-retarding means. It has been found that perforated screw sections wherein each section contains between one and two revolutions serve to both move as well as retard the solids flow through respective chambers.

Each screw section is joined by means of rigid bladelike wiping members secured, preferably spirally, around the periphery thereof and extending between each section. In order to further insure adequate agitation throughout each zone, additional agitating means in the form of centrally positioned rotatable pugs, paddles, or the like are provided in each pyrolysis chamber. It should be noted that the perforations in each screw section allow passage of the pyrolysis vapors therethrough to an entrained solids separator.

In the apparatus illustrated in Figure 1.18a, the pyrolyzer or retort is composed of a generally horizontal cylindrical pyrolysis chamber **60** and a vapor-solids separation chamber **61**. The pyrolysis chamber is provided with inlet means **62** adjacent one end wall thereof for introducing hot ash and inlet means **63** juxtaposed the inlet means **62** for introducing preheated raw shale onto the surface of the hot ash. Within the pyrolysis chamber **60** is a horizontal rotatable conveying and mixing mechanism comprising a conveyor section **65** and a pyrolysis mixing section **66** positioned along shaft **64** which is rotatably driven by motor **85**.

The conveyor section **65** comprises a conveyor screw flight **70** secured to the rotatable shaft **64** along one end thereof so as to be positioned below shale inlet **63** and ash inlet **62**. The pyrolysis mixing section **66** comprises a series of longitudinally spaced pyrolysis mixing chambers **67, 68** and **69**, each chamber being terminated at opposite ends thereof by solids flow retarding members **71, 72, 73** and **74** distributed at spaced intervals along and secured to shaft **64**. The diameter of the conveyor screw **70** should be substantially less than the internal diameter of the conveyor section **65** so as to permit convenient withdrawal of pyrolysis vapors.

The flow-retarding members, as shown in greater detail in Figure 1.18b, are composed of perforated screw sections having a diameter substantially the same as the internal diameter of the mixing section **66**. Each screw section **71, 72, 73** and **74** should contain not less than one or no more than two revolutions, and preferably about 1¼ revolutions. Additionally, each screw section contains a plurality of perforations **75** for alternately passing vapors and solids as the section is rotated. Since each screw section contains between one and two revolutions, the solids flow will be impeded and at the same time conveyed from one pyrolysis chamber to the next.

Each chamber is provided with a plurality of bladelike wiper members **76**, preferably of angular cross section and secured in reellike fashion (see Figure 1.18a) around the periphery of each screw section and extending therebetween for lifting and mixing unpyrolyzed pieces to the upper surface of the ash which is maintained at about ¼ and ½ reactor capacity. In addition to the bladelike members, each chamber is provided with a plurality of pugs or paddlelike mixing elements **77** secured to shaft **64** which serve to agitate the upper portion of the ash-shale mixture. The pyrolysis section may also be provided with an end dam **78** extending upwardly from the bottom of the pyrolysis mixing section **66** to further impede the solids flow.

The vapor-solids separation chamber **61** comprises a cylindrical vertically disposed housing **79** thermally connected to pyrolysis chamber **60**. Within chamber **61** a plurality of cylindrical filtering elements **80**, shown in greater detail in Figure 1.18c, are suspended from tube sheet **81** mounted in the upper portion of the housing for filtering entrained dustlike particles from the pyrolysis vapors.

FIGURE 1.18: HORIZONTAL RETORT WITH SOLID HEAT TRANSFER MEDIUM

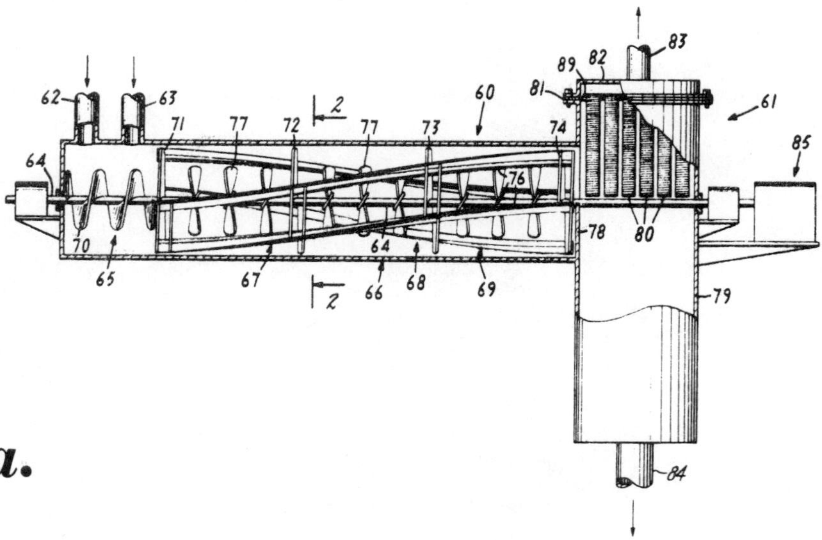

a.

Longitudinal Section of Pyrolysis Apparatus

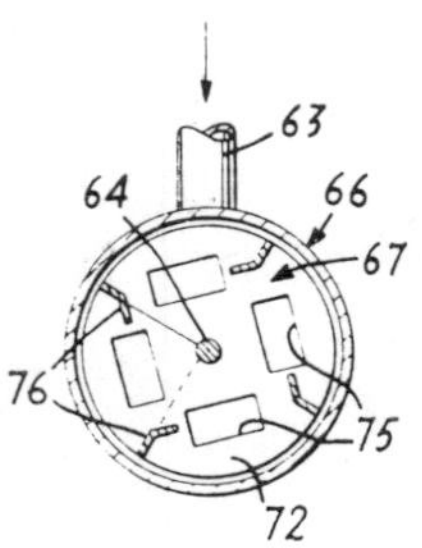

b.

Cross-Sectional View Along Line **2–2** of Figure 1.18a

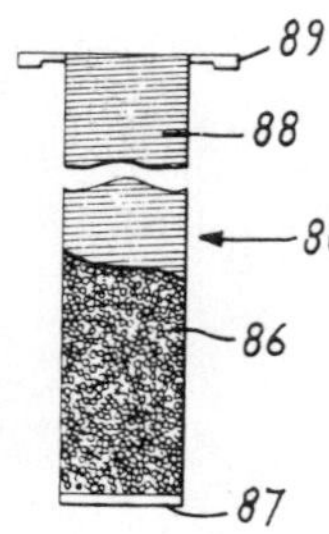

c.

Partial Cross-Section of Filter Element

Source: R.N. Hall, R.H. Beaver, R.G. Vawter and C.J. Mains; U.S. Patent 3,616,266; October 26, 1971

Above the tube sheet **81** is positioned a header **82** for gathering the filtered vapors which are thereafter removed via outlet **83**. The lower portion of housing **79** serves as a hopper for the spent solids-ash mixture from the pyrolysis chamber **60** prior to removal therefrom via solids outlet **84**.

Figure 1.18c illustrates the construction of one of the filter elements **80**. Each filter element **80** comprises a perforated cylindrical support member **86** which is closed at its lower end by a cap **87**. The upper end is open and has secured around the outer periphery a flange **89** which is used to suspend the filter element **80** from the tube sheet **81** as illustrated in Figure 1.18a. The external surface of the perforated support member **86** is entirely covered by a spirally wound continuous wire **88** which is secured to the support member. The perforations are generally in the order of ⅛ to ⅜ inch in diameter. The wire is so wound as to provide a spacing between adjacent winds. The spacing determines the minimum size of the particles to be filtered. It has been found that entrained dustlike shale ash particles may be satisfactorily filtered from the oil vapor by utilizing fibers having 6 to 40 micron spacings between winds. During operation a filter cake of shale ash is formed on the surface of the filter which further aids in the filtering process. The support member **86** and wire **88** are composed of any suitable material (e.g., stainless steel or the like) capable of withstanding pyrolysis temperatures in the order of 750° to 950°F.

Removal of Ash Prior to Retorting

J.H. Barney and F.B. Carlson; U.S. Patent 3,691,056; September 12, 1972; assigned to The Oil Shale Corporation describe a process for the recovery of shale oil from oil shale which contains substantial amounts of carbonate-containing minerals. As the world's petroleum reserves become more depleted and/or nationalized, increasing attention and interest is being directed to the recovery of shale oil from the plentiful reserves of oil shale located in the western states of Colorado and Utah and elsewhere throughout the world.

When oil shale of the type found in the Green River shales of the western United States is pyrolyzed at about 800° to 1000°F, the kerogen decomposes to gas, oil and spent shale. The oil and gas are recovered for downstream processing and upgrading. The spent shale is a waste product comprising coke-like residue which contains large quantities of calcite ($CaCO_3$) and dolomite ($MgCO_3{\cdot}CaCO_3$) having a particle size as small as ¼ inch and smaller. This shale matrix is to be distinguished from other oil-bearing matrixes such as coal which contains minor amounts of ash and tar sands which consist mainly of dense particles of silica.

Green River shale typically contains carbonate minerals in the range of about 35 weight percent dolomite and 15 weight percent calcite. The spent shale also averages about 3 to 6 weight percent fixed organic carbon, depending, of course, on the specific pyrolysis process employed. This amount of fixed carbon in terms of energy constitutes enough heat to sustain a very large retorting facility without resorting to the use of a portion of the oil or gaseous pyrolysis products for supplying the heat requirement. In this regard, many processes have been envisioned which contemplate the use of this available energy source for conducting the pyrolysis reaction, but none of these processes have fully anticipated the gamut of problems inherent in actually recovering and utilizing this energy source.

It has been found that the presence of the combusted spent shale residue, i.e., shale ash, in the pyrolysis zone is detrimental to the economical recovery of shale oil from oil shale containing substantial amounts of carbonate minerals. This is because the shale ash sorbs a significant quantity of the product shale oil during the pyrolysis reaction which is subsequently lost to the recovery system. The sorbed oil is burned during the combustion of the spent shale used to provide the heat for reheating extraneous heat-carrying bodies. This sorption is believed to be peculiar to oil shales containing substantial carbonate mineral concentrations because upon decomposition of the carbonates, the residue or shale ash is converted to an extremely fine (50 to 90 weight percent smaller than 325 mesh or 44 micron), porous and oleophilic material.

If this shale ash is allowed to remain in the heat-carrier circuit, the total hydrocarbons available for recovery will be adversely affected by sorbing more of the oil product than is necessary to furnish the overall process heat requirements. As previously noted, the fixed carbon content of the spent shale alone is generally more than sufficient to provide the pyrolysis heat requirement. As to the endothermic carbonate decomposition reaction, it has been determined that for spent shale combustion temperatures up to about 1400°F the process heat requirement is still in balance. Additionally, because of the minute particle size of shale ash, substantial levels of base sediment contaminate result from vapor entrainment of this fine ash if it is not purged from the circuit. In general, this process provides for the recovery of shale oil from oil shale via solid-to-solid heat transfer techniques while preventing the loss of shale oil by sorption on porous shale ash in the pyrolysis zone. This is achieved by insuring that the heat-carrying bodies recycled to the pyrolysis zone are free entirely or substantially of shale ash. The heat-carriers used may be any small attrition resistant material such as silica, minute alumina beads, or the like. The shale ash is removed from the heat carrier body circuit by attrition and elutriation steps. These steps are preferably carried out in conjunction, however, each step may be performed independently of the other. Thus, in combusting the fixed carbon on the spent shale or shale coke by-product of the pyrolysis of oil shale in a dense phase fluid bed combustion zone containing the cooled heat-carrying bodies, the spent shale is converted to friable particles of shale ash which are attrited to fines by frictional contact in the turbulent dense phase fluid bed combustion zone and the fines elutriated by the hot flue gas by-product of combustion.

Referring to Figure 1.19, oil shale, previously crushed to a particle size of nominal minus ½ inch, is fed from source **1** to the pyrolysis zone (retort) **2**. The shale may be preheated prior to being fed to the retort by passing the cold shale at about 60°F through preheater **3**. Shale preheating may be accomplished by recovering the sensible heat from the discharge flue gas product of spent shale combustion described below.

FIGURE 1.19: REMOVAL OF ASH PRIOR TO RETORTING

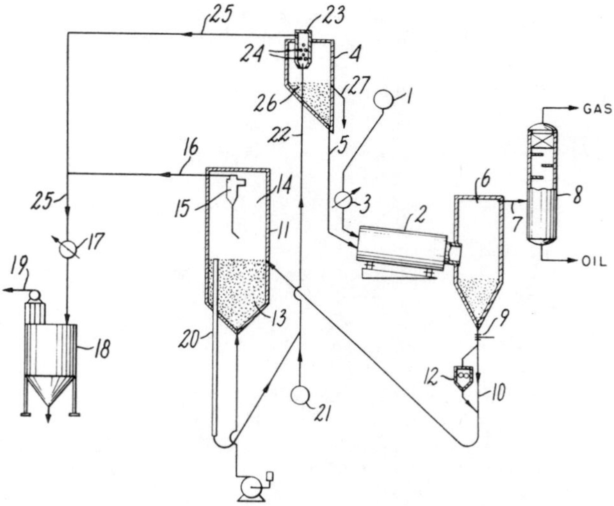

Source: J.H. Barney and F.B. Carlson; U.S. Patent 3,691,056; September 12, 1972

The sensible heat may be recovered either directly by entraining the cold shale in a stream of hot flue gas from which the solid combustion zone products have been removed, or indirectly by passing the flue gas through a heat exchanger of the type resembling a bundle of fin tubes. Regardless of the preheat procedure, it is desirable to preheat the shale to at least 220°F to remove as much of the free moisture as possible and reduce the size of the downstream shale oil recovery equipment.

In one pilot plant test run, the crushed oil shale was preheated to about 230°F in a dilute phase fluid bed and fed at an average rate of about 1,180 pounds per hour into the retort **2** operating at a pressure of 0.75 psig. Hot heat-carrying bodies substantially free of shale ash and having an average temperature of about 1215°F, which are preferably highly attrition resistant silica sand particles having the granular particle size distribution set forth in the table below, are fed from the heat-carrier accumulator **4** to the retort **2** via line **5** at an average rate of about 4,000 pounds per hour. The process contemplates the use of all attrition resistant heat-carrying fluidizable solids having a particle size range in the order of 100 mesh to about 4.0 mesh.

Sieve U.S. No.:	Cumulative weight percent retained on sieve
20	18.2
40	52.7
60	84.9
80	94.0
100	96.8
200	99.6
325	99.8

In the retort **2**, the hot heat-carrying sand particles are mixed with preheated oil shale and conveyed therethrough. The retention time required in the retort is in the order of 2 to 5 minutes. The hot sand particles exchange heat with the preheated oil shale to convert the kerogen content thereof into shale oil vapors at a temperaure of about 950°F and a spent shale residue. The shale oil vapors comprise noncondensable or gaseous hydrocarbons as well as vaporized condensable hydrocarbons. The spent shale or shale coke contains fixed residual carbon in amounts ranging from about 3 to 6%.

The typical carbon content of the Green River oil shale found in western Colorado and processed via solid-to-solid heat exchange techniques is in this range. The fixed carbon content of oil shale usually varies with its origin or location and can be as high as 10% or more by weight. Since the fixed carbon represents an available energy source, all efforts should be exerted to recover or utilize it efficiently.

The solids blend of cooled heat-carrying bodies, i.e., sand, and spent shale particles plus the shale oil vapors pass from the retort **2** into a solids-vapor separator vessel **6**. The shale oil vapors collect in the upper portion of the separator **6** and are removed therefrom via conduit **7** and passed to a vapor recovery zone or fractionator **8** for further downstream processing.

The solids blend of cooled heat-carrying bodies and spent shale particles is withdrawn from the bottom of the separator **6** through the solids flow control valve **9** via line **10** and passed therethrough to a dense phase fluidized bed combustion zone **11** by conventional dense phase transport procedures. If the solids blend contains a substantial quantity of large spent shale particles, for example, minus one-inch to plus ¼-inch size, the particles may be comminuted via crusher **12** prior to entry into the combustion zone **11**. Because spent shale is somewhat friable and easily crushed, the crusher rolls are disposed to impose only moderate pressure on the larger spent shale particles without fracturing the heat-carriers.

In the fluidized bed combustion zone **11** the carbon content of the solids blend is combusted to provide the heat for reheating the heat-carrying bodies therein. The heat-carrying bodies are generally covered with deposited coke, while the spent shale contains coke throughout its fine-grained matrix. If the fuel requirement provided by the carbon content of the solids blend fed to the combustion zone **11** is insufficient to accomplish the desired retorting, additional extraneous fuel may be added to the fluid bed in the form of heavy oil, tar or the like.

The fluidized bed is operated to provide turbulent frictional contact within the bed section **13** and freeboard section **14** to thereby cause severe attrition of the friable spent shale ash into shale ash fines having a particle size preferably less than 200 mesh. The fluidized bed is operated at temperatures ranging from about 1100° to 1650°F and at fluidizing superficial air velocities of from about 2.5 to 6.0 feet per second. The higher combustion temperatures of 1400° to 1650°F are desired because carbonate decomposition of the spent shale more readily occurs, thereby enhancing the breakdown and degree of attrition of shale ash in the combustion zone. A further product of the combustion reaction is hot flue gas. The attrited shale ash fines are elutriated from the reheated heat-carrying bodies by the hot flue gas and withdrawn from the freeboard section **14** via cyclone separator **15** and line **16**. The cyclone separator **15** returns any of the larger heat-carrying bodies which may be jetted out of the fluidized bed back to the bed. The hot flue gas which contains shale ash fines is conveyed via line **25** through a cooler **17** and into a solids removal zone **18**, i.e., bag filter, prior to being discharged to the atmosphere via line **19**.

The reheated heat-carrying bodies and any larger remaining particles of the friable shale ash not removed with the hot flue gas are withdrawn from the dense phase fluid bed combustion zone **11** via standpipe **20**. The reheated heat-carrying bodies and the small remainder of the friable shale ash particles are fed through line **22** by hot air transport gas from source **21** into the heat carrier accumulator **4** having an elutriator **23** equipped with a plurality of attriter bars or baffles **24**. The friable shale ash particles are attrited and comminuted by impacting against the attriter bars **24** and thereby converted into shale ash fines having an average particle size of about 200 mesh (U.S. sieve) or smaller while the attrition resistant hot heat-carrying bodies are not significantly attrited. The elutriator also serves to disperse the shale ash fines so as to significantly improve the elutriation thereof from the heat-carrying bodies.

The suspension of shale ash fines in the hot air transport gas is withdrawn from the elutriator **23** via line **25** which joins line **16** and passes through cooler **17** for cooling prior to subsequent fines-gas separation and disposal. The hot heat-carrying bodies, now entirely or substantially free from shale ash, accumulate in a bed **26** within the base of the heat carrier accumulator **4**. The heat carriers are fed from the accumulator **4** via line **5** at a temperature of from 1100° to 1650°F into the retort **2** to effect pyrolysis of fresh oil shale introduced therein. Excess solids are withdrawn from accumulator **4** via line **27**.

In one test conducted over a 16-hour period, the shale ash fines were allowed to build up in the recycle sand circuit to determine the effect of shale ash fines on the product oil yield. The overall effect was determined by noting the decrease in the fluid bed fuel (natural gas) requirement to maintain an average temperature of about 1400°F. It was observed during this test that the whole oil yield decreased as the test progressed, and that the decrease in oil yield was accompanied by a corresponding decrease in the fluidized bed fuel requirement. The data are presented in the table below. Visual inspection of the recycle sand circuit confirmed that the ash content increased substantially over the test period.

Comparison of Oil Yield and Fluid Bed Fuel Requirement Necessary to Maintain 1400°F Bed Temperature

	Whole oil yield [1]	Natural gas to fluid bed, s.c.f.h.
Test period:		
First 4 hours	68.5	1,030
Second 4 hours	61.0	816
Third 4 hours	64.6	690
Fourth 4 hours	52.2	687

[1] Weight percent of modified Fischer assay.

From the foregoing discussion and the data in the table above, it is evident that product oil was sorbed and burned in the combustion zone. Extraneous fuel was required in all of these pilot plant tests to compensate for heat losses from the small-scale equipment. In another test conducted to determine the effect of removing the shale ash from the recycle sand circuit on the overall product oil yield, a total of 8.9 tons of the same Green River oil shale as used in the above test were preheated to 230°F and processed over a test period of 11 hours. A charge of clean silica sand having the particle size distribution set forth in the previous table was used for this test. The fluid bed combustion zone was operated at a superficial air velocity of 4.2 ft/sec to remove all solids of less than about 200 mesh. Upon completion of the test, it was determined that the recovered oil plus C_4 and heavier gaseous hydrocarbons amounted to 100.5 weight percent of Fischer assay. The total amount of hydrocarbons recovered, which also includes C_1 through C_3 hydrocarbons, oil, and C_4 and heavier gaseous hydrocarbons was 107.2 weight percent of Fischer assay.

IN-GROUND PROCESSES

Solution Mining to Form Cavern

T.N. Beard; U.S. Patent 3,739,851; June 19, 1973; assigned to Shell Oil Company describes a method of producing hydrocarbons and optionally water-soluble minerals from a subterranean or underground oil shale formation containing zone(s) of water-soluble minerals by forming interconnecting cavities or a cavern within the water-soluble mineral-containing oil shale formation by fluid leaching the water-soluble minerals and simultaneously or sequentially flowing a hot fluid into the upper portion of the cavern in order to place the cavern roof including the cavern wall under stress so as to effect oil shale spalling and rubbling and improve recovery of hydrocarbons.

In some subsurface oil shale formations such as in certain areas of the Green River formation in the Colorado and surrounding areas of the United States, such formations also contain water-soluble minerals such as halite, nahcolite, trona, etc. in the form of beds, lenses, nodules, iodes, veins, or the like.

Such minerals can be leached, dissolved and removed by suitable means such as solution-mining, e.g., by contacting the minerals with a solvent such as an aqueous solution which may be neutral, alkaline or acidic and recovering if desired, the mineral-containing solution and at the surface extracting the minerals from the solution and reusing the solution by recycling it back into the formation for extracting more minerals from the underground formation or dispersion of it if necessary. By treating such formations in this manner permeability and voids are created and cavern formation and rubblization of oil shale in the cavern can be induced from which hydrocarbons can be recovered by suitable in-situ thermal means.

As illustrated in Figure 1.20, oil shale formation **12** contains a generally horizontally extensive bed **13** of a continuous deposit of a water-soluble mineral and nodules of minerals **12a**, such as halite, nahcolite or trona. Such mineral-containing areas are frequently encountered near the bottom of a layer of oil shale and may be utilized to form cavities or caverns into which oil shale solids can be moved by caving-in of the oil shale within the formation.

In carrying out the method, the location of zones rich in substantially water-soluble minerals may be determined by geologic investigations based on an analysis of geophysical and geochemical data. For example, the location and extent of such zones man be determined by obtaining and correlating seismic, gamma ray and/or density log data, core analysis data, etc. The borehole of at least one well is extended into a zone containing a soluble mineral and fluid communication between a surface location and at least two spaced points within a substantially continuous and generally horizontally extensive zone rich in soluble mineral is established through the borehole or boreholes or the one or more wells.

Fluid communication is thus established between at least two spaced portions of well boreholes and the water-soluble minerals (as for example, through perforations **22** through **23** of well boreholes **10** and **11**, respectively, and the bed **13** of water-soluble minerals in Figure 1.20a). The well boreholes of Figure 1.20a extend into the soluble mineral bed located within oil shale fromation **12**.

The fluid communication between well boreholes and the bed in water-soluble minerals therebetween may be established by solution-mining a flow channel through the soluble mineral bed and/or by means of conventional hydraulic, electric, and/or explosive fracturing techniques. Where, for example, the subterranean stresses in and around the soluble mineral bed are conducive to the formation of horizontal fractures, the fluid communication between well boreholes and the soluble mineral bed may be established by a conventional hydraulic fracturing technique.

FIGURE 1.20: PRODUCTION OF HYDROCARBONS FROM OIL SHALE FORMATION CONTAINING HALITE

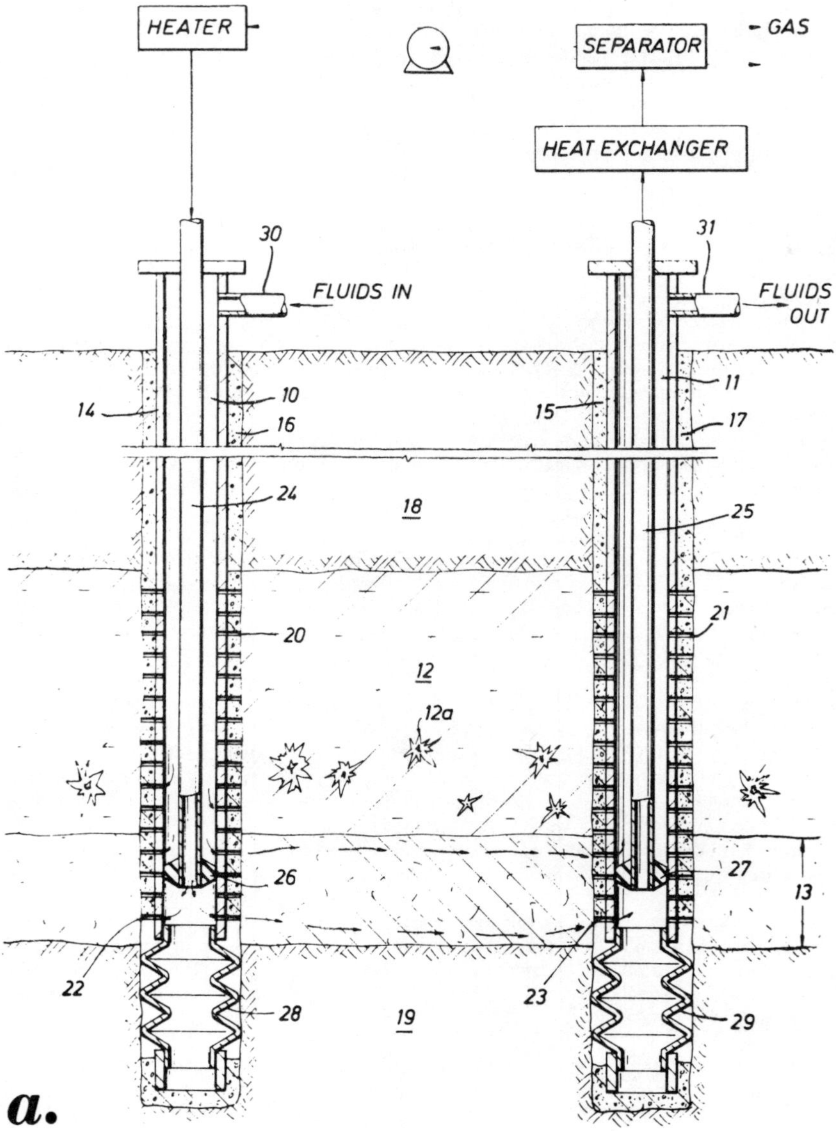

Vertical Sectional View of Process

(continued)

FIGURE 1.20: (continued)

b.

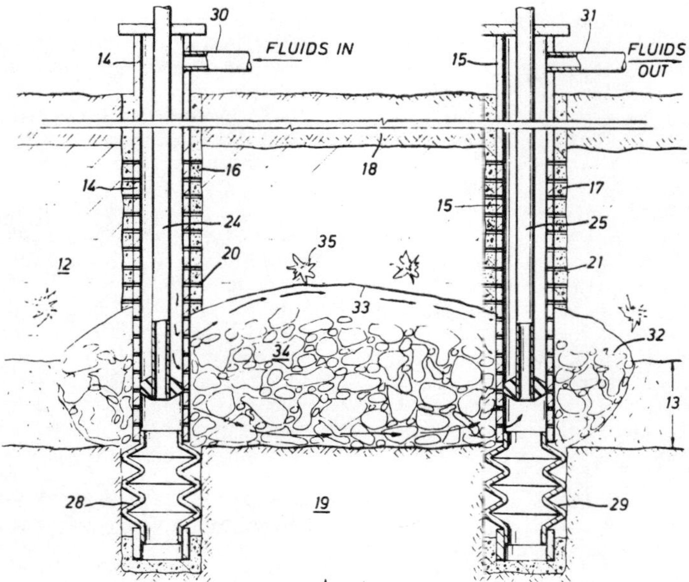

c.

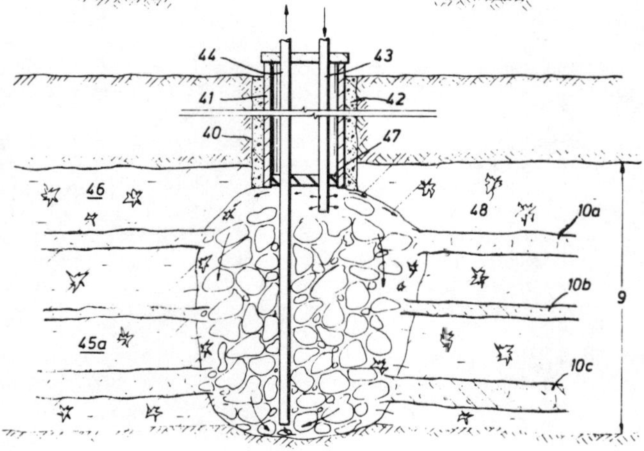

(b) Vertical Sectional View of the Treatment of the Cavern Formed by Leaching or Solution-Mining of Water-Soluble Minerals as Shown in Figure 1.20a

(c) Sectional View of a Formation Being Penetrated by a Single Well

Source: T.N. Beard; U.S. Patent 3,739,851; June 19, 1973

Where the subterranean stresses are conducive to the formation of vertical fractures, special steps are needed to form fractures that will interconnect pairs of spaced wells with each other and the soluble mineral. Such interconnecting fractures may be provided by initially forming parallel vertical fractures in each of a pair of wells. When at least one of the fractures is heated until it becomes closed, by a thermally-induced swelling of the fracture walls, a differently oriented vertical fracture is formed at a higher temperature and pressure. Such fractures are then extended until they intersect, and, the walls of the intersecting fractures are leached, for example, in the zone of their intersection with the soluble minerals. Such interconnected vertical fractures provide flow channels for circulating a soluble mineral-extracting solvent to form the caverns into which shale solids can be displaced prior to or during the pyrolysis of the kerogen. Such flow channels may be formed between a single pair of wells, to form a test pattern which can subsequently be expanded, or may be formed between all members of an extensive pattern of wells.

Alternatively, in similar tectonic environments, such interconnecting fractures may be formed by a combination of steps comprising (a) injecting a fluid through each of a pair of wells into the soluble mineral-rich portion of the oil shale formation at a pressure sufficient to form and extend generally vertical fractures within the formation so that at least one pair of generally vertical fractures are extended within the formation for significant distances from the points of injection; (b) filling the first-formed fracture in one of the wells with a fluid able to solidify in situ to a compression-resistant solid and allowing it to solidify at substantially the pressure at which the first-formed fracture was extended; and (c) injecting a fluid through the well containing the so-plugged fracture at a pressure sufficient to form and extend a second-formed generally horizontal fracture through the oil shale formation until fluid injected through this well is transmitted to the other one of the pair of wells.

Thus, by means of this procedure a low-cost fluid that is capable of solidifying in situ, e.g., a slurry of cement or a solution of a resin, may be used to interconnect at least a pair of wells thereby aiding in solution-mining and shale oil production.

After fluid communication has been established between well boreholes **10** and **11**, aqueous leaching or solution-mining fluid and/or liquid is injected down tubing string **24** and annulus outlet **30** in well borehole **10**, out perforations **20** and **22**, through bed **13** and up tubing string **25** and outlet **31** in well borehole **11** via perforations **21** and **23**.

The aqueous fluid and/or liquid may comprise water and/or steam or aqueous solutions of acid or acid-forming materials and is circulated at pressures either above or below the overburden pressure. The circulating aqueous liquid dissolves the water-soluble minerals and mineral by-products thereof are recovered from the fluid flowing out of well borehole **11**, for example, by conventional evaporation and/or precipitation procedures. Thus, the solution-mining is preferably conducted by injecting hot water or steam or mixture of steam and aqueous liquid into the earth formation while producing fluids from the earth formation at a rate such that most or all of the produced fluid is liquid.

During the solution-mining operation described above, the depths at which the circulating aqueous liquid is injected into and produced from the earth formation are relatively immaterial as long as fluid is flowed through the water-soluble bed.

The solution-mining is continued until the removal of water-soluble material has created a cavern **32** (Figure 1.20b) having a significantly large volume and areal extent and containing exposed oil shale along its roof **33**. Such a cavern should extend over a significant distance, such as at least about 50 ft or more along at least one horizontal direction while extending over an area of at least several hundred square feet and having a volume of at least several hundred cubic feet. Thus, as illustrated in Figure 1.20b, the cavern is formed within the bed extending to substantially the top of underlying formation **19** and the bottom of overlying oil shale formation **12**.

Oil shale **34** (Figure 1.20b) from the roof is caved into the cavern and the cavern roof is migrated up within the oil shale formation by circulating a kerogen-pyrolyzing fluid through the cavern. The circulation of such a fluid is controlled by adjusting the points of injection and production and the temperature and pressure of the circulating fluid so that the roof of the cavern is contacted by kerogen-pyrolyzing fluid at a temperature at which significant amounts of the kerogen in the oil shale **34** are converted to fluid and removed from the cavern **32**. In the initial stage this can be accomplished by injection down annulus outlet **30** and tubing string **24** of well borehole **10**, through the cavern into contact with the roof **33**, and into well borehole **11** and out the annulus outlet **31**.

Such a fluidization of kerogen tends to fracture the oil shale since it both increases the volume of at least one component within the normally impermeable oil shale and removes a significant proportion of solid material from the oil shale. In oil shale, the kerogen tends to be concentrated along layers such as bedding planes, and the effects of such kerogen fluidizing and removing actions tend to be localized, e.g., along the layers.

In the process, the rate of the fracturing, flaking, or spalling of oil shale material into the cavern is increased by the fact that the oil shale being treated is under stresses of compression and/or tension along the roof of the cavern. The caving of the oil shale into the cavern is further enhanced by controlling the circulating kerogen-pyrolyzing fluid so that the roof-contacting portion has a temperature above 450°F and preferably between 550° and 850°F at which kerogen is pyrolyzed at a relatively rapid rate. Since heat is transferred within the oil shale, the heat causes portions of kerogen which are surrounded by impermeable oil shale material to be converted to a liquid or vapor at pressures that fracture the surrounding oil shale material.

Since this material is under stress (tension or compression) the fractured portions separate and cave into the cavern. Such a caving is further enhanced when the roof is contacted with an aqueous fluid, such as steam or hot aqueous liquid, that tends to dissolve and remove water-soluble material **35**, such as entrapped or dispersed halite, nahcolite, or the like, from within the oil shale formation **12**. The dissolving and removing of solid material is beneficial in both enhancing the caving and creating additional void space into which additional amounts of oil shale can be caved.

As the roof of the cavern is migrated upward within the oil shale formation, the depth at which the circulating kerogen-pyrolyzing fluid is injected is preferably raised to keep it near the roof of the cavern. This causes the roof to be contacted by the hottest and most dilute incoming portions of the fluid. By similarly raising the depth of the points at which fluids are produced from the cavern, the circulating fluid is caused to flow along the roof of the cavern as a stream that flows over the top of a body of substantially static fluid within the cavern.

The tendency to form such a flow path is enhanced when the circulating fluid is heated at a surface location or within well borehole **10** prior to its injection into the cavern. This causes the incoming fluid to be hottest and less dense than the fluid in the lower portion of the cavern. Such an effect is further enhanced by circulating a vapor-containing hot fluid, such as steam, hot gas, or a mixture of steam or hot gas and a liquid. The cavern extends substantially to the top of oil shale formation **12**.

The kerogen-pyrolyzing fluid continues to be circulated from well borehole **10** through the cavern of the oil shale formation and out of well borehole **11** in the manner discussed above with respect to Figure 1.20a. Hydrocarbons are continuously recovered from the heated fluid circulating out of well borehole **11** by means well known in the art.

Thus, as illustrated in Figure 1.20a, conventional equipment and techniques, such as heating means, pumping means, separators and heat exchangers may be used for pressurizing, heating, injecting, producing, and separating components of the heated fluid circulating through the cavern within the oil shale formation. The production of the fluid may be aided by downhole pumping means, not shown, or restricted to the extent necessary to maintain the selected pressure within the oil shale formation. The fluid circulated through the cavern to recover hydrocarbons from the oil shale formation may comprise any heated fluid and/or liquid, such as hot gases, hot water, and mixtures thereof.

The solution-mining operation is preferably conducted by circulating hot fluid or aqueous liquid through a well-interconnecting flow path. The flow of a hot fluid, e.g., steam or aqueous fluid, tends to be confined within the water-soluble zone even if intercommunicating fractures have extended into the overlying oil shale. Within the oil shale, the fractures tend to be closed by the thermal swelling of the fracture walls. Within the water-soluble zone, such a closure is prevented by the dissolving of the wall material.

After forming the cavern **32** along the bottom of the oil shale, the roof **33** of the cavern is migrated up within the oil shale by circulating a kerogen-pyrolyzing fluid along the roof of the cavern. The kerogen-pyrolyzing fluid may be the same fluid used in the solution-mining operation which may, to some extent, be concurrent with the roof-migrating operations.

The migration of the cavern roof up within the oil shale formation **12** forms a fracture-permeated zone that may contain a significant amount of recoverable hydrocarbons. Hydrocarbons may be recovered from this zone by substantially any oil shale pyrolyzing procedure such as an in situ combustion, hot fluid heating, or the like type of subterranean retorting procedure. Beyond the areal extent of the cavern, the oil shale **34** tends to remain impermeable and thus tends to confine any fluids which are being circulated through the fracture-permeated zone.

A single well system shown in Figure 1.20c may be utilized by a dual zone completion arrangement such that fluids can be injected at one point of the well and produced from another point of the same well. This fluid communication can be established in one borehole between at least two spaced portions of the well borehole and the water-soluble minerals (as for example, in Figure 1.20c communication is through the tubing string the ends of which are open to the water-soluble minerals and some distance apart).

Thus a single well may be utilized by a dual zone completion arrangement as shown in Figure 1.20c such that fluids can be injected at one point of the well and produced from another point of the same well. In Figure 1.20c, the well bore is **40**, the casing is **41**, the sealant is **47**, within the casing are the injection tubing string **43** and production tubing string **44**, the borehole **40** penetrates oil shale formation **9** with mineral zone(s) **10** or multizones **10b** and **10c** and mineral nodules **48**.

Fracturing pressures are generated within the oil shale formation **9** while lower pressures are maintained within the cavern **45** which is formed within the oil shale formation by the removal of the water-soluble minerals. These pressures are preferably generated by merely circulating hot fluid through cavern **23a**. As the roof of the cavern **45** is heated kerogen is pyrolyzed within the cavern walls and the pressures of the pyrolysis products increase until portions of the walls are spalled into the cavern **45** creating a rubblized zone **24a** and surrounding fracture area **46**.

In summary, then, well boreholes are opened into at least two horizontally spaced points within such a water-soluble zone and are equipped to convey fluids between that zone and at least one surface location. Fluid circulation is established between the spaced points within the soluble zones. An aqueous fluid is circulated through that zone in order to solution-mine a cavern having a roof along which oil shale is exposed.

The cavern is extended to form a void space having horizontal dimensions sufficient to create a tensile stress, due to the tendency of the roof to sag, and to have a volume sufficient to accommodate a significant amount of solid material. Oil shale is caved into the cavern by circulating a kerogen-pyrolyzing fluid through the cavern at a temperature at which kerogen in the oil shale is converted to fluid material that is extracted and removed by the circulating fluid. The depths and rates at which the circulating fluid is injected into and produced from the cavern are adjusted to maintain a flow along the roof as the roof migrates up within the oil shale formation.

Examples: (A) A single injection-production well was designed properly insulated to penetrate a nahcolite-rich oil shale formation and through the injection tubing hot water was injected into the nahcolite-bearing formation for about a week to leach out the nahcolite as an aqueous solution which was recovered above ground via the production tubing. The size of the cavern thus produced was determined by conventional means such as measuring the concentration of the salt in solution and thereafter steam at a temperature above 500°F was injected along the roof of the cavern to effect roof stress resulting in rubbling and spalling of the oil shale into the cavern and cavern enlargement. Steam injection along the cavern roof was continued as well as through the rubbled oil shale bed at a temperature between 550° and 850°F effecting kerogen decomposition to hydrocarbons which were recovered via the production tubing where the hydrocarbons were separated above ground from the condensed steam by conventional means.

(B) The procedure as described in (A) was repeated except that instead of using hot water to leach out the nahcolite steam was used. Thus, once the cavern was formed by steam leaching out the nahcolite, steam injection was continued under controlled conditions so that its flow path was directed along the roof of the cavern placing it under stresses of tension and compression producing results as noted in (A). Thereafter a portion or the entire steam injection flow was directed to flow through the rubbled oil shale to effect kerogen decomposition and hydrocarbon recovery as noted in (A).

Other than steam flow control and direction the (B) process can be regarded as a single process operation in which steam is used as the leaching, rubbling and hydrocarbon recovery agent provided temperature, pressure and flow control of the steam is carefully controlled.

In related work *T.N. Beard; U.S. Patent 3,753,594; August 21, 1973; assigned to Shell Oil Company* describes a method of producing hydrocarbons and optionally halite from a subterranean oil shale formation containing zone(s) of halite. The process involves penetrating the formation with at least one borehole and leaching or dissolving the halite from the formation with a solvent fluid so as to form a cavern(s) and/or interconnected cavities, followed by fracturization and/or rubblization of the oil shale surrounding the caverns or cavities, and thereafter injecting into the fracturized and/or rubblized zones, a pyrolyzing fluid to effect in situ hydrocarbon recovery.

Laterally Expanding Permeabilization

R.C. Ueber, P. Van Meurs and J.R. Brew; U.S. Patent 3,759,328; September 18, 1973; assigned to Shell Oil Company describe an improved method of recovering hydrocarbons and water-soluble carbonates and/or heat-sensitive materials from underground oil shale formations containing substantial amounts of the carbonate and/or minerals by forming a cavern by leaching with an aqueous fluid the carbonates and/or minerals and imparting permeability while effecting rubbling of the oil shale in the treated area. The process involves contacting and flowing a hot aqueous fluid downward from a relatively shallow depth along a vertical interval of the treated oil shale to cause horizontal expansion and recovering from a relatively deep depth an aqueous liquid containing dissolved therein water-soluble carbonates and/or heat-sensitive minerals and subsequently injecting a pyrolyzing fluid or solvent to effect recovery of hydrocarbons from the rubbled oil shale.

The process is in part premised on a discovery that, in a cavern in an oil shale that contains a significant amount of heat-sensitive minerals and/or water-soluble carbonates a hot aqueous fluid can be caused to flow along a path that causes a horizontal expansion of the cavern. In a cavern within such an oil shale, the rate at which a hot aqueous fluid is segregated into layers having increasing densities, has been found to be related to the rate at which heat can be transferred into the walls of the cavern in a manner conducive to the establishing and maintaining of the flow path described above, and the resultant heating and leaching along substantially vertical portions of the walls of such a cavern has been found to cause a horizontal expansion of the cavern.

Figure 1.21a shows a portion of a well borehole **1** which has been drilled through an overburden **2**, comprising successively shallower earth formations, and opened into an oil shale formation **3** that contains a heat-sensitive carbonate mineral.

FIGURE 1.21: LATERALLY EXPANDING OIL SHALE PERMEABILIZATION

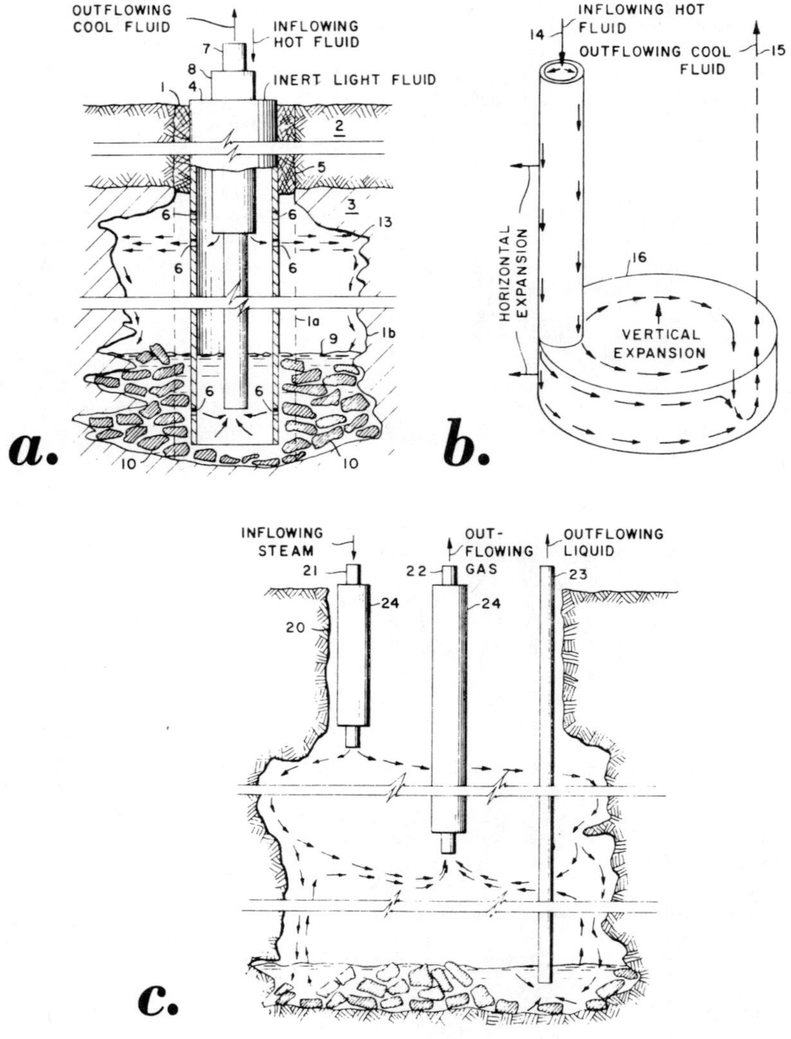

(a) Vertical Section Showing a Subterranean Oil Shale and Downhole Equipment

(b) Schematic Illustration of a Flow Path for Circulating Fluid

(c) Vertical Section Showing an Alternative Arrangement of Downhole Equipment of the Type Shown in Figure 1.21a

Source: R.C. Ueber, P. Van Meurs and J.R. Brew; U.S. Patent 3,759,328; Sept. 18, 1973

The oil shale formation that is placed in fluid communication with a portion of the borehole to be used in practicing the process, should be a formation containing a significant proportion, e.g., greater than 5% by weight, of heat-sensitive carbonate mineral. Borehole

1 is equipped with a string of casing **4**, which is bonded to the surrounding earth formations by cement **5**.

Separate conduits for conveying fluids between a surface location and, respectively, relatively shallow and relatively deep depths within the oil shale are provided by tubing strings **7** and **8**. Alternatively, such conduits may comprise two or more parallel strings of tubing and may be located in two or more well boreholes that intersect or extend into a common cavern within the oil shale. Such conduits can be installed and equipped by means of known procedures and devices and heat insulation is preferably installed around at least those of such conduits that are used for the inflowing of hot fluid.

As indicated by Figure 1.21a, the vertically extensive cavern or opening that is expanded by the process can comprise the borehole of a well that extends into an interval of oil shale that contains heat-sensitive carbonate material. Such an interval preferably has a vertical thickness of at least about 100 feet. In the initial stages, such a borehole may have a generally cylindrical form, such as indicated by the dotted line **1a**, and may comprise a relatively slender, generally vertical cavern within the oil shale. In operating the process with the equipment shown in Figure 1.21a, a hot aqueous fluid is flowed into contact with the wall of the cavern by inflowing hot aqueous gas and/or liquid through the annulus within pipe **8** (i.e., the space between pipes **7** and **8**) and through adjacent perforations **6** at a relatively shallow depth within the carbonate mineral-containing portion of the oil shale.

The inflowing fluid such as hot water and/or steam flows downward along the face of the vertical interval of oil shale (along the wall of the borehole) and decomposes and dissolves the heat-sensitive carbonate mineral material. The dissolving of water-soluble material forms a liquid solution **9**. This solution, which is usually mixed with at least some gas, such as carbon dioxide and gaseous hydrocarbon, is out-flowed through pipe **7**, which extends to a relatively low level within the borehole. The decomposing and dissolving of carbonate mineral components of the oil shale causes the spalling and caving in of particles **10** of the oil shale and causes a generally horizontal expansion of a rubble-containing cavernous zone of permeability within the oil shale.

Where the oil shale being treated contains a significant proportion of a mineral, such as a halite, which is water-soluble in its natural form, the inflowing of hot aqueous fluid can advantageously be preceded by a circulation of aqueous liquid at a relatively low temperature, such as the wellhead temperature, the temperature of the source of the liquid or the like. In such a pretreatment, the circulating liquid may leach out significant portions of distributed layers or particles of the soluble mineral.

This increases the surface area of exposed oil shale and/or weakens the support for layers or chunks of the oil shale. Such a pretreatment circulation can advantageously be continued while the rate of dissolution is high, e.g., as indicated by the proportion of solute in the outflowing liquid. The so-circulated aqueous liquid can then be gradually or rapidly heated to the temperature selected for the inflowing hot aqueous liquid used to decompose heat-sensitive carbonate material, with or without an interruption of the flow through the cavern.

When necessary or desirable the vertical expansion of the cavern can be inhibited by spotting and maintaining a relatively light and cool fluid **13** along the roof of the cavern. Such a fluid is preferably a gas and can be inflowed, or maintained substantially stationary, in and around the annulus within the casing (i.e., the space between pipe **8** and casing **4**) and the upper portion of borehole **1** (below the cement) to extend along the roof of the horizontally expanding cavern as the walls of the cavern move radially outward to and beyond the location shown at **1b**.

The hot aqueous fluid used in the cavern-enlarging procedure is preferably steam, hot aqueous liquid (hot water) or a mixture of such fluids. The hot fluid is preferably inflowed at a temperature, e.g., at least about 250°F, that is significantly higher than the normal temperature of the subterranean oil shale formation. The heat transported by such a hot fluid converts the heat-sensitive carbonate material to carbon dioxide and water vapor within portions of the normally impermeable oil shale matrix. Such a generation of gas causes localized fracturing and/or spalling of the oil shale.

The aqueous liquid component of the inflowing hot fluid dissolves water-soluble mineral material and creates additional solid-free void space. This occurs along most, if not all, of the vertical extent of the flow path used in the process. The spalling and dissolution causes a horizontal expansion of a rubble-containing cavern. The inflowing hot aqueous fluid can comprise superheated, dry, or wet steam, or a mixture of such a steam with substantially any gas vapor or liquid, such as carbon dioxide, phenols, hydrocarbons, alcohols, halogenated hydrocarbons, acids, or the like, or with substantially any aqueous solution, such as an aqueous acid or base or solution or neutral salt. Where the inflowing fluid is substantially completely gaseous it should contain sufficient steam to provide a significant amount of aqueous liquid as it condenses within the cavern.

The inflowing hot aqueous fluid can be heated by means of surface located and/or downhole located, steam generators, water heaters, or the like. Alternatively, or additionally, such heating can be effected or supplemented in an in situ combustion within the oil shale formation. The temperature of the inflowing hot aqueous fluid can range from about 250°F to one sufficient to cause a relatively rapid oil shale pyrolysis, e.g., a temperature of from about 600° to 1000°F.

The inflowing aqueous liquid phase of the hot aqueous fluid dissolves naturally water-soluble minerals such as nahcolite, trona, halite, or the like, and/or water-soluble decomposition products from a heat-sensitive carbonate material, such as nahcolite, etc., to create solid-free space within the oil shale. Various water-soluble minerals, such as nahcolite ($NaHCO_3$), may dissolve prior to any thermal decomposition, if the pressure is sufficiently high at the temperature of the inflowing fluid. Alternatively, such minerals may be partially or wholly decomposed to gaseous fluids and sodium carbonate before dissolution.

Although the portion of oil shale formation which is treated in accordance with the process must contain a significant amount of heat-sensitive carbonate material, it may contain sections, or vertical intervals of as much as several tens of feet thick, which are substantially devoid of heat-sensitive and/or water-soluble minerals. In such heterogeneous regions, the heat-sensitive or soluble minerals are converted or dissolved and removed. Portions of the so-converted oil shale materials become incompetent and break into pieces under the existing local stress field. Such pieces, or chunks, of oil shale mineral materials tend to accumulate on top of ledges of oil shale that contains little or no heat-sensitive or soluble material.

The accumulation of weight from such chunks, together with the existing stress field, cause such ledges to break into pieces and fall to a lower level. The action of converting kerogen into shale oil materials such as gaseous and liquid hydrocarbons enhances such an operation and, where the oil shale is relatively lean with respect to heat-sensitive and soluble materials, the use of hot aqueous fluid heat to a kerogen-pyrolyzing temperature is desirable. Also hydrocarbons can be extracted from the rubbled oil shale by solvent means such as by use of phenols, aromatic solvents, e.g., benzene, xylene, etc.

Due to mechanisms such as those mentioned above, the process provides a generally vertical cavernous zone to grow in a horizontal direction. The rate of growth will vary depending upon the heat-sensitive and water-soluble mineral content of the particular zone. The outer boundary of the zone will generally be very irregular with portions extending several tens of feet further than others. In order to enhance horizontal growth while injecting a hot aqueous fluid that is predominantly liquid, it is generally desirable to maintain most or all of the rubble-containing cavern full of liquid. Alternatively, when the injected hot aqueous fluid is steam, it is generally preferable to keep much of the rubble-containing kerogen filled with steam and/or gas.

A particularly suitable arrangement of flow paths to be used in the process is shown in

Figure 1.21b. At least two horizontally separated wells are opened into a region of oil shale that contains heat-sensitive carbonate mineral and is located immediately above a layer or zone of oil shale or other earth formation material that is rich in water-soluble mineral and/or heat-sensitive carbonate mineral. Such wells are used to form an inflow path **14** and an outflow path **15** that are interconnected by a path extending through an areally extensive cavern **6**.

As indicated by the arrows, hot aqueous fluid is inflowed into contact with oil shale containing heat-sensitive carbonate material at a relatively shallow depth, flowed down along a vertical section of such oil shale, flowed along the roof of a horizontally extensive cavern within such oil shale, and, liquid containing dissolved mineral material is removed from within the horizontally extensive cavern. Such a horizontally extensive cavern can advantageously be formed by means of mechanical fracturing, and/or solution-mining techniques, for example, by one or more of such techniques.

A principal advantage of a flow path of the type shown in Figure 1.21b is the heat economy and the fact that much larger volumes of oil shale can be rubbled per unit time than could be achieved by either a horizontal or vertical rubbling by itself. Relative to horizontal rubbling from a single well, the concurrent vertical and horizontal rubbling is capable of providing much higher oil production rates, particularly in the early stages of the process. Such a flow path can be utilized to produce a relatively coo fluid with much of the produced hydrocarbon and injected fluid being outflowed in the liquid phase.

A flow path of the type shown in Figure 1.21b can be initiated between one or more pairs of wells. Initial communication is preferably achieved by fracturing or dissolving within the layer of water-soluble material until fluid injected in one well can be produced from another. Hot aqueous fluid is then injected at the top of an injected well and fluid is produced from within a generally horizontal cavern or flow path through a production well. The upper portion of the injection well will enlarge laterally and the lower portion or rubble-containing cavern will enlarge vertically so that the permeable zone is expanded both laterally and vertically. It is in the lower region of the rubble-containing cavern that heat improvements are made. With such a flow path, the inflowing fluid preferably has a temperature below one at which the pyrolysis of kerogen is rapid.

Where it is desired to rubble large volumes of oil shale while removing solid materials and preheating the shale for later pyrolysis, such a use of a relatively low temperature results in significant heat economy. If communication between different patterns of injection and production wells is desired the depth of the location from which liquid is produced can be kept relatively deep within the soluble layer so that the circulating fluid will continue to enlarge the areal extent of the dissolved zone. Where communication between different well patterns is not desired, the production point and production rate can be adjusted to leave a substantially saturated liquid solution in the soluble layer in order to prevent its further growth.

Such a versatility with respect to the size and shape of the cavernous zones that are formed before and/or during a recovery of shale oil is a unique advantage of the process. For example, where the oil shale is thick, large amounts of shale oil can be recovered from a series of zones that are vertically extensive but are horizontally spaced so that problems due to subsidence are avoided. For example, wells in a plurality of horizontally separated patterns that each contain one or more wells opening into a layer rich in water-soluble minerals can be operated as described in connection with Figure 1.21b to from horizontally expanding permeable zones and produce shale oil.

The sizes of the permeable zones can be monitored by means of acoustic, electromagnetic or like measurements of the extents of the substantially void space and/or measurements of the volume of fluids that are contained in the caverns. The horizontal expansion of the caverns can be controlled to provide an efficient recovery of oil from nonintersecting, generally vertically extensive zones that are spaced so that undisturbed columns capable of supporting the overburden are left between the depleted zones.

Figure 1.21c shows downhole equipment of the type shown in Figure 1.21a arranged to effect a downhole separation of the gaseous and liquid phases of the fluid being produced. Particularly when the concentration of heat-sensitive carbonate material is relatively high, and/or the temperature of the inflowing hot aqueous fluid is relatively high, a significant amount of gaseous carbon dioxide and water will be formed. However, to the extent that it is feasible, it is desirable to produce a relatively cool liquid phase fluid that contains a significant proportion of produced shale oil hydrocarbon. In the arrangement shown in Figure 1.21c, borehole **20** is equipped with pipe strings **21, 22** and **23**.

Some or all of such pipes are preferably thermally insulated, as indicated by coatings **24** on pipes **21** and **22**. Pipe **21**, through which the hot aqueous fluid is inflowed, opens into the borehole at a relatively shallow depth. Pipe **22** extends to an intermediate depth and is used to outflow fluid that is relatively cool but is predominately gaseous. Pipe **23** extends to a relatively deep depth, is preferably equipped with downhole pumping means, and is used to outflow fluid that is predominately liquid. The vertical section of borehole between the ends of pipes **22** and **23** serves as a downhole gravity of separation chamber.

Steam or a mixture of steam and hot aqueous liquid (hot water) is inflowed through pipe **21**. The inflowing hottest and lightest gas tends to remain above the cooler and heavier gas and in situ generated carbon dioxide. The cooler gases outflow through pipe **22** while the hotter and lighter inflowing gases tend to flow along the walls of the cavern. Where desirable a relatively light and cool gas, such as methane, hydrogen, etc., can be maintained substantially static, or slowly injected, through and around the upper portion of the borehole and cavern.

Once the rubbled oil shale cavern has been established and the heat-sensitive minerals and water-soluble carbonates removed as an aqueous solution, the hydrocarbons (oil) can be recovered by suitable means such as by contacting the rubbled oil shale within the cavern with a pyrolyzing fluid to effect decomposition of the kerogens to hydrocarbon which is removed from the formation. In recovering the hydrocarbons, the pyrolyzing fluid can be injected (Figure 1.21a) via **7** and recovered via tubing **8** vice versa and in a dual system as shown in Figure 1.21c the pyrolyzing fluid such as steam can be injected via tubing string **21** and the hydrocarbons recovered via **22** or the process can be reversed.

Fracturing Using a Pair of Well Boreholes

P.J. Closmann and R.P. Nordgren; U.S. Patent 3,501,201; March 17, 1970; assigned to Shell Oil Company describe a method of recovering shale oil from a normally impermeable subterranean oil shale formation.

The process involves forming generally vertical fractures extending along generally parallel paths from at least a pair of well boreholes extending into a normally impermeable subterranean oil shale formation. Hot fluid is injected through at least one of the well boreholes until flow into at least one of the fractures therein is thermally closed by the swelling-shut of the walls of the fracture. Fluid in at least one borehole in which at least one fracture has been thermally closed is pressurized until at least one new fracture is formed and the steps of injecting hot fluid and the pressurizing of fluid are repeated at successively higher temperatures and pressures until the resultant fractures form a channel interconnecting the well boreholes, through which channel fluid can flow from one borehole to another.

The walls of the fracture interconnecting the boreholes are leached at a controlled temperature until channels provided between the well boreholes are capable of remaining open while hot fluid is circulated between the boreholes. This is accomplished by injecting a leaching fluid into one of the boreholes opening into the well-interconnecting fracture channel and producing fluid from another of the boreholes opening into the fracture channel, with the fluid being injected at a temperature below that at which the last fracture was opened into the well borehole into which the circulating fluid is injected but above that at which the last fracture was thermally closed within the well borehole into which the circulating fluid is injected so that most of the injected fluid is conveyed through the channels interconnecting the boreholes.

Thus, when such a flow path is opened between a pair of wells, the path may be kept open by lowering the temperature of the fluid pumped through the path to a temperature that is less than that at which the last fracture was formed but greater than that at which the next preceding fracture was thermally closed. In addition, if the effective permeability of the flow path between the well boreholes is then increased by circulating through it a fluid that removes solid components from the walls of the channels, the flow path is converted to one through which the flow of fluid from one well borehole to the other remains efficient at whatever temperature is subsequently imparted to the fluid.

The attainment of a significant increase in the effective permeability of such a flow path may be detected by an increase in the rate of flow in response to the injection pressure that was initially required to displace fluid from one well borehole to another. If the temperature of the circulating fluid is too high relative to the rate at which permeability is being increased, this may be detected by a reduction in the flow rate. If the temperature is too low relative to the rate at which the permeability is being increased, this may be detected by a decrease in the rate of outflow from the production well borehole without a corresponding decrease in rate of inflow into the injection well borehole. Where the temperature is too high, the thermal closing of the last opened fracture throttles the flow and where the temperature is too low, a thermal opening of previously closed fractures diverts the flow of the injected fluid.

After all the steps discussed above are carried out, an underground combustion process which develops considerably higher temperatures, may be undertaken. The steps of leaching out part of the kerogen and the rock generally make closure of the fracture paths during combustion very unlikely. In this manner, it is possible to treat a substantial part of the formation by underground combustion.

Nuclear Detonation

A process described by *P.J. Closmann; U.S. Patent 3,578,080; May 11, 1971; assigned to Shell Oil Company* involves exploding a relatively high-energy explosive device within an oil shale formation thereby forming a chimney of rubble within the formation having fractures extending from the chimney through the formation. A plurality of spaced wells are extended into the formation radially outwardly from the chimney and adjacent to at least some of the fractures. Fluid flow paths are formed from the wells through the fractures into the chimney and fluid is circulated through these fluid flow paths and into the chimney at rates creating a pressure drop from the wells to the chimney. Oil shale-reactive properties are imparted to the circulating fluid whereby the fluid reacts with the oil shale moving solid components into void spaces formed within the chimney increasing the permeability of the oil shale formation relative to the permeability of the chimney in regions surrounding the chimney.

A process described by *G.O. Grady; U.S. Patent 3,601,193; August 24, 1971; assigned to Cities Service Oil Co.* involves the recovery of shale oil by in situ retorting of underground formations of oil shale. The process involves a method for initiating and supporting in situ retorting of the oil shale in the chimney of fractured and crumbled shale of high permeability formed by the detonation of a nuclear device positioned in a subsurface oil shale formation.

The process permits and facilitates the establishment of a hot zone substantially across the cross-sectional area of the chimney. This is accomplished by injecting the retorting fluid into the chimney space by means of at least one injection well positioned generally horizontally across the chimney. The retorting fluid is injected through perforations in the injection well liner.

The injection of a retorting fluid into the chimney from a multiplicity of points permits the establishment of a hot zone at a temperature sufficient to support retorting across substantially the entire width of the chimney. By injecting retorting fluid through a series of injection wells thus positioned, the establishment of the desired hot zone substantially across and around the entire cross-sectional area of the chimney may be accomplished.

The support and advancement of the hot zone throughout the chimney face can readily be accomplished by continued injection of the initial retorting fluid or a suitable gas into the injection wells and consequently into and around the cross-sectional area of the chimney. The injection wells may be positioned either in the upper or the lower portion of the chimney, with the distilled shale oil being recovered through production wells in accordance with normal techniques.

In order to initiate and support retorting of the shale across and around the chimney formed by the detonation of a nuclear device in a subsurface in an oil shale formation, at least one injection well is positioned so that it extends generally horizontally across the width of the chimney. The injection well liner contains perforations through which the retorting fluid may be injected into the chimney. The retorting fluid is used to establish a hot zone in the chimney at a temperature capable of supporting retorting of the shale. The projection of the retorting fluid through a series or multiplicity of perforations in the generally horizontally extending injection well liner facilitates the establishment of a hot zone substantially across the width of the chimney.

Carbon Dioxide as Heat Carrier

In a Process described by *H.E. Gilliland; U.S. Patent 3,480,082; November 25, 1969; assigned to Continental Oil Company* oil shale in association with carbonate minerals are heated to a temperature where the hydrocarbon associated with the oil shale is liquefied or gasified by contacting the oil shale with hot carbon dioxide. The oil shale can be contacted in situ or can be mined and contacted in a retort.

When oil shale is associated with carbonate minerals, certain problems are encountered which do not exist with other oil shales. Frequently the carbonated minerals will equal approximately 50% of the mixture of minerals and hydrocarbon material. At temperatures of approximately 1000°F which are required to retort the shale for recovery of hydrocarbons, the carbonates begin to calcine, e.g., decompose, consuming prodigious quantities of heat which will generally range from 600,000 to 1,000,000 Btu/ton of shale. The cost of supplying this heat can range from 2 to 4 times or more than that required for supplying the minimum heat requirement for heating the shale to retorting temperatures.

The carbonates in the shale are also one of the principal agents for bonding the mineral matrix of the shale. Thus, the strength of the shale is deleteriously affected by carbonate decomposition. This loss of strength leads to particle size degradation and shale dust is formed. Particularly in surface retorting, this dust creates numerous operating problems and contributes to atmosphere pollution. To a lesser degree, the dust is also a problem from in situ retorting since the dust is easily carried out through the producing well and also tends to block the flow paths in the subterranean formation.

The decomposition of the carbonate minerals can be substantially reduced if the mineral is heated to calcining temperature in an atmosphere of CO_2. Even at atmospheric pressures $CaCO_3$ decomposition is reduced by a factor between 3 and 4 as compared to the decomposition at calcining temperatures in a nitrogen atmosphere. As the partial pressure of the CO_2 is increased, the amount and rate of decomposition is decreased; therefore, it is preferred to use a pressure of at least 500 psig and generally use a pressure of at least 1,000 psig.

The upper limit on the pressure is limited only by the structural strength of the formation or retorting vessel. Since the retorting temperature is generally around 1000°F, the CO_2 must be heated to some temperature in excess of the 1000°F, the minimum temperature depending upon the formation temperature, the volume of gas per volume of shale to be heated and other heat loss factors which will vary with each given set of conditions. In general, the minimum practical temperature will be about 1250°F and preferably about 1500°F.

The use of CO_2 as the heat carrier has additional advantages over other inert gases. For example, CO_2 has a relatively high heat capacity and therefore is an efficient heat carrier.

The CO_2 is easily liquefied at temperatures below 87.8°F, thus facilitating recovery of the CO_2 for recirculation and handling.

Referring to Figure 1.22, carbon dioxide gas is passed via conduit **1** and well **2** to oil shale formation **3** at a rate of 1,000,000 scfh (standard cubic feet per hour), a temperature of 1500°F and at a pressure of 1,000 psig. The CO_2 heats the shale and vaporizes and/or entrains the hydrocarbons produced by cracking and/or vaporization and liquefaction and is produced from the formation through production well **4** and passed via conduit **5** to heat exchanger **6**. The temperature of the stream entering exchanger **6** is 300°F and has a pressure of 750 psig.

FIGURE 1.22: IN SITU RETORTING PROCESS USING CO_2 AS HEAT CARRIER

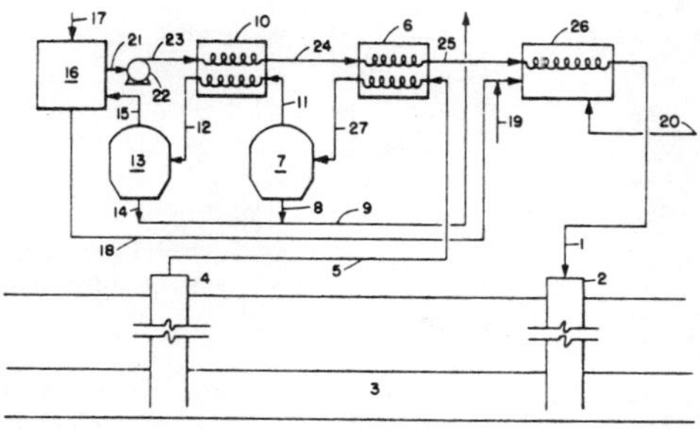

Source: H.E. Gilliland; U.S. Patent 3,480,082; November 25, 1969

The stream is cooled to about 200°F and passes via conduit **27** to separator **7** where a first condensate is removed via conduit **8** and passed to product line **9** to be sent to storage, not shown. The gaseous material is then passed to a second heat exchanger **10** via conduit **11** where it is further cooled to about 100°F and passed via conduit **12** to a second separator **13** where the remaining condensible liquid is separated from the gaseous material and is passed via conduit **14** to conduit **9** and mixed with the liquid from separator **7** to be passed to storage. The mixed product amounts to 1,000 bbl/day oil and 150 bbl/day water.

The water can, of course, be separated from the oil by known decanting or other separating means. The CO_2 and light hydrocarbon gases are then passed at 700 psig to CO_2 liquefaction zone **16** via conduit **15**. Makeup CO_2 is added via conduit **17**. The CO_2 is liquefied and is separated from the gaseous hydrocarbons. These hydrocarbons total 1,400 scf/bbl of shale oil recovered and have an energy value of about 800 Btu/scf. This fuel is passed via conduit **18** to primary heating zone **26**.

Additional fuel is mixed with this recovered fuel via conduit **19**. The liquid CO_2 is passed via conduit **21** and pump **22** at 40°F and 1,200 psig to heat exchanger **10** where it is heated to 140°F and passed via conduit **24** to heat exchanger **6** where the temperature is raised to 240°F. The CO_2 then passes via conduit **25** to primary heater **26** where it is heated by burning the fuel from conduit **9** with air supplied via conduit **20** to a temperature of 1500°F, the pressure having dropped to 1,000 psig and is then passed via conduit **1** to injection well **2**.

Normally, with an oil shale-carbonate deposit, one would expect a net make of CO_2 particularly where the hot gases enter the formation. However, operating according to the method described above, it is necessary to add makeup CO_2 via conduit **17**.

Perforated Pipe

J.H. Medlen; U.S. Patent 3,448,794; June 10, 1969 describes a device for extracting oil from oil-bearing shale having a 4" diameter iron or stainless steel pipe perforated with a series of 1½" holes to allow gaseous vapor from the superheated shale to escape down the pipe for cooling and condensation, the cooling pipe consisting of aluminum tubing of 3 to 5 feet in length with coils of copper tubing around it. A cooling agent of water or alcohol may be pumped through the copper tubing. Gaseous oil vapor condenses in the cooling unit into oil and 40° to 50°F and is collected in a storage tank or oil barrel, as desired. Pumping of gaseous vapors through the perforated tube into the aluminum tube is not necessary, the heavier than air gas being found to flow through the tube by natural gravitational forces.

An advantage of the process is that the device and apparatus is intended to be placed through a drilled hole of a size common to the size of the pipes, into solid shale bends, and may be placed as close together as desired to extract the most oil.

Referring to Figure 1.23a, there is shown a series of perforated pipes **10**, having a cap **12**, and having a series of vapor holes **14**, disposed about the perforated pipe. The perforated pipe may be 4" in diameter and the holes are 1" to 1½" holes. Heating elements **18** connected to electrical conductors **20** provide for the heating and superheating of the shale to produce gaseous vapors which are drawn into the perforated pipe through the vapor holes, and then into a connected solid aluminum pipe **24** around which is provided an iron jacket **26** shown in Figure 1.23a and in part in Figure 1.23b.

Surrounding the cooling pipe **24** is a spirally configured copper tubing **30** through which a refrigerant gas or liquid is pumped from a pump **36** from a refrigerating means **38** to each of the cooling coils **30**, which then returns through a return pipe **40** to the refrigerating means. By means of the cooling effect that is derived and imposed upon the cooling pipes, oil is condensed and is then collected in pipe **44** which connects with a barrel or storage tank **46**. The tank has a vent **48** at the top and an oil outlet valve **50** at the bottom.

FIGURE 1.23: IN SITU OIL EXTRACTION DEVICE

a.

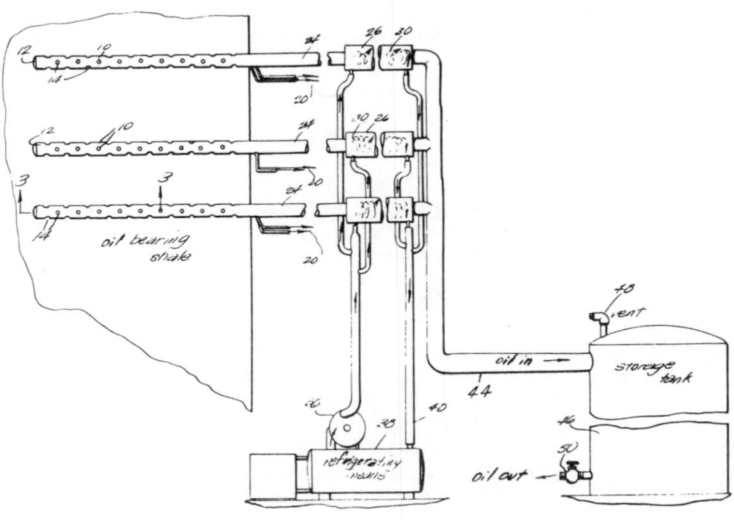

Perforated Pipe Inserted in Oil-Bearing Shale

(continued)

FIGURE 1.23: (continued)

b. 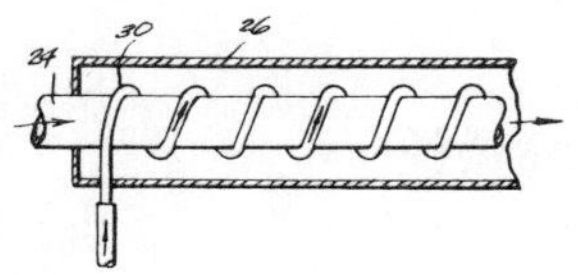

Cross-Sectional View of Cooling Unit

c. 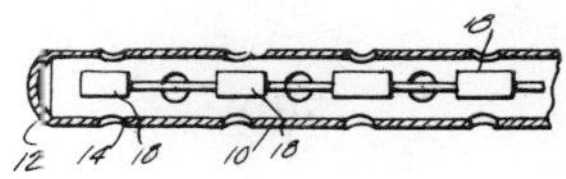

Cross-Sectional View Along Line **3–3** of Figure 1.23a

Source: J.H. Medlen; U.S. Patent 3,448,794; June 10, 1969

In processing the gaseous vapors through the cooling unit, it is found that aluminum tubing of 3 to 5 feet in length disposed within a cooling jacket of iron or stainless steel and filled with insulating material such as glass fibers and in which there is also disposed a spiral of coils of copper tubing around the cooling tube, achieves a maximum result.

The cooling agent may be water or alcohol that passes from the refrigerator pump **36** to the return pipe **40** into the refrigerating means which may include a storage tank. Gaseous oil vapor condenses into usable oil at 40° to 50°F. Pumping of gaseous vapors through the cooling tube and the perforating tube is not necessary due to at least the reason that heavier-than-air gases will flow through the tube by natural gravitational forces, and condensation of vapors in the cooling tube induces a certain pressure to draw the vapors.

Halogenated Organic Compounds

According to a process described by *R.L. Dilgren and G. Drinkard; U.S. Patent 3,695,354; October 3, 1972; assigned to Shell Oil Company* organic materials are recovered from an oil shale by circulating hot fluid containing a halogenated organic compound through a permeable portion of the oil shale and recovering organic materials from the circulated fluid. The organic materials which are produced comprise the components of a conventional shale oil supplemented and/or modified by the presence of halogenated and/or depolymerized derivatives of the relatively high-molecular-weight pyrolysis products of kerogen and/or bitumen.

The process is particularly useful in producing shale oil from a subterranean oil shale that contains water-soluble minerals and/or heat-sensitive carbonate minerals. The presence of such minerals tends to enhance the rate of growth of a permeable zone due to the fluidization of inorganic components of the oil shale. The heat-sensitive carbonates are fluidized by both a thermal conversion to carbon dioxide and water, and a dissolution of water-soluble unreacted portions or conversion products. The hydrogen halides produced by the reaction of the halogenated organic compound used in the process are effective in enhancing the rate of such fluidizations of the inorganic components of the oil shale.

In situations in which it is desirable to mine portions of an oil shale formation and pyrolyze the mined oil shale in a fluid-heated, pressure-tight, surface-located retort, the pyrolytic recovery of petroleum material is improved by incorporating an effective amount of halogenated organic compound in the hot fluid that is circulated through the oil shale. In such a process, chunks or fragments of the oil shale are preferably contacted with a hot aqueous and/or hydrocarbon fluid that contains at least about 2% by weight of at least one halogenated organic compound at a temperature of at least about 500°F under a pressure sufficient to keep a significant proportion of the fluid in the liquid phase.

Where the oil-shale-contacting fluid predominates in hydrocarbon components, e.g., a benzene solution of a halogenated compound, the oil shale is preferably preheated by contacting it with hot aqueous liquid. The contacting of chunks of oil shale with a hot aqueous liquid (i.e., with hot water utilized in preheating the oil shale and/or a hot aqueous solution containing at least one halogenated organic compound) tends to exfoliate the pieces of the oil shale. The exfoliation reduces the tendency for clinkers to be formed during the pyrolysis-extraction operation and reduces the extent to which the oil shale needs to be crushed in order to obtain an efficient recovery of petroleum material.

OTHER PROCESSES

Diluent Soaking Followed by Gradual Heating

According to a process described by *R.S. Brimhall; U.S. Patent 3,697,412; October 10, 1972* crushed shale is provided and a hydrocarbon diluent such as a fuel cut or a stove oil cut is added. From laboratory tests it is indicated that not less than 15 ml of diluent should be employed for every 57 grams of shale. The diluent and oil shale mixture is allowed to remain as a slurry mixture under ambient temperature conditions for approximately 10 minutes of time but in no case less than 7 minutes.

After such soaking has been completed then the oil shale and diluent mixture or slurry is gradually increased as to temperature, but with vapors being maintained in intimate contact therewith and then collected and condensed as a first fraction or first series of fractions. The oil shale, whether or not now in slurry form, is subequently passed through a series of temperature zones of progressively increasing temperatures so that successively higher boiling fractions may be obtained. It is noted in the process that the kerogen is not heated to extremes to produce initially a crude oil for subsequent processing as by distillation. Rather, fractional distillation is commenced at the beginning of the retort cycle of the soaked oil shale.

The first step in the process is to provide a suitable quantity of suitable oil shale for processing. Oil shale, of course, is a sedimentary deposit rich in organic matter and which, under generally used retort processing, yields a shale crude oil when heated in the absence of air. While most shale sediments form oil upon distillation, the industry classifies as oil shale those sediments which produce at least 15 gal of oil per ton. Commercial grades of oil shale will provide from 25 to 50 gal/ton of shale, to extremes of up to 150 gal/ton.

The oil shale to be processed is next crushed. Preferably, the crushing will be such that a majority of the oil shale to be processed will pass through a ¼" screen mesh. Although the process will still be operative, yet it is preferable that the particle size of the crushed oil shale is such that a majority of the same will not pass a 25-screen mesh, as sand granules may tend to settle out in the process.

The next step in the process is to add a hydrocarbon diluent to the crushed oil shale. The term diluent is used in its broadest sense, i.e., the amount of crushed oil shale present in any given volume is reduced thereby, and the diluent liquid may be either reactive or nonreactive relative to the shale. In point of fact, the diluent will constitute a hydrocarbon which enjoys a liquefied state and ambient temperatures, e.g., 70°F ± 35°F. Preferably, the diluent will take the form of a bituminous distillate such as a diesel cut or stove oil. Even gasoline can be employed; howbeit, sealing problems and safety problems will be somewhat complicated thereby. While the ratio of diluent to crushed shale is not critical in practicing the process, optimum results are obtained where there is present at least 15 ml of diluent for every 57 grams of crushed shale.

The next step in the process is extremely important and comprises the soaking of the oil shale in the added diluent at ambient temperature for a period of time. This time is preferably of the order of 10 minutes, but should be at least 7 minutes or more. Experimentation has indicated that the so-provided soaking time is essential in preparing the shale for further processing in order that a maximum recovery may be obtained.

The next step in the process comprises gradually raising the temperature of the oil shale and diluent to vaporize a first fraction in a low temperature zone. Ancillary to this step is to distill the so-vaporized first fraction as by conventional distillation and condenser techniques. As to this first heating step, the same should take place such that the slurry temperature (e.g., 50° to 70°F) is gradually raised from ambient temperature conditions to approximately 350°F. During this step, various gases and vapors will be given off, the nature of the same and their volume depending upon the type of shale being processed.

Extremely important in the process is the gradual temperature rise of the shale and diluent subsequent to the extended soak time of the prior step. It is the combination of these two steps, essentially, that increases yield in the low temperature range, e.g., below 850°F. While the exact chemistry of this phenomenon is currently unknown, yet it is believed that the prior extended presoak plus the gradual raising of the temperature of the so-soaked oil shale, enables carbon, hydrogen and oxygen constitutents to aid in expanding hydrocarbon chains of kerogen constituents, original or thermally modified, so that a maximum yield can be achieved.

This first step of gradually raising the temperature of the oil shale and diluent is followed by corresponding and similar successive steps, respective ones of which are operated in progressively higher temperature regions for obtaining successive fractions. Accordingly, the second temperature rise step may be performed in a vapor temperature region of 350°F to 480°F, the next succeeding step from 480° to 560°F, and the subsequent step from 560° to 760°F.

At some point in this process, subsequent to the commencement of the application of heat, it will be found that a diesel cut or stove oil cut derived from the ascending vapors will be available for feeding back to the hydrocarbon-diluent additive step, to supply diluent to the oil shale being processed. But a small fraction of such cut need be used to be supplied the oil shale.

By the process, either wide or narrow cuts or fractions are possible, this by regulating the temperature ranges involved and adding to or substracting from the heat additive steps. Any number of heat additive steps may be incorporated in the system, and with a selected hydrocarbon distillate, or a portion thereof, being rerouted back to the crushed oil shale entering the system.

Noncombustive Process Using Superheated Steam

O.R. Waltrip; U.S. Patent 3,567,622; March 2, 1971 describes a noncombustive, atmospheric pressure, retorting process for the production of high-grade crude petroleum and petrochemicals from carbonaceous solids. In the process, superheated steam comprises a system carrier fluid and hydrogen donor and the retorting is carried out in the presence of a minor proportion of a middle fraction hydrocarbon oil selectively commingled with the carbonaceous solid in the retort zone or injected into the retort zone with the carrier fluid.

The hydrocarbon oil utilized in conjunction with the noncombustive retorting of the carbonaceous materials comprises a middle fraction, either from an external source or a fraction recycled from the process product, which may be characterized as a No. 2 fuel oil having a boiling point range from about 440° to 675°F.

The hydrogen donor-carrier fluid utilized in the noncombustive retorting process comprises superheated steam supplied to the retort zone in a controlled amount at a temperature within the critical range of about 460° to about 1000°F and wherein the initial flow of steam is at a temperature within the range of about 460° to about 800°F. Steam at a temperature within this critical temperature range is not introduced into the noncombustive retort zone until the carbonaceous materials being processed have been heated by external means to minimum temperature of about 550°F and wherein the temperature of the carbonaceous materials being retorted are not permitted to exceed about 750°F during retorting.

The exemplary apparatus schematically illustrated comprises a noncombustive retort indicated

FIGURE 1.24: NONCOMBUSTIVE RETORTING PROCESS

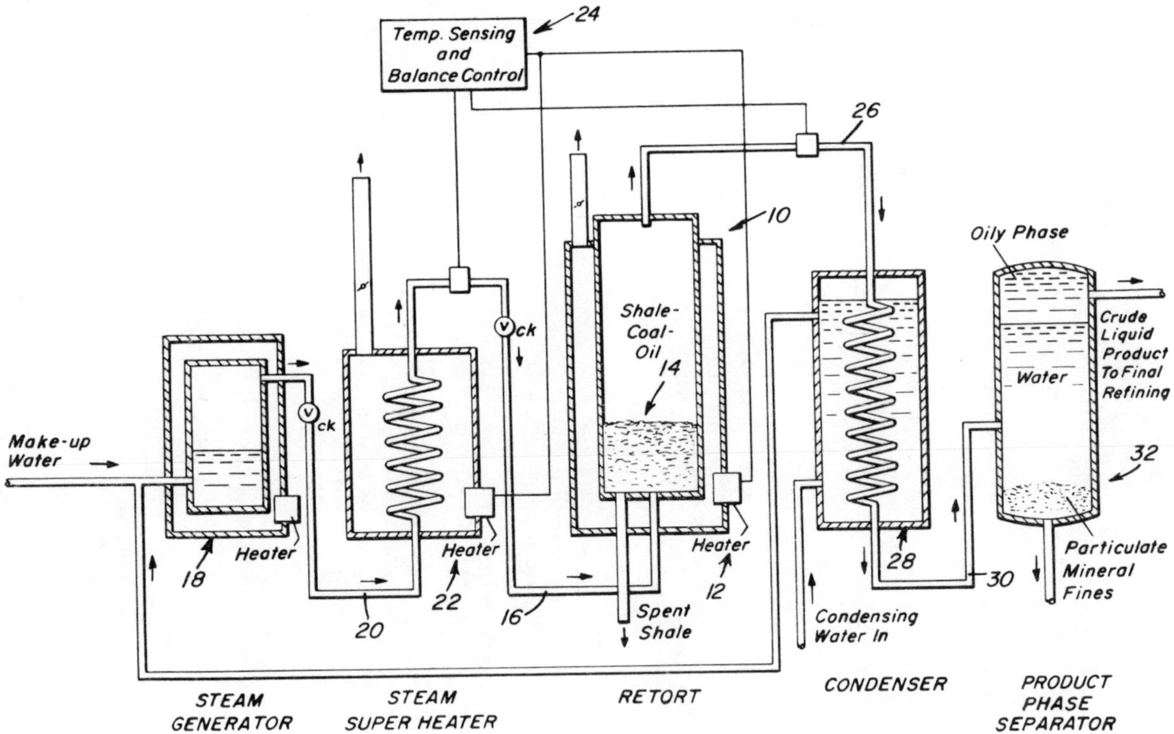

Source: O.R. Waltrip; U.S. Patent 3,567,622; March 2, 1971

generally at **10** maintained substantially at atmospheric pressure and externally heated by a heater means indicated generally at **12** deriving its energy source from any suitable means, even including utilization of process gas subsequent to the initial start-up of the system by the utilization of externally derived fuel. As indicated generally at **14** the batch retort **10** is shown to be charged with particulate carbonaceous material, i.e., oil shale **14** having a particle size between 0 and 6 inches for example, wetted with about 2 to 5% by weight of No. 2 fuel oil.

Steam for introduction into the retort through the conduit **16** is supplied from a wet steam generator **18** provided with a heater means deriving its energy from any suitable source, and wherein water for the steam generator may comprise fresh make-up water or recycle water derived in conjunction with the condensing or fractionating of the product of the process. Wet steam generated within the steam generator is directed through the conduit **20** to a steam superheater means indicated at **22** provided with a heater means, suitably energized, and wherein the superheated steam delivered therefrom through the conduit **16** is continuously monitored to insure that the superheated steam is within the critical temperature range of about 460° to 1000°F.

As indicated, the forward movement of the hydrogen donor-carried fluid, i.e., superheated steam, within the process apparatus is assured by check valves suitably disposed in the conduits **16** and **20**. In addition, a temperature sensing and balance control means **24** sensing the temperature of the superheated steam in the conduit **16** as well as the effluent of the retort discharging through the conduit **26**, as schematically shown, selectively and automatically controls the heater means of both the steam superheater and the retort to insure that superheated steam, within the prescribed temperature range, is not admitted to the retort until such time as the particulate shale has been heated to a minimum temperature of 550°F. In addition, the temperature of the shale, which may be sequentially raised in a stepwise fashion for the off-take of various product fractions, is not permitted to rise above about 750°F.

The effluent discharging from the noncombustive retort through the conduit **26** is directed to a condenser means **28** wherein the condensibles are liquefied, such as by being placed in indirect heat-exchange relation with condensing water, and the condensate directed through conduit **30** to a product phase separator indicated generally at **32** where the product separates into a plurality of distinct phases such as exemplified by the oil oily phase comprising the crude petroleum derived from the shale, and including synthesis products derived from the hydrogenation of hydrogen-deficient hydrocarbons therein.

The water phase includes a number of water-soluble oxygenated organic compounds and may be treated for the separate recovery thereof by suitable conventional means with recycle of the water per se for generation of additional amounts of steam for the treatment of additional carbonaceous material. In addition, means is provided for withdrawal of any particulate mineral fines, i.e., asbestos fines, that may be carried over from the retort by the carrier fluid. The following table is included to specifically show a number of the properties of Rifle, Colorado oil shale, Utah tar sands and Price, Utah oil shale noncombustibly retorted in accordance with the process such as by the exemplary batch process as discussed above.

	Rifle, Colorado oil shale	Utah tar sands	Price, Utah oil shale
Gravity, ° API at 60° F	23.3	35.6	31.0
Sulfur, wt. percent	.24	1.01	.12
Nitrogen, wt. percent	1.44	.073	.083
Viscosity, CST at 100° F	7.8	2.5	3.82
ASTM distillation:			
I.B.P. ° F	152	150	359
10%	360	356	390
20%	424	404	410
30%	468	418	424
50%	560	446	450
70%	(1)	486	507
80%		518	604
90%		610	
Percent distilled	67	90	70
Percent volatiles plus residue	33	10	30

[1] E.P. 634 (at 67%).

From the foregoing table it will be seen that the carrying forth of the process results in the production, for example, of a high-grade crude petroleum characterized by a relatively high API gravity and a relatively low viscosity.

Traveling Grate Shale Retorting

J.H. Haddad and J.G. Mitchell; U.S. Patent 3,483,115; December 9, 1969; assigned to Mobil Oil Corporation describe a method of retorting oil shale in a horizontally moving bed where gaseous material flowing transversely through the shale bed, all in a downflow direction through a plurality of gas contacting zones establishes a restricted kerogen decomposition heat front in the top of the bed and drives the decomposition heat front progressively downwardly as the bed moves horizontally through the retorting section. Gaseous material obtained from an intermediate portion of the shale bed, from which oil mist was recovered, is employed to effect partial cooling of gaseous material recovered from a latter portion of the shale bed. This cooled gas is introduced to an initial portion of the shale bed to condense out entrained oil constituents and to preheat the shale.

Figure 1.25 represents in cross section the process where three gas recovery zones are employed and gases from the second and third zones are mixed prior to being directed to the first zone. Particulate shale which has been ground to a target average size of about 0.75" with a maximum particle size of about 4" is introduced into a hopper **1**.

FIGURE 1.25: TRAVELING GRATE SHALE RETORTING

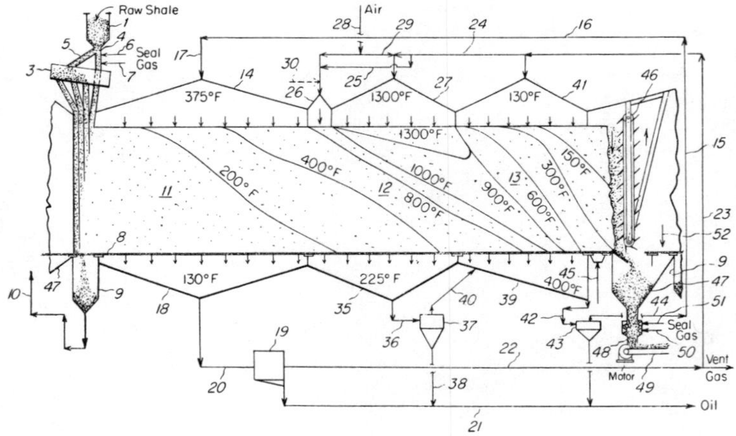

Source: J.H. Haddad and J.G. Mitchell; U.S. Patent 3,483,115; December 9, 1969

The shale is directed from the hopper to classifier **3** through conduits **4** and **5**. In the classifier, the shale particles are classified according to size with the larger particles being directed to the bottom of the bed and the fines being directed to the top of the bed. Seal gas is introduced into conduits **4** and **5** to prevent air from entering the shale bed.

In this example, a double seal gas system is employed in conduits **4** and **5**. The two seal gas flow rates are controlled so that the differential pressure between the two seal gas inlets **6** and **7** is essentially zero. The shale is introduced at a rate so that a relatively thick bed is formed on a horizontally moving perforated grate **8**. The bed thickness on the grate is maintained at about 6 feet.

The small amount of fines which sift through the bed are recovered in hopper **9** and recycled through conduit **10** to the shale inlet hopper **1**. The moving grate **8** causes the shale bed thereon to progress sequentially through zones **11, 12** and **13**. The uncondensed gas from zones **12** and **13** is recycled to gas plenum **14** on top of the bed in zone **11** through conduits **15, 16** and **17**. The uncondensed gas passes downwardly through zone **11** through the grate and into a gas recovery zone **18** located below the grate. The gas in gas recovery zone **18** is maintained at a temperature below about 130°F. At this temperature, most of the shale oil is in the form of a condensed mist. The condensate and remaining gas are directed to an oil recovery step **19** through conduit **20**. Oil recovery step **19** can be, for example, a cyclone separator or an electrostatic precipitator. Condensed oil obtained from oil recovery step **19** is directed through conduit **21** to storage.

Uncondensed gas from recovery zone **19** is directed through conduits **22, 23, 24** and **25** to the top of the shale bed in zones **12** and **13**. The uncondensed gas to zone **12** is first introduced into plenum chambers **26** and **27**. Air is introduced to plenum chambers **26** and **27** through conduits **28** and **29**. At start up, a fuel can be introduced to plenum chamber **26** through conduit **30** to mix with the air and recycled uncondensed vapor under conditions to cause combustion in plenum **26**. After start up it is usually unnecessary to add fuel to maintain combustion in plenum **26**. In plenum **27**, a portion of the uncondensed gas is burned. The heat from the combustion in these two plenums is absorbed by the shale bed directly underneath these zones. By this heating a heat wave at kerogen decomposition temperature is initiated on top of the bed in zone **12**.

The kerogen decomposition zone is caused to move downwardly and toward the shale outlet in a configuration as shown in Figure 1.25. Representative temperatures of the gas introduced to each zone as well as the bed temperature profile is shown in Figure 1.25. The ratio of gas rate to the shale rate in each zone is adjusted to provide the representative temperature profiles shown. The gas in vapor zone **12** is caused to move downwardly through the entire vertical height of the bed. This gas is recovered in plenum zone **35** at a temperature below about 225°F. At this temperature a major portion of the oil has condensed. The uncondensed and condensed oil and gas are directed through conduit **36** to an oil separator **37** where oil is recovered through conduit **28** and uncondensed vapor and gas are directed to plenum chamber **39** through conduit **40**.

In zone **13**, recycled gas from oil recovery step **19** is introduced into plenum chamber **41** on top of the bed at a temperature of about 130°F. This gas is caused to move downwardly through the entire vertical height of the shale bed. In so doing, the top portion of the bed is cooled while the bottom portion of the bed is progressively heated until the kerogen decomposition heat wave reaches the grate. That portion of the bed lying between this point and the shale outlet in a horizontal direction is cooled through the entire vertical height. The off-gas from zone **13** is recovered in plenum chamber **39** at a temperature, when mixed with the uncondensed vapor from oil recovery step **37**, of below about 400°F. At this temperature an additional portion of the higher boiling fraction of the shale oil is condensed.

The uncondensed and condensed shale oil and gas is directed from plenum chamber **39** through conduit **42** to oil recovery step **43** where shale oil is recovered. The uncondensed vapor and gas from oil recovery step **43** is directed through conduits **44, 15, 16** and **17** to the top of zone **11** into the plenum chamber **14**. The gas in plenum chamber **14** is caused to move through the vertical height of the bed in zone **11** in a manner described above. Blowers are provided in the various gas conduits to ensure gas flow downwardly through the bed. The gas flow rates to plenums **14, 27** and **41** are maintained respectively at about 0.765 ton vapor per tons shale, 0.447 ton vapor per tons shale and 0.483 ton vapor per tons shale.

The grate, after progressing through zone **13** is contacted with a small portion of gas introduced through conduit **45** which removes any coke which has formed in the perforations of the grate. This gas moves only through a very small portion of the vertical height and is recovered in plenum chamber **39**. The grate is in sections and above hopper **47** the sections are tipped so that the shale thereon can be removed. The shale is broken up by the clinker breaker **46** and thereafter moves by gravity to hopper **47** and directed outwardly from the retort zone through conduit **48** onto a conveyor **49**.

Conduit **48** is sealed from the atmosphere by a dual-gas seal system by the introduction of seal gas through conduits **50** and **51** similar to that described above for the inlet seal system. The relative sizes of the gas inlet plenums are approximately proportional to the inlet gas flow rates thereto. To afford relatively complete clearing of the grate **8**, it is subjected to a jet of fluid and/or solids through conduit **52**. The material entrained on the grate is directed to the hopper.

In the process, the shale bed can be moved in a closed horizontal path as for example in a circular path or in an open horizontal path as for example in a straight line. In any event, the shale bed is continuously moved through a plurality of vapor contact zones. The vapors in each zone are introduced at a pressure sufficient to ensure vapor flow in one direction through the bed vertical height. The gas pressure differential between the top and bottom of the bed to ensure this flow depends upon a number of factors including the average shale particle size, bed thickness, the gas flow rate used, and the temperature conditions of the bed. In the process, the average shale particle size is maintained below about 2" and preferably below about 0.6". The shale bed thickness is maintained below about 15 feet and preferably between 4 and 8 feet.

In operation, the particulate shale can be directed onto the grate through a feeding device which stratifies the shale according to size with larger shale particles being located on the grate and the fines being located on top of the bed. Means can be provided below the bed and preceding the first vapor recovery zone to recycle the small portion of the fines sifting through the bed back to the top of the bed. The stationary gas plenum chambers above and below the moving retort are sealed from the atmosphere as for example by a liquid seal to contain the gases.

The moving grate moves the shale bed serially through the gas contact zones to effect complete kerogen decomposition. The temperature profile of the spent shale upon discharge from the grate may vary from 150°F at the top of the bed to as high as 600°F at the bottom of the bed. The spent shale is removed from the grate and directed to spent shale container where the shale is mixed and the temperature of the individual particles allowed to equilibrate by conductive heat transfer to reduce maximum particle temperature below 400°F. To assist removal of spent shale from the grate, apparatus can be used to break up the shale bed to assist in shale movement to storage. The spent shale container is also sealed from the atmosphere but is open to the retort vapors.

The cool spent shale is removed from the container and disposed of. The moving grate is then directed to the shale inlet portion of the retort where fresh shale is introduced. After spent shale is removed from the grate and prior to introducing fresh shale thereon, any coke and/or particulate shale in the grate perforation is removed therefrom as for example with high-velocity jet cleaning or by burning.

The vapors directed to each zone are first directed to inlet plenum chambers adjacent the bed. These inlet plenums operate to separate the vapors so that different conditions of vapor temperature and vapor flow can be maintained in various portions of the bed. In addition, the vapor inlet plenums provide relatively uniform vapor flow through the vertical height of a given portion of the shale bed. Each of the plenums can be subdivided into smaller chambers to provide more efficient vapor flow control. Similarly, the vapor outlet plenums can be subdivided into smaller chambers to improve vapor flow control. Compressors are provided in various vapor lines as needed to maintain the desired pressure gradient through the shale bed.

In a process described by *T.E. Ban; U.S. Patent 3,325,395; June 13, 1967; assigned to McDowell-Wellman Engineering Company* oil-bearing material is first obtained or crushed to a workable particle size not exceeding 18" in diameter. This material is then charged to a traveling grate to form a bed. The traveling grate with its burden passes into a first zone where the burden is exposed to heat reducing or neutral gases at a temperature of from 1000° to 1500°F, by passing the gases through the burden in either updraft or downdraft fashion. This elevates the temperature of the burden to an oil-educting temperature of about 800°F, at which point the kerogen breaks down or cracks into an oil vapor which becomes condensed and suspended in the gas stream passing through the bed. This draft or gas is then conducted through a cyclone separator where the oil is separated from the gas. Any suitable means for separating the oil from the gases may be used.

The traveling grate moves to a second zone where the oil-depleted gases are recirculated through the spent burden to aid in cooling the burden, and to preheat the gases to oil-educting temperature for recycling through a new burden in the first zone. Air may be added to the gas stream as to burn part of the burnable constituents contained in the gas stream, and to elevate the temperature of the gas to from 1000° to 1500°F. The spent burden is removed from the traveling grate.

Agglomeration of Fines in Traveling Grate Process

C.A. Rowland and R.D. Frans; U.S. Patent 3,560,369; February 2, 1971 and U.S. Patent 3,560,368; February 2, 1971; both assigned to Allis-Chalmers Manufacturing Company describe a system for recovering oil from oil-bearing shale rock in which the rock is screened to separate fines from larger particles having a dimension of at least about ¼". The system includes a retorting zone in which heated gases, reducing or neutral, are passed through the larger particles and agglomerates of the fines to heat these particles and agglomerates to oil-educting temperature and after which the gases, then containing educted oil, are passed through a zone in which particles and agglomerates are preheated and the gases cooled to condense the oil to small droplets and vapor in the gases. The oil is separated from the gases and into various fractions.

The heaviest fractions of the oil, representing an amount in a range of from 5 to 50% of the total oil educted, are recirculated through the system by using such oil as binder to agglomerate the fines into agglomerates of at least ¼". These agglomerates may be advantageously utilized by charging them to a traveling grate apparatus with the larger particles to form a bed, with the agglomerates forming an intermediate layer between upper and lower layers of the larger particles. The larger particles will, in a preheating condensing zone, trap and prevent escape from the bed, dust and drops of heavy binder oil (which may escape from agglomerates).

With either an upflow or downflow of gas in this preheating condensing zone, heavy binder oil attempting to escape the bed will adhere to and coat particles in the upper or lower layers and dust attempting to escape the bed will adhere to the larger particles coated by the heavy binder oil. Thus, both the heavy binder oil and dust attempting to escape in this preheating condensing zone will be trapped in the bed. In the retorting zone such heavy fraction binder oil retained in the bed will pyrolyze to produce further useful lighter oil, which will leave the bed as vapor, and residual coke which will remain in the bed.

In related work *R.W. Weggel and W.A. Blann; U.S. Patent 3,644,193; February 22, 1972; assigned to Allis-Chalmers Manufacturing Company* describe a system for recovering oil from oil-bearing shale rock in which shale is placed on a traveling grate and transported through a preheating zone; a retorting zone where oil is educted from shale; a combustion zone where the residual carbon in the shale after the oil is removed, is burned; and a cooling zone. A first gas stream, which may be air, is passed through shale that has traveled on the grate at least to where combustion takes place.

This air is thereby heated, and then used to heat a heat-transfer medium such as alumina balls from which a second and oxygen-free gas stream extracts the heat. The second gas stream, heated indirectly by the first gas stream but not contaminated with combustion gases from the first gas stream, passes through the shale in the retorting zone to educt oil from the shale and then through the shale in the preheating zone where the educted oil is condensed and becomes suspended as a stable mist in the second gas stream from which the oil may be mechanically separated.

Electrothermal Pyrolysis

I.S. Salnikov; U.S. Patent 3,377,266; April 9, 1968 describes an electrothermal method and apparatus for the pyrolysis of oil shale to recover shale oil.

Referring to Figure 1.26a, there is shown a vertical furnace or retort **10** of hollow, elongated cylindrical configuration and which is formed to define in downward succession a preheat zone **12**, a distillation zone **13**, and a lower heat exchange zone **14**. In construction, the retort preferably consists of a pair of heavy walled, upper and lower tubular sections **17** and **18** connected in end-to-end relation by means of abutting flanges **19** interconnected by suitable bolts **20**.

To form the preheat zone, the upper tubular section includes an upwardly convergent wall portion **22** for connection in end-to-end relation with a tubular section **24**, of reduced size, again through abutting flanges **25** being interconnected by suitable bolts **26**. The top tubular portion also defines the entrance into the preheat zone from the top of the furnace and for this purpose is provided with a central opening **28** for insertion of a generally funnel-shaped hopper **30** in which is disposed an auger **32**, the latter being driven by an electric motor **33** through gear drive **34** and drive shaft **35**.

In order to separate the tubular section into spaced upper and lower chambers, the lower tubular section has an internal annular shoulder **38** just beneath its point of connection to the upper tubular section which serves to support a downwardly convergent wall portion **40** including a lower tubular portion **42** of reduced size for disposition of a feed screw of auger **43**. Here, the wall portion defines the lower terminal end of the distillation zone and, together with the tubular sections and upper feed screw **32**, provides a sealed chamber for passage of shale at a controlled rate of travel through the preheat zone and distillation zone.

Similarly, the lower end of the tubular section has a convergent wall portion **44** extending downwardly into a lower tubular portion **46** for an ejection screw of auger **47**, and it will be noted that the augers **43** and **47** are driven through a common drive shaft **48** by electric motor **49** through gear train **50**. In general, the various tubular sections and wall portions are heavy-walled and in a conventional manner are surfaced with an inner, refractory lining, not shown, so as to be capable of withstanding the intense heat developed within the furnace during pyrolysis of the shale.

An important feature of the process resides in the construction and arrangement of the heating elements employed in the distillation or pyrolysis zone; and in the preferred from of apparatus a nest or series of electrical resistance heating elements **52** are arranged in spaced parallel relation to one another in the zone with each element being disposed on an axis parallel to the path of flow of the shale material through the retort.

Preferably, each heating element is defined by an elongated, cylindrical rod **54** of metallic material having an outer jacket **55** composed of a suitable fire clay or ceramic material whose expansion and contraction characteristics correspond with that of the material defining the rods. The heating elements are supported in vertical, spaced relation within the distillation zone by means of an open grid or mounting frame **56** having intersecting cross bars **58** and an outer circular ring **59**; and the frame is removably seated in horizontal relation across the distillation zone upon an inwardly projecting annular shoulder **60** at the upper end of the tubular section and is held firmly in place by means of another internal shoulder **62** at the lower extremity of the tubular section, the latter engaging the ring on the mounting plate when the tubular sections are secured in connected-together relation as illustrated.

FIGURE 1.26: ELECTROTHERMAL PYROLYSIS

a.

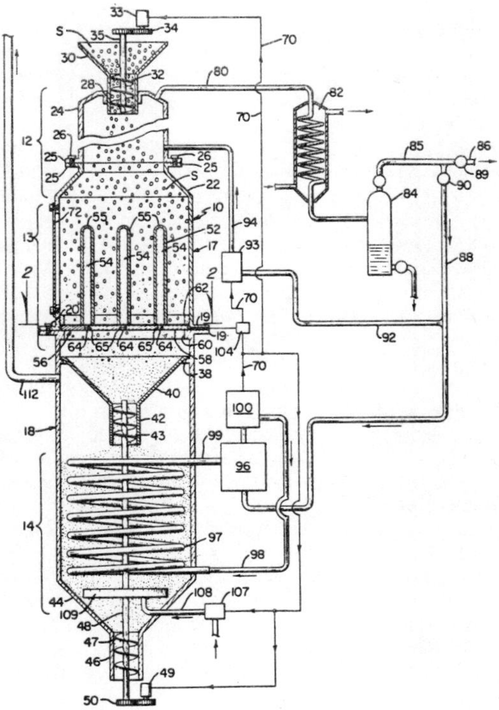

Detail of Apparatus

b.

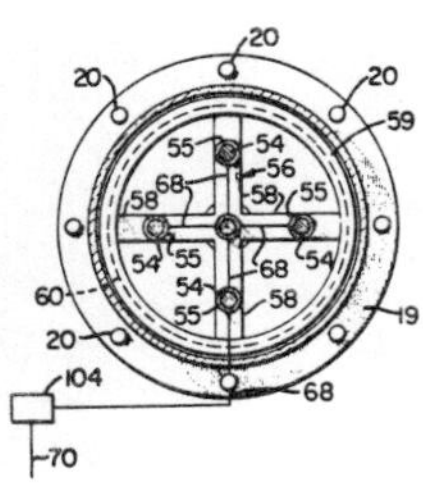

Sectional View Along Line **2–2** of Figure 1.26a

Source: I.S. Salnikov; U.S. Patent 3,377,266; April 9, 1968

Preferably each heating element **52** has a lower threaded end **64** for threaded insertion into a socket **65**, there being a plurality of sockets provided at uniformly spaced intervals and in upwardly facing relation on the mounting frame, as best seen from Figure 1.26b, to support the heating elements in desired spaced relation for upward extension through the distillation zone. Moreover, as represented, electrical wire cable network **68** is imbedded within the mounting frame to establish electrical connection from a main power line **70** to each of the sockets **65** and, in turn, to each of the respective heating elements **52**.

It will be seen that the entire heating assembly including the heating elements and the mounting frame **56** may be removed from the chamber by removing the top tubular section **17**; and a detachable manhole cover **72** is positioned over one open side of the upper tubular section to afford access to the heating assembly so that the heating elements can be individually removed or replaced without disassembling the entire apparatus.

In the process, particles of crushed oil shale **S**, for example, marlstone crushed to a size of ¼" to 2" in diameter are conveyed to the hopper **30** at the top of the furnace by suitable means and caused to descend slowly in succession through the preheat zone **12** and distillation zone **13** by rotation of the auger **32**. The rate of travel of the shale through the upper sealed chamber is controlled also by the lower auger **43** which as shown is independently driven by the drive shaft **48**, and both the augers **32** and **43** are dimensioned to establish sealed relation with the internal wall surfaces of the respective hoppers so as to exclude air from the preheat and distillation zones during pyrolysis of the shale.

It will be noted that the heating elements, being aligned in the direction of movement of the shale material through the distillation zone, will establish direct contact with the material throughout a substantial portion of its travel through the chamber so as to bring about most uniform and thorough heating of the shale, and the outer protective jackets **55** will serve to insulate the rods from direct physical contact with the shale while effectively conducting the heat developed in the rods but without impeding the movement of the shale particles through the distillation zone. This arrangement is further advantageous in that the fire clay jackets, once heated to the elevated temperature required for distillation will not experience rapid temperature changes in heating the shale while being maintained at the high-temperature level by the resistance heating rods.

Thus, by heating the elements to the elevated temperature level necessary to heat the distillation zone to the temperature of distillation of the shale oil-forming gases, this being on the order of 900° to 1000°F, the shale particles will liberate hydrocarbon gases and vapors which for the most part are the shale oil-forming gases and which can be separately collected and recovered outside the furnace or retort in a well-known manner. By excluding air from the distillation zone, the residual or fixed carbon in the shale will pass with the spent shale into the lower heat exchange zone, again at a controlled rate of travel as determined by the speed of rotation of the auger **43**.

The vaporous products of distillation formed will rise in countercurrent fashion from the pyrolysis zone through the preheat zone in heat transfer relation to the incoming shale, then are recovered through gas line **80** for removal through condensor **82** to an accumulator **84** for collection of the condensed shale oil with minor percentages of other condensed hydrocarbon vapors.

The noncondensable gases will tend to rise in the accumulator through line **85** either for separate collection through branch line **86** or for recycling through line **88**, and the proportionate flow between lines **86** and **88** may be suitably controlled by valves **89** and **90** in the respective lines. Of the recycled gases, a portion is drawn through line **92** and delivered by means of a blower, represented at **93**, through line **94** to the preheat zone. In this way, the noncondensable gases will mix with the gaseous products of distillation rising from the distillation zone in order to raise the condensation level of the gases formed by pyrolysis and to minimize condensation of the gases when cooled by the incoming shale.

Another portion of the recycled gases may be drawn through recycle line **88** as a source of fuel either in the heat exchange zone **14** or for a steam power plant **96**. However, the main source of energy for the power plant is derived from the heat exchange zone, and to this end a circulating coil **97** is directed in helical fashion upwardly through the heat exchange zone having a water inlet **98** and a steam outlet **99** leading to the power plant. In this way, water circulated through the coil is increasingly heated as it winds upwardly through the coil in contact with the progressively hotter shale bed moving downwardly through the heat exchange zone, until it is vaporized to form steam for discharge to the power plant.

The steam power plant **96** is most desirably employed to operate a steam turbine or other suitable prime mover for an electric generator **100**, thus converting the heat energy from the heat exchange zone to electric energy. The outlet power l ne **70** leads from the generator to the wire network **68** in the frame **56** for heating the elements **52**, and a rheostat **104** is positioned in the line to regulate the supply of current to the heating elements and consequently to establish close control over the temperature level developed in the distillation zone.

Heat transfer to the water circulated through the heat exchange coil is aided by combustion of the fixed or residual carbon in the shale, this being accomplished by directing air from a blower **107** through line **108** and vent **109** to discharge the air upwardly through the downwardly flowing shale bed and promote combustion of the carbon. Thus, the heat of combustion developed from oxidation of the fixed or residual carbon will substantially increase the heat energy available for conversion of the water to steam supplied to the power plant in order to operate the electrical generator and provide a closed regenerative system for oil shale pyrolysis. The gases formed in the heat exchange zone are recovered through outlet line **112** located above the heat exchange zone.

The rate of travel of the spent shale through the lower heat exchange zone is controlled independently of its rate of travel through the upper zones by regulating the speed of rotation of the augers **43** and **47**, and will permit direct introduction of the spent shale at an elevated temperature from the distillation zone into the heat exchange zone for most effective use in forming steam to generate electricity. Moreover, electrical power line **70** serves as a power source for the gas blower, air blower and electrical motors for the feed screws so that the entire system is self-contained without resorting to outside power sources.

Controlled Nuclear Reactor

M.L. Natland; U.S. Patent 3,109,781; November 5, 1963; assigned to Richfield Oil Corp. describes a process which comprises a unitary system for heating and recovering by-products from materials such as oil shales. The apparatus consists of a retort, the upper end of which is in communication with a fractionation unit and a gaseous heating unit, preferably a gas-cooled nuclear reactor, the outlet of the heating unit being in communication with the lower portion of the retort and preferably the inlet of the heating unit being in communication with the upper portion of the fractionation unit.

Sonic Energy Induced Separations

C.E. Morrell and P.V. Smith, Jr.; U.S. Patent 2,722,498; November 1, 1955; assigned to Esso Research and Engineering Company describe a process which permits the separation of organic chemical constituents from solid inorganic material with which it may be associated. In the process, the pulverized or finely divided inorganic material, containing organic material to be extracted is subjected to the action of a suitable organic solvent while being submitted to the action of compressional waves. This technique results in the extraction of a substantial portion of the organic material contained in the inorganic material treated.

The following examples illustrate the process. In all of the experiments reported, the sedimentary material treated was first reduced to a particle size of about 1 micron by the action of a micropulverizer.

Example 1: A shale sample obtained from a point 3,400 ft above the base of the Green River formation in Duchene County, Utah, was employed in a variety of extraction experiments. In a first series of tests, this shale was subjected to the action of carbon tetrachloride in a Soxhlet extractor. The extraction was conducted over a period of 15 hours employing 50 ml of carbon tetrachloride and 2 grams of shale. It was found that 5% of organic material was removed by this treatment.

Example 2: The shale employed in Example 1 was subjected to this process employing the same solvent, carbon tetrachloride, as employed in Example 1. A piezoelectric compressional wave generator was employed operating at 1,500 kilocycles on an electrical power input of 0.6 kw. Under these conditions, the compressional wave generator produced compressional waves causing cavitation. In exposing a sample of shale and solvent having the weight relation given in Example 1 to these compressional waves for a period of 5 minutes, and on then filtering the shale from the solvent, it was found that 49.8% of organic material had been extracted.

This same experiment was repeated under the identical conditions except that the frequency of the compressional waves was changed to 700 kilocycles. It was again found that 49.8% of organic material was extracted. These data indicate that extraction of organic material from an inorganic material such as shale is relatively ineffective by utilization of solvent action alone. However, when employing a solvent in the presence of compressional waves of sufficient intensity to provide cavitation, the amount of organic material which can be extracted is remarkably increased.

Pipeline Transportation and Recovery

In a process described by *P.E. Titus; U.S. Patent 3,527,692; September 8, 1970; assigned to Shell Oil Company* hydrocarbons are recovered from an oil shale formation by fragmenting oil shale and forming a slurry of the fragmented oil shale which is injected into a pipeline and maintained at a sufficiently high temperature to aid in extracting hydrocarbons from the oil shale and recovering it at the terminal end of the line. The oil shale can be obtained from a subterranean formation by any known means, such as mining, to produce appreciable quantities of large fragments of oil shale.

The fragments of oil shale can be passed to a crusher to reduce the fragments to a size sufficient to be flowed, when mixed with a solvent, within a pipeline. For example, about 15% fines have been produced in crushing Green River shales, this size being sufficient for the operations to be discussed. However, power requirements increase as the size of the oil shale particles decrease; accordingly, for purposes of this process, a preferred particle size distribution should consist of a range of particle sizes from ¼" to fines passing a 325-mesh sieve.

The crushed oil shale may be preheated if desired. However, preheating is merely a way to conserve the heat energy latent in a shale and to counteract the low rate of heat transfer. It does not increase the oil yield or improve its quality.

Preferably, the mined oil shale fragments are passed into the crushing plant where the fragments are crushed and the crushed oil shale is mixed with a solvent to form an oil shale/solvent slurry. The oil formation may, for example, be located in the Green River Formation of the Piceance Creek Basin of Colorado.

The solvent is preferably crude oil, retorted shale oil, or some appropriate fraction of these. It has been found that oil shale bitumen solubility is favored by the presence of aromatics, both cyclics, such as benzene, and heterocyclics, such as pyridine. Therefore, while any fraction of crude oil containing some aromatics would act as a solvent, those fractions rich in aromatics, either cyclics or heterocyclics, are preferred as the most efficient for extraction. It is also preferable to add small quantities of H_2S, as in the order of about 2.2 mol percent, to the solvent. This has been found to accelerate the rates of extraction of the kerogen in the oil shale into the solvent.

The slurry is then heated and passed into a pipeline where it is transported, to a remote refinery area, as for example, to the west coast or to the Gulf coast refineries. The slurry is periodically heated by means well-known in the art to replace heat lost in transit and to maintain the desired temperature range. The pipeline is preferably insulated to minimize heat loss. The slurry may be removed from the pipeline to effect the heating, if desired.

It has been found that heating a slurry of crushed oil shale and solvent for a relatively long period of time, such as periods of up to 18 to 20 days, and at a temperature of approximately 550° to 600°F, results in the extraction of shale oil having 80% weight of Fischer assay with benzene and 90% weight Fischer assay with benzene-H_2S mixtures. This relatively low-temperature process breaks the kerogen within the oil shale down into bitumen, which is then dissolved in the appropriate solvent. Thus, at approximately 550° to 600°F, the insoluble kerogen in the oil shale decomposes to soluble bitumen very similar to conventional crude oils. The material recovered is more like typical refinery crudestocks (i.e., not thermally cracked) with relatively high hydrocarbon content (i.e., relatively low ash, nitrogen, oxygen).

Since the refinery area is generally far enough away from the location of the oil shale formation, as for example, from the oil shale formations in Colorado to west or Gulf coast refineries, efficient utilization is taken of the relatively long processing time required (approximately 18 to 20 days) to decompose the kerogen to bitumen and extract oil therefrom. Thus, the shale oil delivered to the refinery is recovered en route. At the refinery, the extracted shale oil is conventionally processed and the waste material is removed following appropriate heat recovery. The heat put into the slurry may be partially recoverable at the refinery during the distillation process furthering the efficiency of the overall process.

Mining of Deep Thick Oil Shale Deposits

J.M. Whiting; U.S. Patent 3,588,175; June 28, 1971; assigned to Atlantic Richfield Co. describes an oil shale mining method for deep thick deposits which can be referred to as a cut-and-fill method or more simply, as blasthole stoping.

In the process, a mining zone is selected and divided vertically by horizontal haulage levels into a preselected number of production levels. A plurality of subordinate level networks is established above each haulage level to divide each horizontal production level into production blocks of desirable size.

Muck raises are driven to create outlets for broken ore at the bottom of each block. Ore remaining at each subordinate level is removed to form horizontal slots or voids. A vertical raise is driven at the longitudinal center of each block and then a slot or void is slashed for the width of the block. Volumes of each block are sequentially blasted from the central vertical slot to form a stope of increasing dimensions. Lastly, the broken ore is drawn and removed via the underlying haulage level. Thin weak pillars are left between adjacent stopes to retain fill and as roof support to temporarily control caving. After suitable bulkheads are constructed, the empty stopes are filled with spent shale and other waste material.

Production is carried out at several horizontal levels at the same time with the active stopes vertically staggered from level to level at such an angle as to minimize stress and strain on adjacent support pillars. Backfilling is accomplished as soon as possible so that each stope remains empty a minimum amount of time. As the development proceeds from level to level, the effect is to create an inverted pyramid-shaped mass of consolidating and subsiding fill.

Advantages of the process include the following. It is possible to safely recover 50 to 80% or more of the oil reserves from oil shale deposits; surface subsidence and attendant restoration problems are minimized; breakdown or weakening of strata and overburden adjacent to the mining limits is prevented or minimized; ground pressures and rock deformation at the periphery of the mined out region are minimized; surface waste disposal requirements are minimized; and hazards such as violent air displacement, rock falls, shock and dust, caused by sudden breakdown of many large open stopes are eliminated.

Laser Beam Heating

In a process described by *M.R. Yant; U.S. Patent 3,652,447; March 28, 1972* the bottom layers of a large body of oil shale are heated by pulsed laser beams. The heated shale when contacted by air drawn into an enclosure, causes eduction of gaseous hydrocarbons that rise upwardly through the shale to a gas collection space. The rising gas and air heat the upper layers of the crushed shale in order to retort the entire body of shale.

Liquid products gravitate downwardly to the bottom and are drawn off while the gas is drawn upwardly by vacuum pressure pulses synchronized to occur between the pulsations of the laser beams. Because of the synchronized vacuum pressure pulses, voids are intermittently produced in the body of crushed shale to maintain its transmissivity with respect to the laser beams which would otherwise decrease as the shale is retorted under the heat produced. The heat generated by the laser beams is also distributed throughout to cause retorting of all portions of the crushed shale body.

Figure 1.27a illustrates the apparatus of the process generally referred to by reference numeral **10** designed to carry out the oil shale recovery process. The apparatus includes a relatively large, elongated enclosure generally referred to by reference numeral **12** which has a base or foundation **14** on which sidewall sections **16** are mounted to form wall assemblies on opposite longitudinal sides of the foundation. The sidewall sections are interconnected by a roof portion **18.** The roof portion as well as the sidewall sections are made of a suitable, opaque material such as steel plate. The sidewall assemblies formed by the sidewall sections are interconnected at the opposite ends by radiation transmissive, end wall portions which are made of a material such as glass including an entry end wall portion **20** at one longitudinal end and an exit wall portion **22** at the opposite longitudinal end as shown in Figure 1.27c.

The enclosure is filled up to a predetermined height with a large body of hydrocarbonaceous solids such as crushed oil shale **24** as more clearly seen in Figures 1.27b and 1.27c supported on top of the foundation by a fire clay floor **26.** A gas collection space **28** is formed above the body of crushed oil shale. As shown in Figure 1.27a, the roof portion at the ends, is supported on opaque end wall sections **30** above the glass wall portions, the end wall sections having openings mounting hinged access doors **32** through which the oil shale is loaded into the enclosure.

Combustion supporting fluid such as air is supplied to the body of oil shale from any suitable source through an air intake conduit **34** which mounts a control valve **36** to regulate the quantity of air drawn in. Liquid hydrocarbon products or shale oils are withdrawn from the body of crushed solids through a conduit **38** which extends upwardly through the floor into the enclosure. A cutoff valve **40** is mounted in the conduit **38** in order to control the outflow of the liquid.

A suction conduit **42** is interconnected with a plurality of exhaust pipes **44** that extend upwardly from the roof portion. Thus, gaseous products within the gas collection space **28** at various locations above the body of solids **24** are withdrawn from the enclosure through the conduit to which a vacuum pump assembly **46** is connected.

As shown in Figures 1.27a and 1.27c, a pair of laser beam generators **48** are mounted externally of the enclosure adjacent one longitudinal end. The laser beam generators are of the pulsed ruby type arranged to direct a pair of pulsed laser beams **50** through the bottom layer of the body of solids. The beams thus enter the enclosure through one end wall portion in horizontally spaced relation to each other, exiting from the other horizontally spaced end wall portion. The beams while passing through the body of solids generate the requisite amount of heat for retorting purposes.

Referring to Figure 1.27d, it will be observed that each of the laser beams is emitted from a conventional type of pulsed ruby laser generator. In order to obtain efficient and thorough retorting of the entire body of solids, the operation of the vacuum pump is controlled by the synchronizing control **52** so as to time the negative pressure pulses applied to the gas collection space between the pulsations of the laser beams being pulsed at a predetermined rate. In this manner, the transmissivity of the body of solids with respect to the laser beams are maintained as the gaseous products rise upwardly into the space **28** and the liquid products are drawn off through the conduit **38**. The negative pressure pulses also draw in a regulated quantity of air through conduit **34** necessary to support the retorting process. Also, the gaseous products may be conducted through a condenser **54** as shown in Figure 1.27d and then delivered by the vacuum pump to a storage container **56**.

By way of example, the capacity of the retort enclosure when the body of solids is filled to the top of the plate glass wall portions **20** and **22** will be 480 cu ft leaving the gas collection space which will be filled with gas from the pyrolytic action of the laser beams. The gas is drawn to the condenser by a vacuum pump of 3 hp capacity with a 30 to 35 cfm rating. The negative pressure pulses of the vacuum pump will be synchronized in continuous out-of-phase relation to the pulsations of the laser beams of 5 pulses per minute. A commercially available rube laser generator of known construction and operation may be utilized. The glass end walls portions will not only transmit the laser beams but will also permit visual inspection of the interior of the enclosure.

FIGURE 1.27: LASER HEATING PROCESS

a.

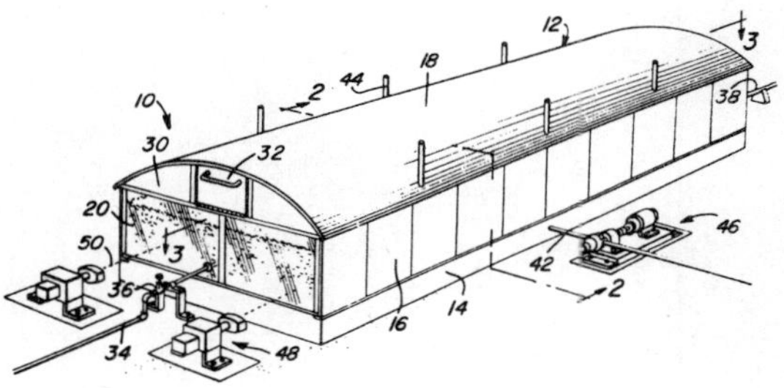

Overview of Housing and Process Equipment

b.

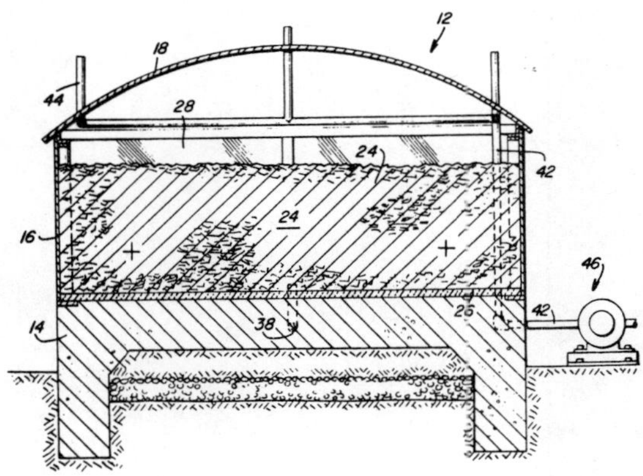

Sectional View Along Line **2–2** of Figure 1.27a

(continued)

FIGURE 1.27: (continued)

c.

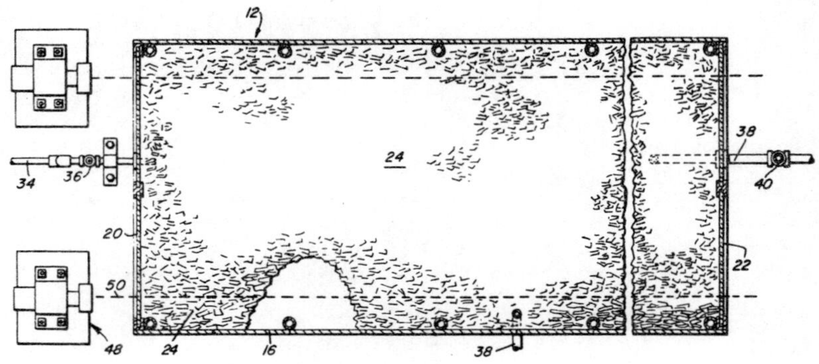

View Along Line **3–3** of Figure 1.27a

d.

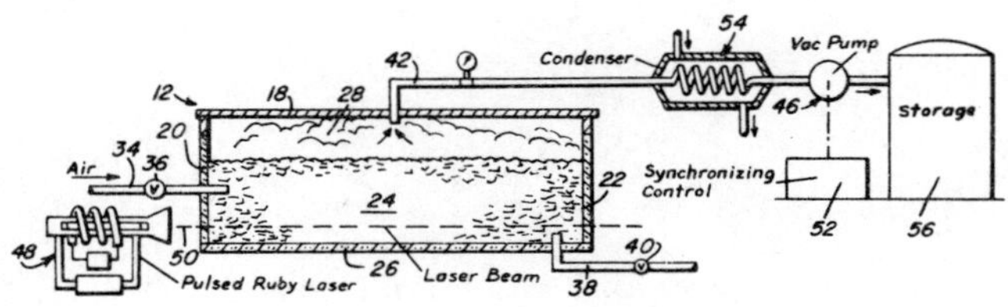

Sectional View of Apparatus

Source: M.R. Yant; U.S. Patent 3,652,447; March 28, 1972

OIL SHALE REFINING PROCESSES

HYDROGENATION

Activated Carbon-Alkali Metal Catalyst

A.N. Wennerberg and A.W. Frazier; U.S. Patent 3,715,303; February 6, 1973; assigned to Standard Oil Company describe a process for substantially upgrading the quality of fossil fuels containing polynuclear aromatics for such uses as low sulfur fuel, fluidized cracking units (FCU) feed, etc. According to the process, polynuclear aromatics, either in fossil fuels or otherwise, are contacted, preferably as an intimate mixture of catalyst and feed, at elevated temperatures and pressures and in the presence of hydrogen with a catalyst comprising activated carbon and an alkali metal component or an alkaline earth metal component or both. The amount of metal component is sufficient to promote cracking of polynuclear aromatics. In some instances it may also be desirable to include metallic components from Groups VI and VIII of the Periodic Table such as, for example, nickel, tungsten, cobalt, molybdenum, iron or their oxides or their sulfides.

Polynuclear aromatics having a number average molecular weight of about 400 or greater can most advantageously be treated according to the process. Fossil fuels such as tar sands, shale oils, coals, or hydrocarbons such as resids contain such heavy polynuclear aromatics. Advantageously, the coal is dispersed in a carrier such as recycle oil or a coal extract is dissolved in a suitable solvent such as Tetralin.

The solid fuels are high in polynuclear aromatics, especially the coal which can contain as high as 50 to 80 weight percent polynuclear aromatics. The liquid fuels, especially resids, include polynuclear aromatics called asphaltenes and resins. The asphaltene concentration in, for example, a vacuum resid ranges between 0.1 and 30.0 weight percent depending on its crude oil source, and the resin concentration ranges between 25 and 70 weight percent.

The resins are low molecular weight materials that are heptane soluble and are absorbed on the surface of silica gel when a heptane solution of resins and oils is passed over a column of the silica gel. The asphaltenes have high molecular weights such as, for example, from 1,000 to 15,000 and are insoluble in hot heptane.

The structure of asphaltenes can be schematically illustrated as follows. (Actual structures can be very complex and can include sulfur, nitrogen and various metals.)

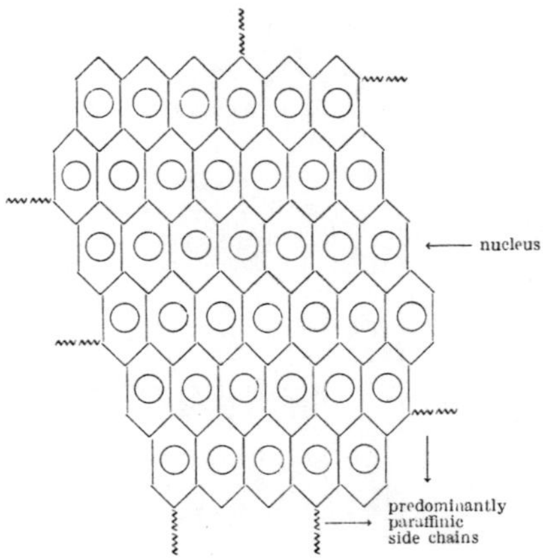

The catalyst used appears to have the ability to split the nucleus (central polynuclear array) of the asphaltene molecule. A typical feed which one would treat according to the process would be a resid including from 0.1 to 25 weight percent asphaltenes based on the weight of the resid. According to the process this feed can be converted into a light, amber-colored, free-flowing oil which contains very low concentration of metals and asphaltenes. In contrast, when the same feed is hydrotreated using a conventional acidic catalyst comprising alumina impregnated with cobalt and/or molybdenum, the feed is converted to a product having a black color. This indicates that many of the asphaltenes nuclei, though modified, remain essentially intact. It is believed that the presence of the alkali or alkaline metal component, which renders the carbon catalyst basic, is critical in obtaining the improved hydrocarbon product of the process.

It appears that the catalyst is somewhat selective. Data indicate that the polynuclear aromatics, when treated according to the process, are converted to lower molecular weight aromatics including substituent alkyl groups. It is desirable to maintain a high concentration of resins in the reaction mixture, since resins help maintain the asphaltenes in solution. Thus the asphaltenes will not precipitate and clog the reactor. The data indicate that the process reduces the molecular weight of asphaltenes by a factor of about 10, and the molecular weight of resins by a factor of about 3.

Consequently, resin concentration in the reaction mixture is always relatively high, minimizing loss of asphaltene solubility. The tendency toward reactor plugging is thus greatly reduced. And, surprisingly, the catalyst cracks or otherwise modifies the metal-containing compounds in the resid, and the metals are deposited on the catalyst without any noticeable adverse effects during the time of the experiments. High molecular weight sulfur-containing compounds are also cracked in the process. The catalyst, however, appears to have little effect on low molecular weight sulfur compounds commonly found in resids such as benzthiophene and dibenzthiophene. Hence, there can be virtually complete demetalization and partial desulfurization of the resid using the catalyst.

Specifically, the catalyst comprises activated carbon having a surface area in the range of from 200 to 2,500 m^2/g of carbon. This catalyst can be in either powder or granular form. In the granular form the preferred surface area ranges from 800 to 2,000 m^2/g of carbon. This activated carbon is impregnated with the metal component which is preferably in the form of a hydroxide, sulfide or oxide. Preferred alkali metals are potassium, sodium, lithium, rubidium and cesium. Preferred alkaline earth metals are magnesium, calcium, strontium and barium. The concentration of metal typically varies from 0.1 to 50 weight percent based on the weight of carbon.

To prepare a catalyst which includes 30 weight percent potassium hydroxide based on the weight of carbon, first prepare at 70°F 100 grams of a methanol solution including 20 grams of potassium hydroxide dissolved in 80 grams of methanol. Seventy grams of activated carbon is then mixed with this solution and the methanol is evaporated. A preferred carbon designated as SGL active granular carbon, mesh size 10 to 20, has a surface area of about 800 m^2.

Figure 2.1 is a flow diagram schematically illustrating the hydrotreating process. Resid from source **10** passes through heater **12** and is then mixed with hydrogen from source **14**. The mixture of resid and hydrogen then percolates down through a bed of alkalized carbon catalyst in reactor **16**. In the reactor the resid is cracked into lower molecular weight hydrocarbons, and metals in the resid are deposited on the catalyst.

Some desulfurization also occurs to produce hydrogen sulfide. The effluent from reactor **16** flows into separator **18** where unused hydrogen, gaseous hydrocarbons and hydrogen sulfide are removed. The hydrogen mixture is recycled to the reactor via scrubber **20** which removes the hydrogen sulfide, and the liquid effluent from the separator flows via line **19** into an atmospheric fractionating tower **22**.

Distillate fuel and gasoline are removed from the top of tower **22** and 650+°F boiling materials are withdrawn from the bottom of the tower. Since these 650+°F materials are of such a high grade quality, containing virtually no asphaltenes or metals they can be fed directly to fluidized catalytic cracking unit **24**.

FIGURE 2.1: HYDROTREATMENT OF FOSSIL FUELS

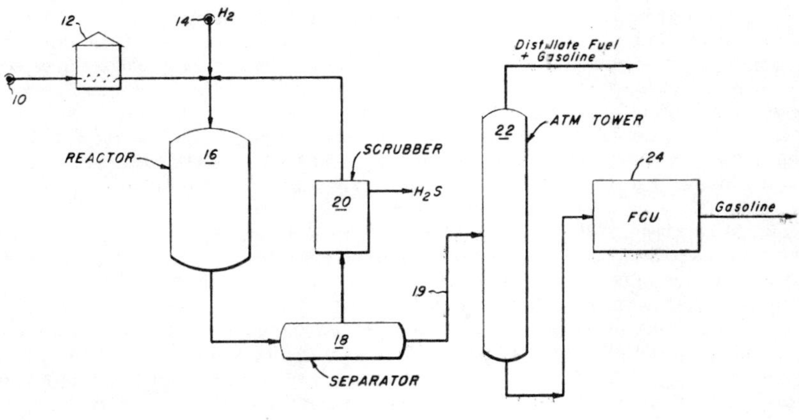

Source: A.N. Wennerberg and A.W. Frazier; U.S. Patent 3,715,303; February 6, 1973

The following table sets forth in detail the composition of the catalyst, the resid feed and the effluent from the reactor, and the conditions in the reactor. The data in the table show that the product has a low metals content and a low asphaltene (heptane insolubles) level.

The carbon residue is only slightly higher than that for a typical FCU feed. The process has removed about 60% of the sulfur. Cracking evaluation of the 650+°F fraction indicate this product has catalytic cracking quality properties similar to a high-sulfur virgin gas oil.

	Reactor 16	
Operating conditions:		
Pressure, p.s.i	3,500	
Temperature, ° F	805	
Volumes of feed/hr. per volume of catalyst	0.34	
	Feed, West Texas 650 +° F. resid	Product of Reactor 16, C_6+ liquid recovery, 95 wt. percent of total product
Process stream properties:		
Gravity, ° API	15.7	23.9
Sulfur, wt. percent	3.3	1.40
Hot heptane insolubles, wt. percent	2.4	0.03
Metals content, p.p.m.:		
Nickel	14	0.4
Vanadium	23	0.1
Carbon residue, wt. percent	8.3	1.9

NOTE.—Catalyst composition—10% KOH 10-20 mesh SGL carbon.

Hydrotorting Using Hydrogen and Water

W.G. Schlinger, D.R. Jesse and J.P. Tassoney; U.S. Patent 3,617,469; November 2, 1971; assigned to Texaco Inc. describe a continuous process for preparing maximum yields of shale oil of reduced nitrogen and sulfur content from raw shale under relatively reduced pressure. It has been found that raw shale can be readily converted to shale oil and relatively kerogen-free, dry powdered shale by injecting a slurry of raw oil shale in shale oil with hydrogen (5,000 to 20,000 scf of hydrogen per ton of raw shale) and water (0.01 to 0.6 tons of water per ton of raw shale) under pressure, and immediately introducing the mixture into an externally fired tubular retort under conditions of turbulent flow.

Within a period of from one-quarter to 3 minutes at an outlet temperature of 850° to 950°F and at a pressure in the range of 300 to 1,000 psig and preferably at a critical pressure of 475 to 525 psig, hydrogenation takes place with no addition of a supplementary catalyst. Shale oil is produced having a substantially reduced nitrogen and sulfur content and with increased yields of 116 volume percent of the Fischer Assay. Furthermore, if desired, still greater yields of shale oil may be obtained (in some instances as much as 135% of the Fischer Assay) by submitting solids-free prehydrogenated gaseous effluent from the tubular reaction zone to further hydrogenation in a separate catalytic hydrogenation zone.

Addition of hydrogen to the slurry and the hydrogenation of the pyrolysis products of the kerogen improves the yield of the product shale oil and provides the product with a greater amount of the desirable middle distillate material, while the formation of heavy polymers, unsaturated hydrocarbons and carbonaceous residues which characterize known processes are suppressed. Injecting water into the slurry before the tubular retort was found to have several benefits. The velocity through the tubular retort, the turbulent flow, and the heat transfer coefficient of the mixture in the retort are all increased. Thus, rapid heat transfer is brought about which allows conversion of the kerogen to crude shale oil in the retort coils at residence times of about one-quarter to 3 minutes.

Furthermore, vaporization of the water in the coils tends to disintegrate the shale particles and facilitates atomization of the shale oil. Also, coking of the slurry may be minimized or eliminated at a substantially reduced hydrogen consumption. Other advantages of injecting the water under pressure into the shale oil slurry just prior to introducing the slurry into the tubular reaction include:

(1) greater concentrations of raw shale may be incorporated in pumpable oil shale slurries,
(2) less water is required in this process than when water is added to the shale in the slurry mixing tank,
(3) clogging of the retort tubing is prevented,
(4) better control of the amount of water added, and

(5) it was found that water addition reduces the endothermic decomposition of inorganic carbonates in the shale to form CO_2, thereby preventing the undesirable reaction between CO_2 and hydrogen to form H_2O and CO.

Thus by water injection, there is a savings of energy in the form of heat used for carbonate decomposition as well as a reduction of hydrogen consumption in the tubular retort.

Referring to Figure 2.2, particles of raw shale in line **1** and heavy shale oil in line **2** are introduced into mixing tank **3** where they are mixed by agitator **4**, forming a raw shale-shale oil slurry. This slurry is passed from the bottom of mixing tank **3** through valve **5** and into the suction end of screw pump **6**. At a temperature in the range of 100° to 500°F, the slurry is pumped through line **7** to a conventional gas-liquid contactor **8**, which may be in the form of a venturi mixer. Recycle hydrogen from line **9** and makeup hydrogen from the line **10** are mixed in line **11** and injected into the accelerated slurry stream at the throat of the venturi **8**. Recycle water in line **12** is similarly injected into the slurry. The pressure of each of the streams in lines **11** and **12** exceeds the system line pressure by 25 to 200 psi.

The intimate mixture of hydrogen gas, water, raw oil shale particles, and heavy shale oil at a temperature below the vaporization temperature of water leaving contactor **8** is directed through line **13** into externally heated tubular retort **14** situated immediately after contactor **8**. Under conditions of high turbulence in tubular retort **14**, the mixture is raised within seconds to a temperature in the range of about 700° to 1100°F and disintegration and pyrolysis of the raw shale, vaporization of the shale oil and water, and hydrogenation of the kerogen and shale oil all take place simultaneously. No supplementary catalyst need be added to the materials in the tubular retort to promote the reactions.

A hot gaseous effluent stream comprising shale oil vapor, unreacted hydrogen, water vapor, H_2S, NH_3, CO_2, CO and shale dust in the form of a fine dry powder of 200 to 325 mesh, leaves tubular retort **14** through line **15** and is discharged into gas-solids separator **16**. Spent shale, substantially free from any hydrocarbonaceous residue, falls to the bottom of chamber **16** and is removed from the system through line **17**. To prevent plugging with spent shale and heat loss, the gas-solids separation chamber **16**, the overhead transfer lines, line **15** from the tubular reactor, and exposed flanges and pipe joints are insulated to maintain the gaseous stream at a temperature of 850° to 950°F.

Since this example of the process involves the production of shale oil by hydrotorting in tubular retort **14** only, that is, with no subsequent hydrogenation in a catalytic reactor **19**, valves **20, 21** and **22** are closed and bypass valve **23** is opened. Hot gaseous effluent from separator **16** comprising shale oil vapor, water vapor, H_2, and minor amounts of NH_3, H_2S, CO_2, CO and CH_4 is then directed to cooler **24** by way of lines **18, 25, 26** and **27**. The water and shale oil that are condensed out in cooler **24** pass through line **28** and into gas-liquid separator **29**.

Unreacted hydrogen and other uncondensed gases are removed from the top of separator **29** and are passed through line **30** into compressor **31**. Compressed hydrogen-rich gas from lines **32** and **9** are mixed in line **11** with makeup from line **10**. If necessary this hydrogen-rich mixture may be heated to a temperature in the range of 100° to 500°F before it is introduced into contactor **8** by means of one of the many heat exchangers in the system.

Shale oil-water mixture is withdrawn from the bottom of gas-liquid separator **29** and is passed through line **33** into shale oil-water separator **34**, where the lighter shale oil separates out and floats on a water layer which contains dissolved H_2S, NH_3 and CO_2. The water layer is removed at the bottom of separator **34** through line **35** to standard water purifier **36** where H_2S, NH_3 and CO_2 are removed through line **37** and are directed to a standard NH_3, H_2S and CO_2 recovery system. Purified water is removed through line **38** and a portion may be recycled to contactor **8** by means of pump **39** through lines **40** and **12**, in the manner previously described. Surplus water is discharged from the system through line **41**.

FIGURE 2.2: CONTINUOUS HYDROTORTING OF RAW OIL SHALE

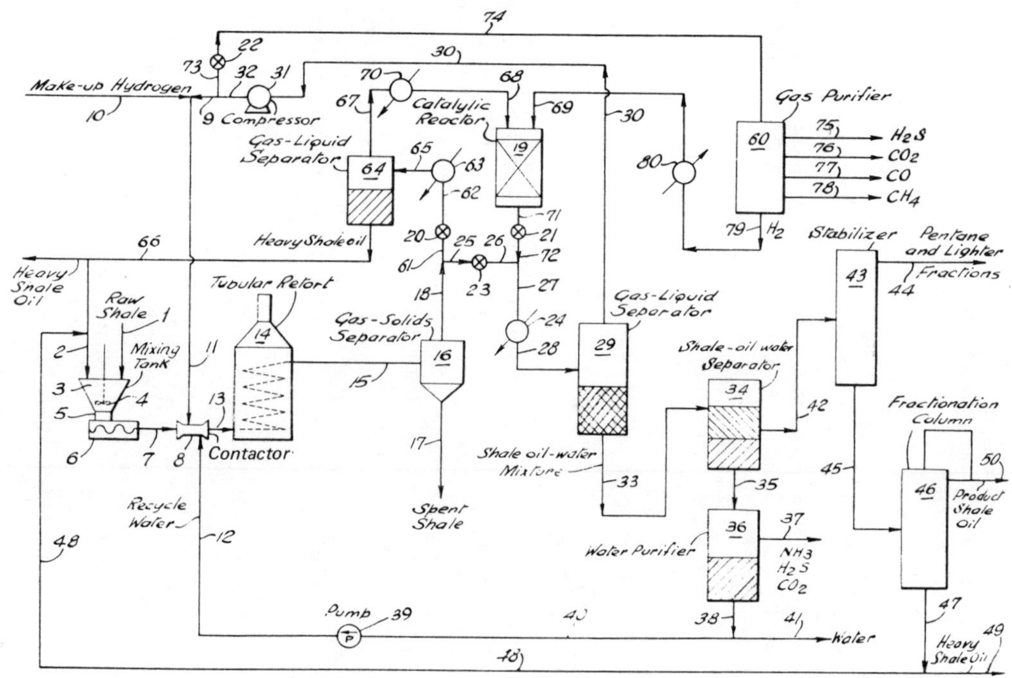

Source: W.G. Schlinger, D.R. Jesse and J.P. Tassoney; U.S. Patent 3,617,469; November 2, 1971

Since most oil shale deposits are located in arid regions, one significant advantage of the process is that there is no net consumption of water; but in fact, an excess of water may be produced.

The crude shale oil layer in separator **34** is withdrawn through line **42** and is introduced into stabilizer **43** where by fractionation, pentane and lighter hydrocarbon fractions are separated and are passed out the top through line **44**. Crude shale oil is withdrawn from the bottom of stabilizer **43** through line **45** and is introduced into fractionation column **46**. A portion of the heavy shale oil bottoms from column **46** at a temperature of 600° to 900°F is recycled through lines **47, 48** and **2** into mixing tank **3** for making the raw shale-shale oil slurry feed to the process, in the manner previously described. Generally, no heavy shale oil recycle pump is necessary since the system pressure will move the oil to the mix tank. The remainder of the heavy shale oil is removed from the system through line **49**. Product shale oil is removed from the system through line **50**.

A second example of the process involves two separate hydrogenation steps: first, the raw shale-shale oil slurry is hydrogenated in the noncatalytic tubular retort as described previously; and second, the prehydrogenated shale oil vapors from the first step are hydrogenated in a fixed or fluid bed hydrogenation catalytic reactor **19**. Also, a gas purification system is integrated into the system to supply pure hydrogen to the catalytic reactor and to prevent the buildup of gaseous impurities in the recycle hydrogen stream.

Catalytic reactor **19** and gas purifier **60** are introduced into the system by closing bypass valve **23** and opening valves **20, 21** and **22**. The hot gaseous stream from gas-solids separator **16** is then passed through lines **18, 61** and **62** into cooler **63** where heavy shale oil and tars condense out and pass with the uncondensed gases into gas-liquid separator **64** by way of line **65**.

Separator **64** also serves as a catch pot for all shale dust that may have passed through gas-solids separator **16**. Heavy shale oil and tar, which are the least valuable portion of the product shale oil may then be removed from the system through line **66** at the bottom of separator **64**. This also protects the catalyst in reactor **19** from contamination. However, if desired all or a portion of the material in line **66** may be introduced into mixing tank **3** by way of line **2** for slurrying with the raw shale.

The gaseous stream leaving from the top of gas-liquid separator **64** is passed through lines **67** and **68** and into the catalytic reactor **19** where hydrogenation takes place. This hydrogenation step is facilitated by a hot stream of pure hydrogen which is introduced into reactor **19** by way of line **69**. The preferred mol ratio of hydrogen (from line **69**) to gaseous feed to the catalytic chamber (from line **68**) is within the range of 0.0 to 0.3. Heat exchanger **70** is provided to help maintain the gas stream at the inlet to catalytic reactor **19** at the proper temperature to effect maximum denitrification and desulfurization.

If it is desired to operate the catalytic reactor at a higher pressure than the pressure in tubular retort **14**, then a pump may be inserted in line **68**. The effluent from catalytic reactor **19** is discharged through lines **71, 72** and **27** into cooler **24** where the treated shale oil and water are condensed out. Except for the gas purification unit, the remainder of the system involves shale oil-water separator **34**, water purifier **36**, stabilizer **43**, and fractionation column **46**. The operation and function of these units are the same as that which has already been described.

To prevent the buildup of gaseous impurities in the system, a portion of the uncondensed gases from gas-liquid separator **29** comprising unreacted hydrogen and traces of H_2S, CO_2 CO and CH_4 is purified, and the hydrogen is returned to the system. For example, a portion of compressed recycle gas in line **32** is passed through lines **73** and **74** into a standard gas purifier **60** where H_2S, CO_2, CO and CH_4 are separated from pure H_2 and leave respectively by way of lines **75, 76, 77** and **78** for recovery. A standard gas purifier utilizing refrigeration and chemical absorption may be employed to effect separation of the gases, e.g., U.S. Patent 3,001,373.

Hydrogen leaves gas purifier **60** by way of line **79** and is passed into heater **80** where it is raised to a temperature in the range of 800° to 900°F before it is introduced through line **69** into catalytic reactor **19**, in the manner previously described. If desired, high purity makeup hydrogen from an external source may be introduced into the system by mixing with the H_2 in line **79**. The preferred mol ratio of gases recycled in line **73** to gases passed through line **11** is in the range of 0.1 to 1.0.

Example: Colorado oil shale having a Fischer Assay of 31.2 gallons of shale oil per ton of raw oil shale and 2.9 gallons of H_2O per ton of raw oil shale is crushed to -8 mesh and mixed with heavy shale oil to form a slurry comprising 75.6 weight percent of raw shale. Immediately after water and hydrogen are injected into the slurry under pressure, the mixture is hydrogenated in a 1" SCH 40 pipe x 530' long noncatalytic tubular retort.

Operating conditions and results of runs in accordance with the first example of the process as previously described are summarized in the following table, column 1. In column 2, there is shown a summary of the conditions and results of double hydrogenation, first in the noncatalytic tubular retort and second over a Co-Mo hydrogenation catalyst.

It appears, from a comparison of columns 1 and 2, that the second method is the preferred procedure because it provides high yields of product shale oil having a higher °API and characterization factor, improved distillation characteristics and considerably less sulfur, nitrogen and carbon residue. Further, in comparison with the Fischer Assay, the data for both examples show a substantial increase in product-oil yields; an improvement in API gravity, pour point and yield of distillate; and a reduction in the nitrogen and the sulfur content in the shale oil product.

Operating Conditions and Results	1	2	Fischer assay
Operating conditions:			
Pressure, p.s.i.g.:			
Tubular retort	500	500	....
Catalyst chamber	N	480	....
Temperature,° F.:			
Tubular retort	925	925	....
Catalyst chamber	N	800	....
Catalyst	N	Co/Mo	....
Retorting period, seconds	19.6	19.6	....
Turbulence level	2,800	2,800	....
Input materials:			
Shale: Raw shale charged, lbs./hr	980	980	....
Hydrogen:			
Feed purity	99.8	99.8	....
Consumption, s.c.f./bbl. oil:			
Total	1,977	2,199	....
In tubular retort	1,977	1,977	....
In catalyst bed	N	222	....
Feed rate to contactor, s.c.f./ton raw shale.	13,800	13,800	....
Make-up rate, s.c.f./ton raw shale	1,709	2,170	....
Temperature at inlet to contactor, ° F.	155	155	....
Pressure at inlet to contactor, p.s.i.g.	625	625	....
Velocity at inlet to tubular retort, ft./sec.	9.00	9.00	....
Water:			
Feed rate to contactor, lbs./ton raw shale.	520	520	....
Temperature at inlet to contactor, ° F.	100	100	....
Pressure at inlet to contactor, p.s.i.g.	625	625	....
Raw oil shale-shale oil slurry: Temperature at inlet to tubular retort, ° F	140	140	....
Gaseous effluent from retort: Velocity at inlet to gas-solids separator, ft./sec	33.6	33.6	....
Output materials:			
Product shale oil:			
Gravity,° API	24.0	30.0	24.1
Viscosity, SSU at 122° F	55	41	50
Pour point,° F	60	55	75
Sulfur, wt. percent	0.64	0.24	0.98
Nitrogen, wt. percent	1.49	0.75	1.80
Conradson carbon, wt. percent	4.10	0.10	2.3
Characterization factor	11.5	11.6	11.4
ASTM distillation,° F.:			
IBP	180	165	192
10%	308	283	336
30%	482	450	518
50%	630	590	655
70%	702	675	705
80%	724	701	....
Heavy shale oil (line 49):			
Gravity,° API	15.0	16.0	N
Viscosity, SSU at 122° F	1,000	900	N
Pour point,° F	125	120	N
Sulfur, wt. percent	0.63	0.25	N
Nitrogen, wt. percent	2.00	1.25	N
Conradson carbon, wt. percent	8.10	.30	N

(continued)

Operating Conditions and Results	1	2	Fischer assay
Recovery:			
Product shale oil:			
Gals/ton of raw shale	36.3	38.0	31.2
Percent Fischer assay	116	121	100
Heavy shale oil: Gals/ton of raw shale	None	None	N
Water:			
Gals/ton of raw shale	4.4	4.9	2.9
Percent Fischer assay	152	169	100
Spent shale:			
Lbs./ton of raw shale	1,653	1,653	1,670
Carbonaceous residue, wt. percent	2.98	2.98	5.0
Pentane and lighter fractions, lbs./ton of raw shale	35.2	38.7	N
Removal of spent shale in gas solids separator, wt. percent	100.0	100.0	N

NOTE.—N = Not applicable.

In related work *W.C. Schlinger, D.R. Jesse and J.P. Tassoney; U.S. Patent 3,617,470; Nov. 2, 1971; assigned to Texaco Inc.* describe a process which combines tubular hydrotorting of a raw shale-shale oil slurry and the centrifugal separation of entrained spent shale from the gaseous effluent stream.

Hydrotorting Using Synthesis Gas and Water Injection

W.G. Schlinger, D.R. Jesse and J.P. Tassoney; U.S. Patent 3,565,784; February 23, 1971; assigned to Texaco Inc. describe a continuous process for recovering shale oil from a slurry of raw oil shale in shale oil. Water and hot unquenched synthesis gas from the reaction zone of a partial oxidation generator are injected into the raw oil shale-shale oil slurry under pressure and the mixture is immediately introduced into a noncatalytic tubular retort maintained at a temperature in the range of 850° to 950°F and at a pressure in the range of 300 to 1,000 psig, and preferably at 500 psig for maximum yields of shale oil having a minimum nitrogen content. Substantially all of the hydrogen and a large fraction of the heat required in the tubular retort are provided by the synthesis gas.

In the tubular retort under conditions of turbulent flow, the raw shale is completely stripped of kerogen in about ¼ to 3 minutes (preferably less than a minute), and by simultaneous pyrolysis and hydrogenation the kerogen is converted to a gaseous effluent from which shale oil is separated. Simultaneously, hydrogen is generated in the tubular retort by the exothermic water-gas shift reaction, whereby CO in the synthesis gas reacts with H_2O the spent shale acting as a shift catalyst. Pure hydrogen and the prehydrogenated shale oil are introduced into a catalytic reactor to produce denitrogenated and desulfurized shale oil at yields of about 125% of the Fischer Assay. Feed to the synthesis gas generator comprises a portion of the heavy shale oil and steam produced by the process and in this respect the process is self-sustaining.

In a related process, described by *W.G. Schlinger, D.R. Jesse and J.P. Tassoney; U.S. Patent 3,617,471; November 2, 1971; assigned to Texaco Inc.* hydrogen-rich gas, e.g., synthesis gas, and H_2O are injected into oil shale at a comparatively moderate pressure in the range of 300 to 1,000 psig and at a temperature in the range of 850° to 950°F, for producing high-quality shale oil in yields that exceed the Fischer Assay. The addition of H_2O reduces the hydrogen consumption and heat load required for a given yield of shale oil. The process can be self-sustaining in that the shale oil and water produced may be used for making the synthesis gas used in the hydrogenation and denitrification of the shale oil.

W.G. Schlinger, D.R. Jesse and J.P. Tassoney; U.S. Patent 3,617,472; November 2, 1971; assigned to Texaco Inc. also describe a process for recovering shale oil from oil shale by retorting with synthesis gas, i.e., a mixture of carbon monoxide and hydrogen, generated by partial combustion of by-product gas with oxygen, where part of the heat required for retorting is provided by the hot synthesis gas, and additional hydrogen is produced in the oil shale retort by the water-gas shift reaction, the shale acting as a catalyst, and the process being self-sufficient in requiring no external source of water.

Use of Coarse and Fine Catalysts

N.C. Stewart; U.S. Patent 3,635,943; January 18, 1972; assigned to Cities Service Research and Development Company describes a method for treating oils in a simple reactor. By employing two differently sized particulate catalysts, two generally separate expanded catalyst beds are established in the single reactor. Basically, the process contemplates feeding regenerated fine catalyst together with the hydrocarbon oil feed and a hydrogen containing gas into a hydrotreating vessel, and passing the mixture upwardly through an expanded coarse catalyst bed in the reactor. Above the level of the coarse catalyst bed the fine catalyst, hydrocarbon reactant and gaseous mixture, passing upwardly into the upper portion of the reactor establishes an expanded fine catalyst bed.

A portion of the hydrocarbon reactant liquid is recycled back to the bottom of the reactor. Recycling effectively adds to the flow of fluids through the lower catalyst bed and the resultant velocity is sufficient to maintain the lower bed in an expanded state. The upward flow above the lower catalyst bed is reduced to a velocity which enables fine particulate catalyst to concentrate and form a fine catalyst bed in the upper portion of the reactor. Effluent is withdrawn from the reactor above the level of the fine catalyst bed.

The particulate coarse and fine catalysts may be of any suitable catalyst composition. Suitable catalyst compositions include a large number of materials such as cobalt, molybdenum, iron, nickel, tungsten, platinum, palladium, naturally occurring clays, as well as combinations of the same. For the particular application described here, a cobalt-molybdenum catalyst supported on alumina is the preferred composition.

The coarse particulate catalyst is in the form of either pellets or extrudates in the size range of from 1/32" to 3/8" in diameter. After becoming inactivated the coarse catalyst may be regenerated by calcining in an air or oxygen atmosphere, and then is either attrited or more preferably pulverized to a size below 600μ, preferably in the size range from 300 to 600μ for use as fine catalyst in the process.

The feed may be any suitable hydrocarbon liquid for which a desired treatment is hydrogenation. For this reason naphtha and other oils boiling below 450°F are not regarded as suitable. Preferably the feed is a liquid hydrocarbon boiling above 450°F, and more preferably above 650°F. Heavy hydrocarbon oils boiling above 950°F such as vacuum or atmospheric residuums are preferred feeds, being particularly suitable for refining by hydrotreating. The term hydrotreating is intended to cover various reactions such as hydrocracking, hydrogenation, hydrodesulfurization and hydrodenitrogenation, all of which are obtainable in varying degrees by utilizing the process described herein.

Impregnation of Shale with Gasoline

A process described by *C.D. Shields; U.S. Patent 3,503,868; March 31, 1970* comprises impregnating mined and crushed geological matrix, e.g., oil shale, with a hydrogen-containing solvent for the kerogen and maintaining the matrix in contact with the solvent for a period of time sufficient for the kerogen to be impregnated by the solvent. The solvent-impregnated matrix is then charged into a rotatable high pressure reactor which is provided with heating means and which contains crusher balls, preferably of heat conductive material, e.g., steel. The reactor is provided with a quick-opening vent valve which is connected to a line leading to a high pressure petroleum fractionating tower. The reactor is closed, heat in the range used for mixed phase petroleum cracking operations is applied and the reactor is rotated.

The crusher balls grind the matrix into smaller particles, which are more readily accessible to the solvent and to the reactant agent. The crusher balls also serve conductively to heat the matrix. Pressure is permitted to rise in the closed reactor, causing increased solubilization of the heated kerogen by the hot solvent. The hydrogen from the solvent reacts with the kerogen to convert it into compounds volatile at 400° to 500°F. After a suitable grinding, heating, solvating, and converting period, a significant portion of the kerogen is made volatile.

The vent valve of the reactor is opened suddenly to release the pressure. The sudden release of pressure causes the solvent in the matrix particles to expand almost explosively thereby pulverizing the ground particles into still finer particles. Solvent and low boiling kerogen products boil out of the reactor. The rapidly boiling materials entrain some higher boiling kerogen compounds and carry them out of the reactor. After the pressure in the reactor has dropped to a set limit, the vent valve is closed and more solvent is charged into the reactor.

A reactive chemical agent, which may be a hydrogenating agent, an alkylating agent, a cracking agent, or other petroleum converting agent is also charged into the reactor. Heating and rotating of the reactor and its contents are continued for a suitable reacting, cracking, converting and solvating period while pressure again builds up in the reactor. At the end of this period, the pressure is again suddenly released from the reactor into the fractionating tower.

The operating cycle of charging with solvent and with reactive chemical agent, rotating, reacting and sudden pressure releasing is repeated until substantially all the kerogen economically feasible to recover has been recovered from the matrix. The spent matrix is then discharged from the reactor and the unit prepared for the next charge of crushed solvent-impregnated shale oil matrix. The products vented into the fractionating tower are condensed and separated by distillation. Recovery of oil from the starting material by this process is about 130% of Fischer Assay.

Solvent Treating Before Hydrogenation to Prevent Plugging

A process described by *R.A. Hanson; U.S. Patent 3,052,620; September 4, 1962; assigned to Union Oil Company* involves a method for treating residual shale oils prior to hydrogenation in such a manner that the hydrogenation unit will not become plugged with gunk. The process involves solvent treating residual shale oils prior to the step of preheating the shale oil in the presence of hydrogen preparatory to catalytic hydrogenation. A preferred form of the process involves a combination of water washing and acetone treating residual shale oils prior to preheating the shale oil in the presence of hydrogen preparatory to catalytic hydrogenation. With this procedure, a total of 2 to 6% by weight is rejected by the solvent.

Example: Crude shale oil produced by the downflow process of U.S. Patent 2,501,153 was mixed with 3 volumes of acetone and agitated while being heated to 135°F over a period of about 0.5 hour. The mixture was then permitted to settle and cool for about 16 hours, at which time it had cooled to about 80°F. The mixture was filtered through No. 40 Whatman filter paper at this temperature, leaving about 4% of the original weight of the crude oil as a solid residue on the filter paper.

The filtrate was vacuum distilled to remove the acetone, and the remaining oil, which constituted about 96% by weight of the original crude, was tested in a filter bomb test in comparison with the crude oil feed. It was found that the original crude shale oil feed plugged the filter so quickly that substantially none of the oil could be filtered, whereas all of the acetone treated oil passed through the filter with no evidence of plugging and no perceptible precipitate. Photomicrographs of the feed and product were also taken, and whereas the feed showed evidences of many solid or semisolid particles of various sizes suspended in the oil, the product was substantially free of any particles.

When particles are referred to here it must be understood that these are not ordinary solid suspensions capable of removal by simple settling or filtration. These crude oils referred to have in many cases been stored for months and even maintained at temperatures of 160°F under a blanket of helium; and while there was in some cases a settling of a small amount of semisolid material, the photomicrographs remained substantially the same, as did the plugging tendency in the bomb test and in actual pilot plant runs. Even filtration after such long standing had no substantial effect on the appearance in the photomicrographs or on the plugging tendency. Yet the acetone treatment, which appears to remove primarily wax-like materials of relatively low melting point, not only clarified the oil but removed all tendency toward plugging or gunk formation during the subsequent hydrogenation. Since wax can be readily hydrogenated under the above conditions without causing any plugging of the hydrogenation reactor or preheater, the effect of the acetone treatment is quite surprising.

It is also interesting that when the crude shale oil feed is vacuum distilled to take 80% overhead, acetone treatment of the distillate as above resulted in rejection of substantially the same amount of waxy solid material, and had almost exactly the same effect of reducing pour point, i.e., a reduction of about 40°F; yet this distillate is free from plugging tendency in the filter bomb and hydrogenation tests, so that acetone treatment is of little benefit to the distillate in this respect.

The acetone-treated crude oil described above was found to be substantially as corrosive as the crude shale oil feed, i.e., both exhibited corrosion of carbon steel at 600°F amounting to about 0.068" per year. One volume of the product was mixed with one volume of water at room temperature and the mixture was agitated while heating to 180°F over a period of 1 hour and allowed to settle for 1 hour at which time it had cooled to about 150°F. The oil was decanted from the water and recovered to the extent of about 99.5% by weight.

It was again tested for corrosivity and found to be substantially noncorrosive, i.e., its corrosion rate for carbon steel was only about 0.003" per year, which is well below the tolerable rate of about 0.008" per year. The water-washed oil remained equally free from plugging tendency in the filter bomb test, and was suitable for catalytic hydrogenation without plugging of the preheater or catalyst bed. Several modifications of the process are also described.

Conservation and Reuse of Hydrogen

A process described by *M.G. Huntington; U.S. Patent 3,106,521; October 8, 1963; assigned to Huntington Chemical Corporation* is concerned with the conservation and maximum recombination and reuse of the hydrogen which is initially contained in the hydrocarbonaceous constituent of oil shale and further concerns itself with hydrofining and hydrocracking of the condensable volatile matter in order to achieve maximum production of the more useful light oils while not requiring an outside source of hydrogen. Raw crushed shale of various sizes may be fed vertically downward through an internally heated retort in measured amounts over gyrating shelf feeders. The raw shale of various sizes is concurrently contacted and heated by a stream of hot hydrogen and the primary volatile matter is distilled. This primary volatile matter is taken off from a duct in the vertical retort and the solid particles of shale continue to be fed downwardly.

A moderate amount of unheated hydrogen is introduced to flow countercurrently through the spent, broken shale to effect the final stripping of volatile matter and to achieve at least partial heat recovery. Safety gas locks continuously purged with noncombustible gas are provided at both ends of the hydrogen contacting area of the retort to insure against ignition of the thermal carrier hydrogen. Heat recovery and preheating sections are provided outside of the safety gas locks at the bottom and top of the retort respectively.

The primary volatile matter offtake is connected to a multiple bed catalyzer and then to a fractionator and the entire stream passes through both vessels at system pressure. From the fractionator the heavy bottoms are recycled for contact coking in a portion of an auxiliary apparatus such as a coal still, the middle oils are passed through a high pressure hydrocracker and the overhead stream is passed to a condenser. After being scrubbed, uncondensable gases from the condenser are thermally cracked and reheated in an auxiliary apparatus and then are recycled to the oil shale retort. All of the heat supplied for heating the oil shale retort is supplied from an auxiliary source.

This auxiliary source for supplying all the heat may be a coal still in combination with the oil refining processes described. This coal still not only furnishes all of the heat but also may be the auxiliary apparatus which provides the means (hot char) for cracking hydrocarbon gases into hydrogen at substantial pressures and in substantial quantities and thus furnishes the thermal carrier gas to accomplish the destructive distillation of the oil shale.

The heavy oil bottoms from the fractionator are also recycled into the coal still where coking and redistillation are performed in contact with incandescent char. The flue gas produced from the coal still may be utilized as the noncombustible gas for the safety gas locks of the oil shale retort, and for heat recovery and preheating.

Thus, this process produces high quality petroleum products without requiring a source of hydrogen and without requiring internal firing of the oil shale retort. Furthermore, by utilizing a system in which the thermal carrier fluid is essentially hydrogen and where the partial pressure of hydrogen has a tendency to increase the yield of condensable liquids and to decrease the splitting off of permanent gases during the pyrolysis and distillation of the kerogen, also, the apparent yield of intermediate semirefined shale oil is substantially greater than 100% of that indicated by the modified Fischer Assay. A complete description of the process is presented in the patent.

OTHER PROCESSES

Screw Conveyor Process

L.C. Gutberlet; U.S. Patent 3,658,654; April 25, 1972; assigned to Standard Oil Company describes an apparatus for the pyrolysis of solids containing carbonaceous materials including a cylindrical substantially horizontal pyrolysis vessel together with solids input and removal means communicating with the vessel. Solids are conveyed through the pyrolysis vessel by means of an auger-type conveyor.

Heating means are provided for heating solids between the input and removal means. Gas input and withdrawal means permit the input and removal of gas from the pyrolysis vessel, and gas delivery means are provided for delivering gas to subsequent processing equipment. The gas is recycled back to the gas input means via gas recycle means. Stripping gas delivery means are also provided to deliver a stripping gas to the pyrolysis vessel downstream of the gas withdrawal means.

Referring to Figure 2.3, the apparatus provides an elongated, substantially horizontal pyrolysis vessel **10**, which is preferably cylindrical in shape. The interior of the pyrolysis vessel contains an auger-type conveyor **12**, which is driven by a motor **14**. As will be seen from the subsequent description, a particular advantage of the process is that many of the parts, such as the motor, are outside the pyrolysis vessel, and therefore are not exposed to pyrolysis temperatures.

A raw shale container **16** communicates with the pyrolysis vessel at an upstream portion thereof through solids input means. The raw shale container is filled with particulate solids containing carbonaceous material, in this instance raw oil shale **17**. Of course, other solids containing carbonaceous materials may be processed in the apparatus of the process, as will be understood by those skilled in the art. The oil shale **17** is delivered to the pyrolysis vessel through a solids input conduit **18** under the influence of gravity.

Downstream of the input conduit, also communicating with the interior of the pyrolysis vessel, is a solids removal conduit **22**. The solids removal conduit provides communication between the pyrolysis vessel and a spent shale container **26**, which collects spent shale **27**. It will thus be seen that, when the auger **12** is in operation, particulate solids will be delivered from the raw shale container **16** through the pyrolysis vessel **10** to the spent shale container **26**. Since the entire system is preferably maintained under pressure, it will be necessary that the raw shale container **16** and spent shale container **26** be closed to the atmosphere. Suitable means, may be provided for introducing and removing particulate materials from these containers **16, 26** without the necessity of depressurizing the system. Between the solids input conduit **18** and the solids removal conduit **22**, there are heating means **28** for heating the solids containing carbonaceous materials in the pyrolysis vessel **10**. The type of heating means employed is not important, so long as it is capable of producing pyrolysis temperatures within the pyrolysis vessel. When the apparatus is employed as part of a petroleum refinery, numerous sources of fuel are available, so that the heating means may comprise a furnace. On the other hand, electrical heating coils surrounding the pyrolysis vessel could also be employed.

Figure 2.3 also illustrates gas input and withdrawal means, as well as stripping gas input means and gas recycle means. Gas is maintained in the entire system under pressure from a gas source, preferably a hydrogen source, designated by reference numeral **30**. This gas is delivered to the pyrolysis vessel through a gas input line **32**. In the preferred case, a portion of the gas input line is surrounded by heating means in order to preheat the gas before it is introduced into the pyrolysis vessel. The gas input line has a coiled portion **34** within the heating means **28**. Of course, separate preheating means on the gas input line may be employed.

Gas is withdrawn from the pyrolysis vessel at a gas withdrawal line **36**, which also serves as gas delivery means for delivering the gas removed from the pyrolysis vessel to subsequent processing equipment designated by reference numeral **38**. This subsequent processing equipment **38** may simply be a condenser to separate vaporized carbonaceous materials from the gas, but preferably comprises hydrotreating or hydrodesulfurizing equipment where the carbonaceous material is upgraded and separated from the hydrogen stream. The hydrogen stream is then returned to the gas input line **32** through a recirculation line **40**. It will thus be seen that the gas input line, the pyrolysis vessel, the gas withdrawal line **36**, the subsequent processing equipment **38**, and the recirculation line **40** form a closed loop. To the extent that any hydrogen is consumed during the pyrolysis and subsequent processing, it is replaced from the hydrogen source **30**.

FIGURE 2.3: SCREW-CONVEYING RETORTING APPARATUS

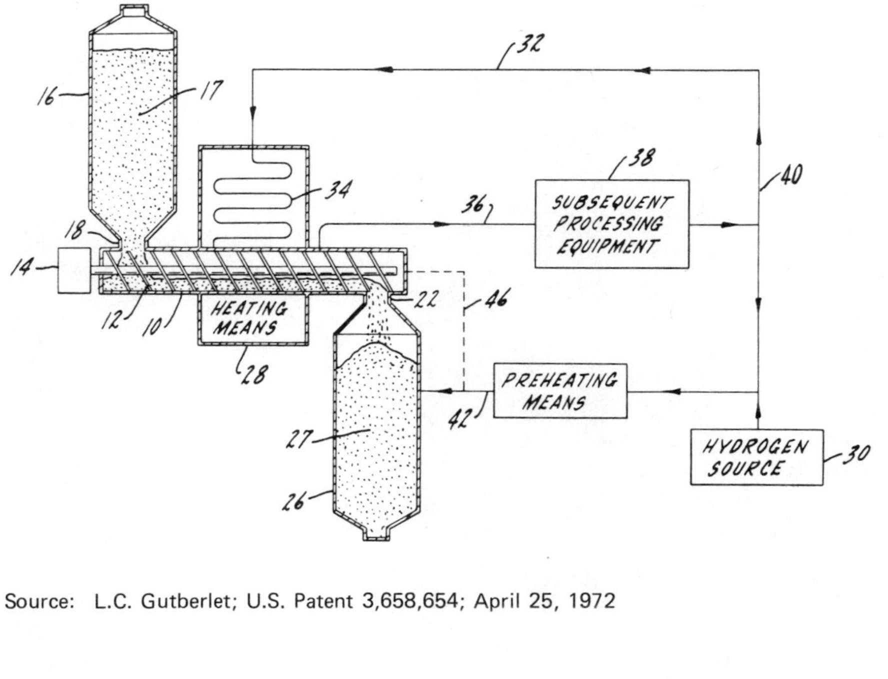

Source: L.C. Gutberlet; U.S. Patent 3,658,654; April 25, 1972

As illustrated, a stripping gas input line **42** also communicates with the interior of the pyrolysis vessel **10**. Two examples are shown in the drawing. In the first case, the stripping gas input line **42** communicates with the interior of the spent shale container **26**. Since the solids in the spent shale container **26** will be hot, the stripping gas will be heated by these solids so that it will vaporize residual carbonaceous material carried by the solids subsequent to pyrolysis. However, if sufficient heating is not provided by the solids in the spent shale container **26**, auxiliary preheating means **44** on the stripping gas input line **42** may be provided.

As shown by a phantom line **46**, the stripping gas input line **42** may optionally communicate directly with the end of the pyrolysis vessel, bypassing the spent shale container **26**. In any event, the stripping gas should preferably flow in a direction opposite to the flow between the hydrogen gas input and the withdrawal lines **32, 36**, respectively, in order to prevent the collection of gas and vaporized carbonaceous materials at the end of the pyrolysis vessel. The stripping gas may be any gas that does not react with the hydrogen or carbonaceous materials in the pyrolysis vessel. In the preferred embodiment, the stripping gas is hydrogen, and is drawn from the hydrogen source **30**.

The runs described in the following examples were performed in an apparatus constructed in a manner similar to that shown in Figure 2.3, where the optional stripping gas inlet line shown by the phantom line **46** was used. The pyrolysis vessel was constructed of 2" diameter stainless steel pipe, and hydrogen entering through the gas input line **32** was preheated through the last 5' of piping, and entered the pyrolysis vessel immediately downstream of the raw shale container. The pyrolysis vessel was electrically heated from the gas input line to the solids removal conduit **22**. Temperatures were measured opposite the gas input line, the gas withdrawal line **36**, and midway between these two points. The auger **12** was normally rotated at a speed of 0.5 rpm, producing a shale residence time in the pyrolysis vessel of 46 minutes. The raw shale storage container **16** was filled with 14 to 28 mesh oil shale No. 4 having the following Fischer assay: Water yield, 1.38 weight percent; oil yield, 18.70 weight percent; and oil gravity, 25.0°API. The operating pressure was 500 psig.

Example 1: Seven runs were made at varying temperatures. The temperatures shown in the following table represent the average wall temperature in the downstream half of the pyrolysis zone.

Run number	1	2	3	4	5	6	7
Pyrolysis temp. (° F.)	890	922	925	930	930	932	934
Shale feed rate (lb./day)	38.1	35.6	38.1	38.8	40.0	38.1	37.2
Gas flow rates (thousands of SCF/ton of feed):							
Recycle gas	363	382	353	362	340	349	369
Stripping gas	5.2	7.3	6.3	6.9	7.[illegible]	5.5	7.4
Total carbonaceous matter recovered (wt. percent based on raw shale)	22.6	27.3	28.4	26.2	25.[illegible]	29.0	26.3
Product distribution (wt. percent):							
Carbon oxides	1.9	4.3	4.4	5.1	4.2	4.6	5.6
Dry gas	3.4	4.9	5.2	5.1	5.3	5.8	7.6
Shale oil	83.2	84.6	84.8	82.0	83.0	83.1	78.0
Coke	11.5	6.2	5.6	7.8	6.3	6.5	8.8
Shale oil gravity (° API)	26.0	24.0	22.6	23.8	23.0	24.0	27.3

The above table shows an acceptable product is produced under any of the reaction conditions shown. However, at the temperatures above 900°F a substantial reduction in the amount of coke is achieved, although larger amounts of carbon oxides are produced. The shale oil production in each of the runs was acceptable.

Example 2: A series of additional runs were made, in which the vapors of the pyrolysis product were conducted directly to a hydrogenation reactor where they were hydrogenated over a conventional cobalt-molybdena-alumina catalyst. The reaction conditions and results are shown in the following table.

Run number	8	9	10
Retorting conditions:			
Maximum temperature, °F	932	940	940
Shale feed rate, lb/day	36.8	39.1	40.0
Recycle gas rate, scf/ton x 10^{-3}	375	343	336
Stripping gas rate, scf/ton x 10^{-3}	6.8	6.4	5.9
Hydrogenation conditions:			
Average temperature, °F	710	835	725
H_2/oil ratio, scf/bbl x 10^{-3}	281	268	272
Space velocity (wt/wt/hr)	0.39	0.52	0.46
Space velocity (vol/vol/hr)	0.35	0.46	0.40
Throughput (vol oil/vol catalyst)	1.33	3.00	4.42
Total carbonaceous matter recovered (wt % based on raw shale)	22.7	27.1	23.4
Product distribution, wt %:			
Carbon oxides	3.0	1.6	3.2
Dry gas	10.3	11.4	10.0
Shale oil	76.5	79.7	79.7
Shale coke	10.2	7.3	7.1
Oil inspections:			
Gravity, °API	33.1	37.5	32.2
Sulfur, wt %	0.16	0.015	0.12
Nitrogen, wt %	1.40	0.91	1.49

As shown in the above table, in each instance, a high yield of shale oil is obtained.

Combined Retort and Cracking Process

F.D. Hoffert; U.S. Patent 3,483,116; December 9, 1969; assigned to Hydrocarbon Research, Inc. describes a method by which shale in lump form may be subjected to retort or distillation conditions and where the volatile products from the retorting step, may be cracked to more desirable products within the same zone in which the retorting takes place. The shale fines produced from the retorting operation may be used as the heat transfer medium for both the distillation or volatilization step and the cracking step.

The process comprises passing lump shale downwardly through a contact zone containing a fluidized bed of hot shale fines. The temperature and inventory of the fluidized bed is maintained by introducing superheated fines at the top of the bed and removing cooler fines at the bottom. Thus, there is an overall downward flow of fines in the bed resulting in a temperature gradient vertically through the bed, the highest temperature being at the top and the coolest temperature being at the bottom.

As the lump shale travels downwardly through the bed, it is finally heated hot enough whereby the hydrocarbons contained therein begin to retort or distill out of the shale. Generally, the retorting doesn't begin until the lump shale is in the lower portion of the contact zone as it takes a finite time to heat it up to the required retorting temperature.

The vaporized materials travel upwardly into the upper superheated portion of the fluidized bed. The temperature in this upper portion is sufficiently high to result in substantial cracking of the distilled material into lower boiling and more valuable products which are removed from the upper part of the contact zone and cooled and fractionated in the usual manner. Thus, the bed is separated by function into two zones, a lower retorting zone where the lump shale is at its highest temperature and the fines are at their lowest temperature, and an upper cracking zone where the shale is at its lowest temperature and the fines are at their highest temperature. The bed of fines is normally fluidized by an external inert gas, however, the distilled vapors may be used to aid in the fluidization.

Thus, the process allows one to successfully and effectively remove the distillable hydrocarbon-containing materials from shale without having to grind the shale to a fine powder. Additionally, the process makes use of the fines which inevitably result even from a relatively crude grinding step to provide heat, not only for the distillation or retorting of the volatile constituents from the shale, but also for the subsequent thermal cracking of the distilled materials. The distilled lump shale is removed at the bottom of the contact zone and the fines are removed separately from the lump shale and may be reheated and recirculated to the upper portion of the zone.

Referring to Figure 2.4, granulated or lump shale having a nominal diameter size between ½" to 4" is introduced from hopper **10** through valve **12** and line **14** into the contact zone **16** which contains a fluidized bed of hot shale fines. The lump shale moves downwardly through the bed where it is heated by the fluid bed to distill the hydrocarbons and exits through stripper leg **20**. Exit valve **22** controls the overall rate of flow of the lump shale through the zone. The hot distilled lump shale enters cooling chamber **48** through line **46** where it is cooled by the passage of air. The cooled distilled shale is then removed from the system through valve **52**.

FIGURE 2.4: USE OF SHALE FINES AS HEAT TRANSFER MEDIUM IN CRACKING PROCESS

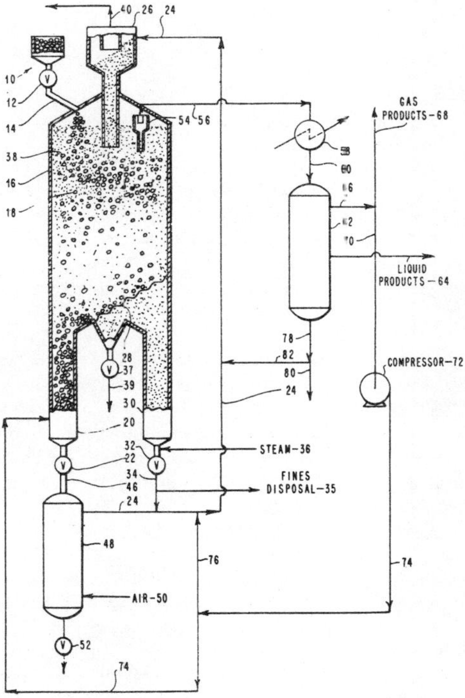

Source: F.D. Hoffert; U.S. Patent 3,483,116; December 9, 1969

The fluidized bed is maintained by introducing superheated shale fines to the chamber through line **24** and gas-solids separator **26**. The rate of passage of the fines through the chamber is a function of the heat transfer requirements of the system. The exiting fines are separated from the larger lumps of shale by screen **28**. The fines enter stripper leg **30**, are steam stripped using steam through line **36** to remove any adhering shale distillate oil and are then removed from the system through valve **32** and line **34**. Fines disposal may take place either through valve **37** and line **39** or through line **35**.

The bulk of the fines is introduced to line **24** where it is mixed with hot air from cooling zone **48**. The air and fines with some residual oils adhering thereon, are passed upwardly through line **24** which constitutes a combustion zone, where the residual oils are burned in order to achieve superheating of the fines. The superheated fines are then recycled through line **24** into separator **26**. The combustion or flue gas is separated from the fines in separator **26** and exits through line **40**. The superheated fines entering the top of the contact zone through separator **26** provide the heat for the distillation of the hydrocarbon materials from the shale. It has been found that such fines have highly efficient heat transfer characteristics and, thus, provide a superior means for providing or transferring the heat source for the shale oil distillation.

The fines in bed **38** at the top of the contact zone are at a relatively high temperature. Thus, there is sufficient heat provided for the required cracking process. It is also noted, however, that a thermal gradient exists vertically through the reaction zone, since at the top of the zone, the lump shale is at its lowest temperature and the fines are at their highest temperature. As the lump shale progresses downwardly through the zone, it becomes hotter as it picks up heat from the fines. The fines thus become cooler. At a relatively low point in the contact zone, the lump shale is at its highest temperature and the fines are then relatively cool. The differential, however, at this low point between the lump shale and the fines is small, whereas, at the top of the reaction zone, the differential was quite high. Thus, the distillation of the hydrocarbons from the shale takes place at the lower portion of the contact zone, since it is at this point that the lump shale is the hottest.

The vapors then proceed upwardly through the lower portion of the bed, which constitutes the distillation zone and into the upper portion of the fluidized bed which constitutes the cracking zone. The vaporized hydrocarbons then exit the contact zone **16** through separator **54** where any fines are returned to the fluidized bed. The fines-free vapors removed from the separator **54** through line **56** are cooled in cooler **58** and enter fractionator **62** through line **60**. Fractionator **62**, represents any known type of fractionation system which is normally used in the separation of multiboiling component mixtures. It may consist of either a single unit or units in combination including both atmospheric and subatmospheric thermal distillations.

From the fractionator **62**, the various products or product fractions of the shale oil are removed. The liquid products are removed through line **64** and the gaseous products are removed through line **66** and fed to distribution through line **68**. A portion of the gaseous products may be taken through line **70**, compressed by compressor **72** and fed through line **74** into the stripper leg **20**. This relatively high velocity compressed hydrocarbon gas serves a two-fold purpose. First, it removes the fines from the lump shale in and entering stripper leg **20** and assures that the distilled lump shale entering valve **22** is essentially dust-free. Second, the gas serves as a fluidizing means for the fines bed **38**.

A unique aspect of the process, embodied in the fluid bed seal in the dip leg of separator **26**, serves to prevent combustion gases from entering the distillation or cracking zone. As shown, the dip leg extends below the top surface or interface of the fluidized bed. A controlled differential pressure exists between the gases in the contact area and those leaving separator **26** through line **40**, the pressure generally being lower in the separator such that the level of the fluidized bed in the dip leg of separator **26** is higher than in the contact zone. This additional depth of the bed, combined with the relatively small diameter of the dip leg compared to the reaction zone and the downward flow of solids being concentrated by the separator, acts to prevent the gases in the contact zone from escaping through the dip leg and, conversely, to seal the combustion gases in the separator from the contact zone.

The bottoms products from fractionator **62** are removed through line **78** and may be sent to further downstream processing through line **80**. If desired, however, a portion of the bottoms material may be introduced to the combustion zone in line **24** through line **82**. The combustion of this bottoms material may then be used to incrementally supply the heat for the fines.

Generally, the conditions under which the process is carried out, would be a distillation zone temperature from 800° to 1000°F and a system pressure from about atmospheric to about 100 psig.

Example: This example gives specific process conditions which have been used in applying the process to the production of oil from African and Australian shales.

Shale contact time retort, min	20
Distillation temperature, °F	950
Cracking temperature, °F	1050
Lump shale feed rate, lb/hr/ft^3 of retort	2,400
Air circulation, scfh/ft^3 of retort	16,400
Fines circulation rate (300°F at), lb/hr/ft^3 of retort	20,600
Heat release in retort, Btu/ft^3/hr	51,000
Volatile matter in feed, %	40
Total oil product, bbl/ton of feed shale	2

In a process described by *L.W. Fish and G.F. Pappas; U.S. Patent 3,093,571; June 11, 1963; assigned to Esso Research and Engineering Company* both retorting and quality improving conversion of the liberated hydrocarbons are effected within a single combination system. A two-zone hydrocarbon conversion process, e.g., a fluid coker and burner, are integrated with a retorting zone in a manner such that a single burner supplies requisite heat for both the retorting and hydrocarbon conversion steps.

More particularly, carbon-coated fine particles are withdrawn from the hydrocarbon reaction stage and passed to the burner where they are subjected to combustion. A portion of the particles heated in this manner are passed to the reaction zone and a second solids stream passed to retorting zone thereby supplying requisite thermal energy. Liberated hydrocarbons pass directly from the retort to the hydrocarbon conversion stage where they are commingled with conversion vapors. A heavy ends portion of the total product vapors is condensed and circulated back to the reaction zone.

Simultaneous Separation and Cracking

L.P. Evans; U.S. Patent 3,281,349; October 25, 1966; assigned to Mobil Oil Corporation describes a process for the simultaneous separation and catalytic cracking of shale oil from crushed shale. The process comprises utilizing functionally integrated zones, including a retorting-cracking zone, a solids classifying zone, a catalyst regeneration zone and a shale fines separation zone. Within the retorting-cracking zone shale oil is separated from crushed shale particles upon being commingled with heated catalyst particles, and the resulting separated shale oil is cracked catalytically to more desired products. During this simultaneous retorting and cracking step organic matter is distilled from the shale particles; concurrently the cracking reaction deposits carbonaceous matter upon the solid catalyst particles and thereby reduces catalyst activity.

Separation of organically depleted shale particles from the deactivated catalyst is effected in the solids classifying zone. Thereafter carbonaceous matter is burned from the catalyst in the regeneration zone; the catalyst is thereby heated and reactivated. After removal of entrained shale fines in the shale fines separation zone, the heated catalyst is recycled to the retorting and cracking zone for reuse.

Simultaneous Retorting and Hydrogenation

A process described by *du B. Eastman and W.G. Schlinger; U.S. Patent 3,117,072; Jan. 7, 1964; and U.S. Patent 3,044,948; July 17, 1962; both assigned to Texaco Inc.* relates to simultaneous retorting of oil shale to recover shale oil and hydrogenation of the recovered shale oil. The process results in the production of a recovered oil of reduced viscosity, lower sulfur and nitrogen content, and improved refinability as compared with oil recovered from shale by conventional retorting in the presence of relatively inert gases. A fluid mixture or slurry, of hydrogen, shale oil, and particles of oil shale is passed through an elongated tubular reactor of relatively great length in comparison with its cross-sectional area.

The volume of flow of feed material in the reaction zone is maintained at a rate such that velocities sufficient to completely disperse the shale particles in the oil-hydrogen composite fluid and obtain highly turbulent flow conditions is ensured. Such velocities may be readily obtained by passing the slurry feed mixture through a tube of relatively small internal diameter, for example, a one-half to one inch tubular reactor. Under these conditions, the hydrogen diffusion distance is greatly reduced and the allowable rate of reaction correspondingly increased.

By the use of highly turbulent flow conditions, the hydrogenation reaction may be accelerated so that conversion of heavy shale oil to lighter products is accelerated and the shale oil product is considerably upgraded by increasing the yield of lower boiling products without the concomitant formation of high boiling polymers or coke.

Oil feed rate, hydrogen recycle rate, reaction tube diameter, and operating conditions of temperature and pressure all tend to affect velocity of flow and turbulence. Temperatures of 700° to 1500°F may be employed. A preferred range of temperatures is from 800° to 1000°F. Pressures of 1,000 to 20,000 psig may be employed, although pressures of 1,500 to 10,000 psig are preferred. Hydrogen feed rates of 1,000 to 100,000 scf/bbl of slurry feed may be employed, but feed rates of 2,000 to 50,000 scf/bbl are preferred. Hydrogen may be supplied in relatively pure form or in concentrations as low as 25 volume percent. Reaction times of 20 to 300 seconds are preferred.

Pyrolysis and Hydrogenation in Single Reaction Zone

G.B. Hoekstra; U.S. Patent 3,565,751; February 23, 1971; assigned to Standard Oil Co. describes a method for pyrolyzing oil shale and simultaneously hydrogenating the shale oil in the presence of a hydrogenation catalyst, and in a single reaction zone. This simultaneous pyrolysis and hydrogenation insures maximum yield of high quality product.

In carrying out the process, oil shale is introduced into a reaction zone which contains a conventional, fluidizable hydrogenation catalyst. The oil shale is heated in the reaction zone to pyrolysis temperatures, whereby to drive shale oil from the oil shale and to produce finely divided particles of spent shale as a result of the removal of the kerogen. Hydrogen is introduced into a lower portion of the reaction zone during the heating, under a pressure of 200 to 2,000 psig and at an upward velocity sufficient to fluidize the catalyst and to carry the finely divided spent shale particles and shale oil upwardly out of the reaction zone. Because of the presence of the hydrogen and a hydrogenation catalyst, the shale oil is hydrogenated prior to removal from the reaction zone.

A process described by *M.G. Huntington; U.S. Patent 3,475,317; October 28, 1969* involves the destructive distillation of hydrocarbonaceous materials. The apparatus and process include a means for feeding at least a solid supported catalyst and a crushed hydrocarbonaceous raw material through a lock and to a vertical reactor capable of withstanding temperatures of up to 2000°F and pressures to 30 atmospheres. The reactor contains a plurality of adjacent vertically positioned distillation zones, gaseous intake and offtake means for each zone, means for providing a gaseous diffusion barrier between zones, and a combination AC agglomerator and widening gaseous paths in each zone in front of the offtake means.

The reactor is further provided with a second lock to allow for the removal of the solid material which is subsequently passed to a system which separates effluent solids. Preheated hydrogen is fed into each zone, entrains the destructively distilled and stabilized product and is passed through a plurality of dephlegmators and catalyst beds. The catalyst in the reactor may be rejuvenated and recycled to the original feed means and a solid heating medium may be utilized in the reactor and continuously recycled. Complete details of process equipment are provided.

Heat Transfer Balls and Controlled Cracking

T.D. Nevens; U.S. Patent 3,034,979; May 15, 1962; assigned to The Oil Shale Corp. describes a process which comprises the steps of pyrolyzing oil shale to produce effluent oil vapors and gases, cooling the effluent to a predetermined temperature to condense the heavier fractions but not the lighter fractions, thermally cracking the heavier fractions within a certain predetermined range of temperature, and condensing the uncracked lighter fractions as well as the cracked heavier fractions to produce a composite oil product. The composite oil product has a pour point and viscosity comparable to, or slightly less than, that of shale oil produced after pyrolysis and cracking of all the effluent oil vapors and gases from the pyrolysis, but the loss of oil is substantially less than in conventional processes for a given pour point.

Referring to Figure 2.5a, a solid material, such as oil shale and generally spherical solid heat-carrying bodies enter the pyrolysis zone or rotatable drum **10** via lines **12** and **14**, respectively. The oil shale usually enters the drum **10** at a low temperature, e.g., 50°F, although, in some cases, it is desirable to preheat the oil shale to a temperature of 300°F to as high as 600°F. The heat-carrying bodies or balls are made of a hard heat resistant material such as alumina, or other ceramic material, or steel and preferably have a diameter somewhat larger than the average mesh size of the incoming fresh or preheated oil shale. For example, the ball diameter usually lies within a range of ½" to 1", whereas the average oil shale mesh size ranges between ¼" to ½".

The heat-carrying bodies or balls preferably enter the drum at a temperature lying within the range of 1000° to 1400°F, and are preferably intermixed in parallel flow, with the oil shale. Counterflow of balls and oil shale is also employed, although parallel flow is preferred. The ratio of the balls to oil shale generally ranges between 1:1 and 3:1 depending upon the nature of the oil shale being processed, the type of heat-carrying bodies being employed, and the rate of heat transfer sought. It should be noted that while other modes of heating and pyrolyzing the oil shale can be employed, e.g., fluidized bed processes employing heated gases, the preferred method and means for pyrolyzation (and also for cracking) utilizes the generally spherical solid heat-carrying bodies, above described.

Upon mixing the balls with oil shale in the rotating drum, pyrolysis of the oil shale occurs and the effluent oil vapors and gases leave the drum and enter conduit **16** passing through chamber **17** as indicated by the arrow **18**. The effluent oil vapors and gases have a temperature preferably lying between 750° to 950°F. The effluent oil vapors and gases are passed into a fractionator or partial condenser **20** and cooled to a temperature lying preferably between 500° and 800°F. That fraction of the oil vapors and gases which remains a vapor at 500° to 800°F, e.g., gas, gas oil, kerosene and gasoline, is, of course, not condensed, and is passed into condenser **22** via lines **24** and **26**. The condensed heavier fraction, which comprises approximately 10 to 50% of the collectible oil product passes along conduits **24** and **28** and then into a reservoir or tank **29** which serves as a gas seal. The heavier fraction is then fed into the thermal cracker **30** via standpipe **32**, and is cracked.

Because the reduction in temperature of the effluent oil vapors and gases entering the partial condenser **20** is approximately only 50° to 450°F the coolant for the condenser is usually ambient air, rather than water, which is extremely scarce in areas where the majority of the oil shale deposits are found. The air enters the condenser **20** via line **21**, passes through condenser **20** around a series of tubes **23** in the condenser and exits via line **25**.

After pyrolysis, the carbonaceous residue passes, together with the balls, into the chamber **17**, as indicated by the arrow **42**. The carbonaceous residue passes downwardly into the inclined passageway **44**, and is separated from the larger balls by passing through the openings in the inclined screen **46**. In this connection, the carbonaceous residue is considerably reduced in size over the initial oil shale mesh size due to the grinding and crushing action of the hard balls on the oil shale in the pyrolysis drum **10**. Separation of carbonaceous residue from the balls is thus readily effected by the screening means **46**.

FIGURE 2.5: PROCESS FOR THE PRODUCTION OF LOW TEMPERATURE PUMPABLE OIL FROM OIL SHALE

a.

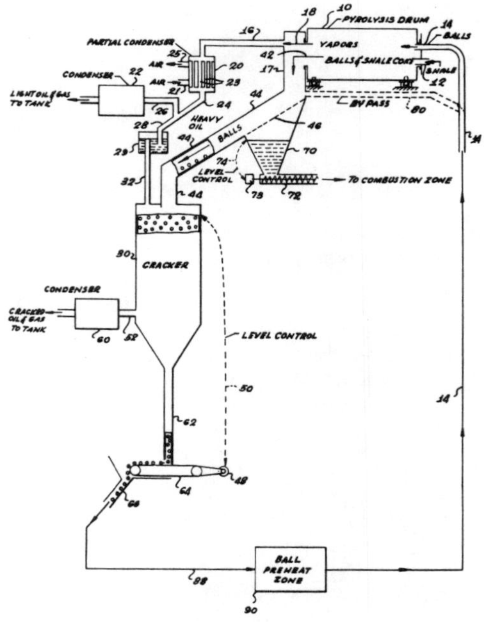

Details of Pyrolysis and Refining Process

b.

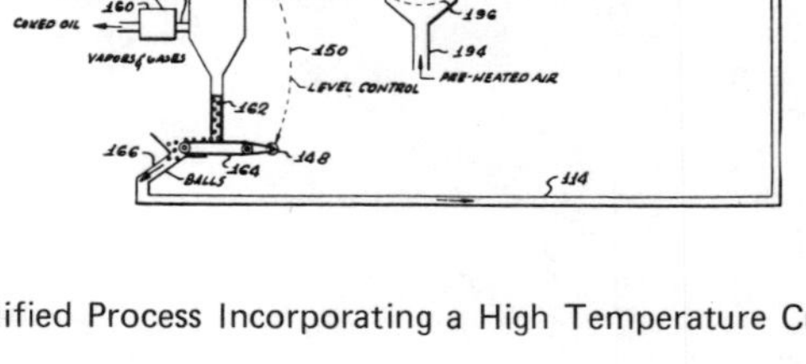

Modified Process Incorporating a High Temperature Cracking Step

(continued)

FIGURE 2.5: (continued)

c.

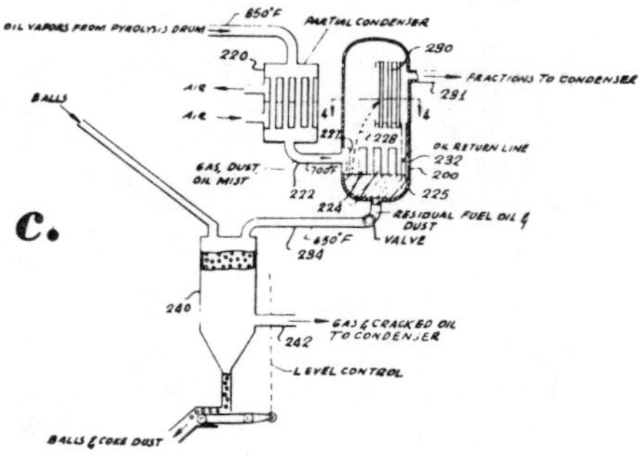

Use of Dust Removal Equipment

d.

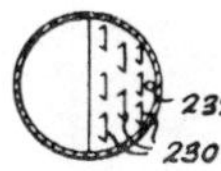

View Along Line **4–4** of Figure 2.5c

Source: T.D. Nevens; U.S. Patent 3,034,979; May 15, 1962

The balls, after being separated from the carbonaceous residue, pass downwardly along the passageway **44** and enter the thermal cracker **30**, the ball temperature, after the pyrolysis, lying preferably between 900° and 1100°F. The hot balls within the cracker **30** are maintained at a constant level by suitable conventional level control means (shown schematically by the numeral **48** and dotted line **50**).

The heavier liquid oil fractions enter the cracker from pipe **32** and pass through suitable distribution means, e.g., a sprayer in the cracker. The heavier oil fraction contacts the hot downwardly moving balls for a predetermined period of time, preferably ranging between one second to two minutes, and a cracking of the heavier oil fractions results. The balls and oils can contact each other in coflow, as shown, or in counterflow, or crossflow, if desired. The liquid oil fractions contain some dust upon entering the cracker, the dust being initially carried with the pyrolyzed oil vapors and gases to the partial condenser **20**.

Because the balls have deposited thereon a certain amount of coke on their surface, the exact amount depending on the conditions under which the cracker is operated, dust adheres to the balls and is carried out with the balls. The dust is removed either during the ball reheating phase of the process, to be described, or is removed during pyrolysis of fresh oil shale in the pyrolysis drum **10**.

The cracked oil vapors and gases leave the cracker via the line **52** at a temperature of approximately 825° to 1100°F and are sent to a conventional condenser **60** and cooled to a temperature of approximately 100°F, usually by means of water. The uncracked noncondensed oil and vapors, entering condenser **22** are also cooled by water to a temperature of approximately 100°F. The oil condensed in both condensers is then fed to a common tank or reservoir to form a composite oil product for transportation by pipeline to market or to further refining steps.

The composite oil product formed by following the above process is advantageous in that, for Colorado oil shale, its pour point generally lies in the neighborhood of 30° to 60°F; and what is also extremely important, the loss of oil in the process typically amounts to approximately only 5 to 15% computed on the basis of a 100% oil recovery. In prior processes, where cracking of all effluent oil vapors and gases, i.e., both light and heavy ends, occurs, for a given pour point of the resulting shale oil, the loss of oil in the form of noncondensables and coke increases appreciably.

After the balls have passed through the cracker **30**, they are returned to the pyrolysis drum **10** for the pyrolysis of additional solid material after being reheated to a temperature of between 1000° and 1400°F. The preferable method of providing heat for the reheating of the balls is to combust the carbonaceous residue remaining after pyrolysis, and to then transfer the requisite amount of heat from the products of combustion of the carbonaceous residue to the balls. Thus, the balls pass from the cracker via an outlet conduit **62**, and endless conveyor belt **64**, and conduit **66** and **88** to a ball heating zone **90** (shown schematically in Figure 2.5a) for heating. As mentioned, the balls have coke deposited on the surface, and this coke is combusted to furnish additional heat for the heating of the balls.

The carbonaceous residue, separated from the balls by screen **46**, falls into chamber **70**. The level in the chamber is controlled by level control means **73** and **74**. The carbonaceous residue is then conveyed, by conveyor **72**, and other means to an appropriate combustion zone, such as a fluidized combustion zone. The hot products of combustion (which may be both solid entrained particulate products and gaseous products) then impart a substantial portion of their heat to the balls in the ball heating zone **90**. The reheated balls, at a temperature of 1000° to 1400°F, then enter conduit **14** and are sent to the pyrolysis drum **10** for pyrolysis of additional solid material. A specific method and apparatus for reheating the balls by means of shale coke is shown in Figure 2.5b and is preferably employed also in conjunction with the process of Figure 2.5a.

Mention has previously been made of the fact that the solid material entering the pyrolysis drum may either be cold, e.g., 50°F or preheated. If preheated solid material is to be introduced into drum **10**, a preferred mode of preheating involves mixing fresh oil shale with hotter balls in either parallel or counterflow, as is described in U.S. Patent 3,008,894. The hot balls employed for the solid material preheating step are preferably those taken directly from the outlet conduit **66** of the cracker. After the solid material preheating, the balls are sent to the ball heating step, just described.

Attention is drawn to the fact that condensation of part of the effluent oil vapors and gases from the pyrolysis zone **10** takes place prior to the cracking of the heavier fraction. While it is possible to first condense all of the effluent oil vapors and gases, and then partially fractionate prior to the cracking of the heavier oil fraction, it will be understood that such a process necessitates an extra condensation step, which in turn, would require additional equipment as well as the use of valuable water. A specific example of the process with reference to Figure 2.5a is set forth below.

One ton of 25 gal/ton of Colorado oil shale of approximately a ¼" mesh size enters the drum **10**, via line **12**, along with 2 tons of ⅝" aluminum oxide-containing ceramic balls, the balls entering the drum via line **14**. The oil shale has an average inlet temperature of 50°F and the balls have an average inlet temperature of 1300°F.

192 lb of oil vapors and 30 lb of gases are produced during the pyrolysis, the oil vapors and gases passing via chamber **17**, and conduit **16** to a partial condenser **20**. The partial condenser **20** cools the oil vapors and gases to 600°F, 60 lb of oil being condensed and sent to the cracker. These heavier condensed fractions are sent to the cracker. There, they contact balls that have come directly from the pyrolysis drum **10** via line **44**, which have a temperature of approximately 1050°F. The period of cracking is approximately 30 seconds, the cracked oil vapors and gases being cooled to 100°F and being condensed in condenser **60**.

The lighter fractions pass from the partial condenser **20** into the condenser **22** and are cooled to approximately 100°F and condensed. Both the condensed lighter fractions and condensed heavier fractions, when combined total 172 lb, only 20 lb or 10% of the maximum oil recovery being lost as noncondensable gases and coke. The pour point of the composite product is 40°F. Comparing these results with a process where all conditions are equivalent to that just described except that all the vapors are cracked, the pour point was 50°F and the loss of oil was 40 lb or 20% of the total oil collectible. The balls from the cracker **30** (at 1000°F) were then reheated to a temperature of 1300°F by directly contacting them with the hot products of combustion of the shale coke. The balls are then sent to the pyrolysis drum to contact fresh incoming oil shale.

The average inlet temperature of the balls entering the cracking zone **30**, in Figure 2.5a lies between 900° and 1100°F. It is sometimes desirable to increase the average inlet ball temperature to the cracker so that a high cracking rate is maintained. To this end, a predetermined portion of balls sent from the ball heating zone **90** is bypassed around the pyrolysis drum **10** and enters the lower temperatured ball stream in conduit **44**, as is indicated by the conduit **80**, shown in dotted line. These bypassed balls may have a temperature as high as 1400°F upon entering the cracking zone, and can appreciably raise the average inlet temperature of the balls.

The amount of total ball temperature increase that is brought about by the use of the bypass feature is optimally approximately 100° to 150°F. Thus, the temperature range of the balls entering the cracking zone **30** of Figure 2.5a can vary from a low of about 900°F to a high of about 1250°F. It will be noted that the total ball to shale ratio will increase in proportion to the amount of balls passing through the bypass conduit **80**.

If a thermal cracking of the heavier fractions at still higher temperatures than previously described with reference to Figure 2.5a (with or without ball bypass) is desired, such a cracking step can be accomplished by referring to Figure 2.5b. The previously described advantages of Figure 2.5a processes with regard to lower pour point and minimal loss of oil are also present in the high temperature cracking process to be described. A preferred method and means of reheating the balls is also shown in Figure 2.5b.

In the process of Figure 2.5b, the balls leave the pyrolysis drum **110** and are reheated prior to the cracking step, rather than after the cracking step, as in Figure 2.5a. Inasmuch as the Figures 2.5a and 2.5b examples are quite similar, similar pieces of equipment and conduits, etc., are designated by 100 plus the numeral designation given it in Figure 2.5a. Thus, the cracker in Figure 2.5a is designated **30** and in Figure 2.5b the cracker is designated **130**.

The balls leaving the pyrolysis drum, after giving up some of their heat therein, have a temperature usually ranging between 900° and 1100°F. Instead of being sent to the cracker at this temperature, the balls first pass through a ball heating zone **190** and are there reheated, in a manner to be specifically described, to a temperature lying between 1200° and 1500°F. These hot balls are then conveyed by a moving gate **184** from the heating zone **190** to the cracker via line **186**, where the high temperature cracking takes place.

It will be noted that the inlet ball temperature in the cracking zone **30** in Figure 2.5a ranges between 900° and 1250°F depending upon whether ball bypass is employed or not. Since in Figure 2.5b, the ball inlet temperature to the cracking zone **190** ranges from 1200° to 1500°F, the variation in ball inlet temperature ranges between 900° and 1500°F. A specific means of providing heat for reheating the balls is also shown in Figure 2.5b. For example, shale coke, the residue produced during pyrolysis of such solid materials as oil shale, is sent from conveyor **172**, along line **182**, in a stream of entraining air, to a fluidized combustion zone **192**. Preheated air enters the zone **192** via line **194**, and passes through air distribution plate **196** which imparts to the air passing through a velocity higher than the settling velocity of the shale coke particles to thereby prevent the shale coke particles from settling through openings in the plate.

The shale coke is combusted in the presence of the air, and the hot products of combustion, the gaseous and entrained solids, pass into the ball heating zone **190** via line **197**. The hot products of combustion contact the balls in crossflow, and leave the zone **190** via exhaust pipe **198**. Referring especially to Figures 2.5c and 2.5d, the plant of either Figure 2.5a or Figure 2.5b is advantageously modified by the inclusion of a dust removal and demistifying apparatus designated by the numeral **200**. The oil vapors and gas from the pyrolysis drum **10** or **110** (e.g., at a temperature of 850°F), enter the partial condenser **220**, which is similar in construction to the partial condenser **20** or **120** of Figures 2.5a and 2.5b respectively. The oil vapors and gases usually carry some dust from the pyrolysis drum. The oil vapors and gases are reduced in temperature (e.g., to 700°F) in the partial condenser **220** by means of an air coolant, and oil mist, gas and dust pass into the demistifier **200** via line **222**.

The oil mist, gas and dust enter passages **224** in the demistifier **200** initially causing some oil and dust collection. The upper ends of the passages **224** are closed, and the oil mist and dust are forced downwardly into a reservoir of oil, designated by the numeral **225**. The oil mist and dust thus agglomerate into larger particles and move upwardly, through pipes **227**, as indicated schematically by a dotted line **228**, and strike a second set of staggered baffles **230** (best seen in Figure 2.5d). This causes the oil mist and dust to be substantially completely knocked out of the gas and the lighter fractions proceed to the condenser **22** or **122** via line **231** substantially free of the heavier oils in the form of mist, or any appreciable amount of dust. The residual fuel oil collected from the oil mist returns to the bottom of the demistifier **200** via oil return line **232**, and along with the dust, is sent to the cracker **240** (which is similar to cracker **30** or **130**) along conduit **234** for cracking. The dust removed from the oil vapors and gases in the demistifier **200** passes through the cracker **240** with the balls, while the heavier oils fractions are cracked and sent via line **242** to a condenser, such as condenser **60** or **160**.

It can thus be seen that the demistifier **200** and ball cracker **240** both act to prevent dust from moving out with the lighter oil fractions, and cracked heavier fractions, respectively. The need for additional dust removal equipment such as cyclones, is thus substantially reduced. It will be seen that the demistifier **200**, or similar apparatus, will appreciably add to the efficiency of this process where any substantial amount of oil mist is present in the outlet line **222** from the partial condenser **220**.

W.J. Culbertson, Jr.; U.S. Patent 3,008,894; November 14, 1961; assigned to The Oil Shale Corporation describes a similar process in which the combined pyrolyzing and thermal cracking continuous process employs one or more streams of hot heat transfer bodies for the furnishing of substantially all of the heat requirements of the process. It is found that the use of such circulating heat bodies in the combined pyrolyzing and cracking process is the primary reason for the substantial reduction in clogging and overheating of the cracking equipment, due to coke and tar formation, and is the primary reason for the substantial removal of dust from the oil vapors and gases produced in the process.

In related work, *T.D. Nevens; U.S. Patent 3,018,243; January 23, 1962; assigned to The Oil Shale Corporation* describes a process which comprises the steps of pyrolyzing oil shale to produce effluent oil vapors and gases, cooling the effluent to a predetermined temperature to condense the heavier fractions but not the lighter fractions, thermally cracking the heavier fractions within a certain predetermined period of residence time, by recycling at least a portion of the heavier fractions to the zone where the pyrolysis takes place, either indirectly, i.e., via one of the several material streams entering therein or directly to the zone of pyrolysis itself. The uncracked lighter fractions, as well as the thus cracked heavier fractions, are then condensed to produce a composite oil product which has a pour point and viscosity comparable to, or slightly less than that of shale oil produced after the pyrolysis and cracking of all the effluent oil vapors and gases from the pyrolysis, the loss of oil being substantially less than in conventional processes.

A process described by *R.M. Otis; U.S. Patent 3,058,903; October 16, 1962; assigned to The Oil Shale Corporation* comprises preheating oil shale by means of hot solid bodies in a first zone, the preheated oil shale then being pyrolyzed by means of hotter solid bodies in a second zone. The effluent oil vapors and gases resulting from pyrolysis are then sent to the first zone, for contact with a mixture of additional fresh oil shale and partially cooled solid bodies, which results in the condensation of heavy ends of the effluent oil vapors and gases on the oil shale. The light ends are not condensed in the preheater and pass therefrom, as vapor, to be condensed at a later time or sent to further processing.

The condensed heavy ends are sent, with the oil shale, to the pyrolysis zone where they are cracked by reheated solid bodies. At the same time, the solid bodies cause pyrolysis of the freshly preheated oil shale, and crushing to render the pyrolysis more efficient. The pyrolyzed and cracked vapors are then sent to the preheating zone. The heavy ends of the vapors are condensed by the oil shale in the preheating zone, and the lighter oil product, both cracked and noncracked, pass out as vapor. The oil product has a low pour point, and is suitable for pumping at temperatures in the neighborhood of 30° to 60°F.

Zeolite Catalysts

H.C. Maryland; U.S. Patent 3,654,141; April 4, 1972; assigned to Atlantic Richfield Co. describes a process for catalytically cracking of oils derived from oil shale which contain high amounts of nitrogen, i.e., over about 1%. The catalytic cracking is conducted at temperatures of 975° up to 1200°F. At these temperatures the adverse effects of nitrogen poisoning of the catalyst are substantially negated.

A test was made in a laboratory catalytic cracking unit which contained a small fluidized bed reactor with provisions for continuously feeding oil and the catalyst, the oil being a 625° to 925°F boiling range distillate shale oil (without hydrogenation) containing 2.2 weight percent nitrogen. The cracking temperature was 1000°F, the shale oil feed rate was 5 weight units per hour per weight unit of catalyst in the reactor hot zone, i.e., 5 WHSV, and the catalyst was fed at the rate of 16 weight units per hour per weight unit of shale oil, i.e., a 16 to 1 catalyst to oil ratio. The catalyst used was a rare earth (including cerium) exchanged crystalline aluminosilicate-containing catalyst, containing 0.5 to 0.9 weight percent Na, calculated as the free metal, about 0.6 weight percent rare earth, calculated as the free metal, a silica to alumina mol ratio of 4.5 to 1, a pore radius of about 13 angstroms, and containing about 10% crystalline aluminosilicate mixed with an amorphous $SiO_2 \cdot Al_2O_3$ cracking base, part of which is derived from clay.

The product distribution showed about 11 weight percent coke on the feed and substantial yields of 430°F EP gasoline and 430° to 600°F distillate cuts The 600°F plus cut contained 4 ppm Ni, 6 ppm V and 1 ppm Fe which is satisfactory for most purposes. The gasoline contained 0.85 weight percent nitrogen and the 430° to 600°F cut contained 1.43 weight percent nitrogen. These nitrogen levels indicate that the gasoline could be improved by denitrogenation by hydrogen treatment and then catalytic reforming to increase its octane rating.

R.L. Clampitt, U.S. Patent 3,551,322; December 29, 1970; assigned to Phillips Petroleum Company describes a destructive distillation process for recovering hydrocarbons from oil-yielding solids where liquid hydrocarbons and noncondensible gas are recovered, the noncondensible gas is separated from the liquid distillation product and passed to a hydrocarbon conversion zone where the hydrocarbon contained in the noncondensible gas is converted, the converted hydrocarbon product is then separated and recovered from the noncondensible gas.

One form of the process comprises catalytically cracking the noncondensible gas stream over a silica-alumina catalyst at from 950° to 1150°F, condensing the cracked product, recovering the liquid product from the noncondensible gas stream, and recycling the gas stream to the retorting zone to cool residue solids and act as heat transfer media.

Pour Point Depressant

D.K. Wunderlich and J.H. Frankovich; U.S. Patent 3,532,618; October 6, 1970; assigned to Sinclair Oil Corporation describe a process for producing a pour point depressant for shale oil which comprises mixing a deasphalting solvent, under deasphalting conditions, for instance, at a temperature of up to about 250°F at a pressure up to 500 psig, with a shale oil which has had its viscosity and pour point lowered as a result of contact with hydrogen under hydrovisbreaking conditions. The mixture of shale oil and deasphalting solvent is then separated into deasphalted shale oil and a heavier portion containing the pour point depressant for shale oil. The following example illustrates the process.

Example: Raw shale oil was passed upflow, in contact with H_2 under hydrovisbreaking conditions, through a 3/8", schedule 40 pipe held at 825°F in an electrically heated furnace. Feedstock properties, reaction conditions and product inspections are shown below.

	Raw Shale Oil	Hydrovisbreaking
Reactor temperature, °F		825
Pressure, psig		1,000
Gas ratio, scf/bbl		2,000
Oil feed rate, g/hr		300
Cold oil time, min		10
Wt percent liquid recovery	100	94.0
	Feed	**Product**
Liquid inspections:		
Gravity, °API	18.9	19.7
Pour point, °F	+85	-45
Viscosity, KV at 122°F	31.01	14.15
Percent Ramsbottom carbon	4.5	5.46
Wt percent hydrogen	11.27	11.11
	Raw Shale Oil	**Hydrovisbreaking**
ASTM (vac) distillation, °F:		
IBP	393	375
5%	457	421
10	505	454
20	590	521
30	670	587
40	741	662
50	802	738
60	856	809
70	920	895
80	990	1017
Trap, ml		17.0

The shale oil having a pour point of -45°F produced by the hydrovisbreaking process was mixed with an equal volume of normal pentane. The mixture was centrifuged and a heavy viscous layer amounting to 2.5 wt % of the shale oil was centrifuged from the lighter deasphalted shale oil layer. This heavy viscous layer contained the pour point depressant. The deasphalting solvent in the deasphalted shale oil layer was evaporated. The resulting oil had a pour point of 0°F. The removal of the viscous layer increased the pour point of the shale oil of lowered viscosity and pour point from the initial value of -45° to 0°F. The viscous layer obtained from the deasphalting step and containing the pour point depressant was mixed thoroughly with a volume of raw shale oil having a pour point of +85°F which was equal to the volume of shale oil of reduced pour point and viscosity from which the heavy viscous layer was obtained. The resulting mixture of raw shale oil and the heavy viscous layer containing the pour point depressant had a pour point of 0°F. Thus, the pour point depressant was effective in reducing the pour point of raw shale oil from +85° to 0°F.

Visbreaking in Presence of Hydrogen for Pour Point Reduction

J.H. Frankovich and D.K. Wunderlich; U.S. Patent 3,738,931; June 12, 1973; assigned to Atlantic Richfield Company found that visbreaking of synthetic crude oil in the presence of hydrogen while utilizing controlled hydrogen consumption rates helps stabilize the reduced pour point of an oil product obtained by combining visbroken oil with visbroken and hydrogenated oil. The resulting combination of oils is a product of reduced pour point which is suitable for storage and transportation without concern about the product becoming so viscous that it cannot be handled.

By this process, after visbreaking in the presence of hydrogen a major portion of the oil can be separated by a holding type of separation treatment instead of distillation. Suitable separation conditions can be achieved while keeping the temperature and pressure the same as the visbreaking conditions. The elevated temperature and pressure of visbreaking can be maintained in the separation step. This is helpful because the vaporized oil removed by the separation step can then be subjected to hydrogenation without loss of heat or pressure between the visbreaking and hydrogenating steps. This conservation of heat and pressure is a significant process advantage.

Figure 2.6 shows synthetic crude oil in pipe **1** passing into visbreaking zone **2**. Hydrogen is added to the oil by way of pipe **3** so that hydrogen is present when the oil is subjected to the elevated temperatures and pressures maintained in zone **2**.

FIGURE 2.6: TREATMENT OF SYNTHETIC CRUDE OIL FOR POUR POINT REDUCTION

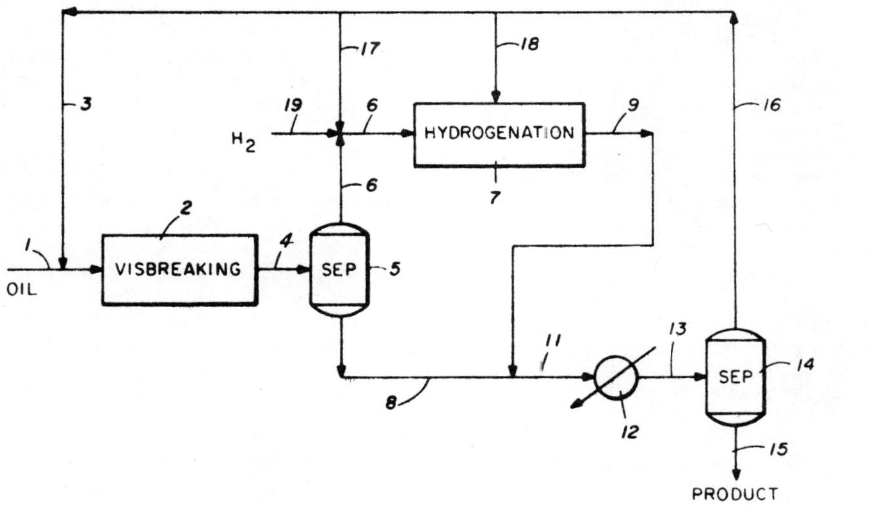

Source: J.H. Frankovich and D.K. Wunderlich; U.S. Patent 3,738,931; June 12, 1973

The visbroken oil is removed from zone **2** by way of pipe **4** and passed to and held in first separation zone **5** where the temperature and pressure conditions are maintained close to those of visbreaking zone **2** except for normal temperature and pressure drops due to pipe **4**. During the holding in zone **5** separation occurs of vapors which can constitute a major portion (i.e., at least about 50 weight percent based on the total weight of the liquid oil in pipe **1**) of the liquid feed oil. The oil vapors in first separation zone **5** are removed by way of pipe **6** and passed into hydrogenation zone **7**.

The unvaporized visbroken oil present in zone **5** is removed substantially in the liquid state by way of pipe **8**. The liquid in pipe **4** has a pour point at least 20°F below that of the oil in pipe **1** and preferably has a pour point below 0°F. In addition, the pour point of the liquid in pipe **4** is stabilized so that it will not increase significantly upon long standing or handling at ambient atmospheric temperatures or even at more elevated temperatures.

The amount of vaporized hydrocarbon or oil removed in pipe **6** can be controlled by the amount of hydrogen added by way of pipe **3**. Generally, the more hydrogen added by way of pipe **3**, the more hydrocarbon or oil vapor removed by pipe **6**. The hydrogenated vaporized oil is removed from zone **7** by way of pipe **9** and all or part can be combined with all or part of the material in pipe **8**. The combined materials in pipe **11** are a pour point depressed and stabilized oil.

The combined materials in pipe **11** are at an elevated temperature which is lowered by way of cooling means **12**. The combined materials, after cooling, are still at an elevated pressure when passed by way of pipe **13** to a second separation zone **14**. In zone **14** the temperature and pressure are maintained substantially the same and gas such as hydrogen gas allowed to separate from the liquid oil to degasify the liquid oil product. The product is removed by way of pipe **15**. Recovered hydrogen gas is returned by way of pipe **16** to the process. The hydrogen gas in pipe **16** is reused in pipe **1** by way of pipe **3** and/or in pipe **6** by way of pipe **17** and/or as a cooling or quenching gas for zone **7** by way of pipe **18**. Makeup hydrogen can be added as necessary by way of pipe **19**.

The product in pipe **15** is a liquid oil product suitable for storage, transportation, further refining, and the like, since the product is readily flowable due to the hydrogenation step and since the pour point is depressed in a stable manner so that the reduced pour point will remain as such during the subsequent handling. Previously it had been found that the pour point of visbroken oil could increase with time, but visbreaking with controlled hydrogen consumption according to this process reduces and stabilizes the pour point of the oil in pipe **15** so that the pour point will not substantially increase thereafter.

By the use of zone **5** in accordance with this process, the temperature and pressure conditions in zone **5** can be maintained substantially the same as the conditions in zone **2**. Thus, a maximum amount of heat and pressure is conserved while still separating the material to be hydrogenated from the liquid in pipe **8**. This is to be distinguished from the situation where a conventional distillation zone is utilized. With the use of a distillation zone the temperature and pressure are decreased substantially from that of visbreaking zone **2**. Therefore, a substantial loss of heat and pressure results whereas with the use of separation zone **5** maximum conservation of heat and pressure for hydrogenation zone **7** is achieved while still achieving separation of the oil desired for hydrogenation. Th synthetic crude oil employed can be any crude oil other than naturally occurring crude oil, e.g., oil obtained from oil shale, tar sands, coal and the like.

The visbreaking conditions are a temperature of from 600° to 1000°F, preferably from 650° to 850°F; a pressure of at least 500 psig, preferably from 500 to 1,500 psig; and a time of at least about 3 minutes, preferably from 3 to 30 minutes. The amount of hydrogen necessarily consumed during visbreaking in order to obtain a stabilized reduced pour point is from 50 to 200 standard cubic feet (scf) per barrel of oil being visbroken.

The vaporized oil in pipe **6** is at an elevated temperature and pressure closely approximating those of visbreaking zone **2** when it reaches hydrogenation zone **7**. The oil in pipe **6** is, therefore heated and pressured sufficiently so that only hydrogen need be added to obtain the desired degree of hydrogenation. Reheating and/or repressuring of the vaporized oil for hydrogenation zone **7** is eliminated. This is an advantage over the use of a distillation zone since repressuring and/or reheating would be necessary with such a fractionation zone.

The hydrogenation operation **7** is carried out in a conventional manner. For example, the zone can contain known hydrogenation catalysts such as a combination of a Group VI metal (e.g., chromium, molybdenum, and/or tungsten) with a Group VIII metal (such as iron, nickel, and/or cobalt). A suitable catalyst is cobalt molybdate. Hydrogenation is carried out at a temperature of at least 650°F, preferably from 650° to 900°F, a pressure of at least about 1,000 psig, preferably from 1,000 to 3,000 psig, a liquid hourly space velocity from 0.1 to 2 vol/hr/vol, and a molecular hydrogen addition rate of at least 2,000, preferably from 2,000 to 10,000 scf/bbl of oil to be hydrogenated.

All or part of the pour point depressant in pipe **8** can be combined with all or part of the hydrogenated oil of pipe **10**. The combination of materials in pipe **11** is at least partially liquefied by cooling substantially to ambient temperature by cooling means **12**. The cooled liquid oil in pipe **13** is then passed to second separation zone **14** where at least hydrogen gas and/or hydrogen containing gases are allowed to separate from the liquid thereby stabilizing the liquid by having degasified same and at the same time recovering hydrogen or hydrogen containing gas for reuse in one or more of zones **2, 5** and **7**.

The product in pipe **15** has a pour point at least about 20°F below the pour point of the feed oil in pipe **1**. The reduced pour point will remain stable during subsequent handling, etc. so that one purchasing this product can rely upon the pour point of the product as purchased remaining substantially unchanged.

Example: A full range shale oil obtained by retorting Colorado oil shale was fractionated and the 380+°F (including bottoms) fraction was tested as the feed oil. The feed oil had a pour point of +80°F. Two visbreaking runs were made, one utilizing hydrogen according to this process and the other utilizing nitrogen so that the conditions of the second run could be kept comparable to the first (hydrogen) run.

In the first run the visbreaking temperature and pressure were 825°F and 1,000 psig. The visbreaking time was 10 minutes and the amount of hydrogen consumed during visbreaking was 50 scf/bbl of feed oil. The pour point of the visbroken oil from the first run was -45°F.

The second run was carried out in the absence of added hydrogen utilizing an 825°F temperature, a 1,000 psig, total pressure maintained by the addition of nitrogen and a 10 minute visbreaking time. The visbroken oil had a pour point of -30°F. Samples of visbroken oils from both the first and second runs were aged 9 months in a sealed container at room temperature (about 80°F) and the pour point of the samples then remeasured. The pour point of the sample from the first (hydrogen) run remained the same at -45°F whereas the pour point of the sample from the second run had increased a total of 35° to +5°F.

It can be seen from the above that by carrying out the visbreaking step with controlled hydrogen consumption in accordance with this process, pour point reduction is stabilized so that the reduced pour point remains substantially unchanged. After separation and catalytic hydrogenation of the vaporized oil over cobalt molybdate catalyst, the unvaporized liquid oil from flash zone 5 and the hydrogenated vaporized oil from zone 7 were recombined and had a pour point 30°F below the pour point of the original feed oil.

Thermal Treatment to Reduce Pour Point

W.J. Culbertson, Jr., T.D. Nevens and W.D. Schnackenberg; U.S. Patent 3,284,336; Nov. 8, 1966; assigned to The Oil Shale Corporation describe a method for treating a heavy fraction separated under substantially noncracking conditions from a crude oil derived by thermal treatment of solid carbonaceous material, to produce the heavy fraction and at least one light fraction. This method comprises heat treating the separated heavy fraction alone above about 600°F and below the point of incipient thermal decomposition of the heavy fraction for a period of time which is inversely proportional to the temperature to produce a product which, when combined with at least part of a light fraction, results in an oil having a pour point lower than that of the original crude oil; the heat treatment producing substantially no noncondensible hydrocarbons and substantially no elemental carbon.

While the precise effect of the heat treatment of the heavy high molecular weight fraction containing the waxes, asphaltenes, and the like is not completely clear, it is theorized that some portion of the heavy fraction undergoes a molecular modification and/or rearrangement which results in a marked improvement in pour point and reduction in thixotropy after combination with lighter hydrocarbon oil fractions.

The molecular weight of the heavier hydrocarbons is not appreciably affected by any such alteration in molecular structure. Moreover, the pour point of the heavy fraction which has been treated in accordance with the process in not appreciably changed from that of the heavy fraction prior to treatment. The temperature and conditions utilized avoid appreciable formation of light gases or of coke. These substances represent undesirable loss in yield in processing oils, and in the process are not formed even as by-products. The pour points were determined by ASTM D97-47. Pour points were taken to the nearest 5°F. The distillations were conducted according to ASTM D1160-52T.

Example 1: In order to demonstrate the effectiveness of the method in lowering the pour point of oils derived by thermal treatment of solid carbonaceous materials by thermal means, runs were carried out on samples of a crude Colorado shale oil having a pour point of 80°F where each sample of the oil initially was fractionated to provide a light fraction consisting of 70% by volume of the crude oil and a residue. Portions of the residue then were heated to and maintained at 650°, 700° and 750°F, for about 1 hour. No discernible amount of noncondensibles or elemental carbon were produced. At the end of the heating period each treated residue was cooled to 70°F, tested for pour point properties, and recombined with its corresponding, previously removed light fraction. Each oil resulting from the recombination was then tested for its pour point characteristics. The results of the various runs are set forth in the following table.

Run	Temp., °F.	Coke and Gas Losses	Residue Pour Point, °F.		Final Oil Pour Point, °F.
			Before Soaking	After Soaking	
1	650	None	80	80	50
2	700	do	80	80	15
3	750	do	80	80	5

Example 2: The general procedure of Example 1 was repeated in a series of runs where various residues were heat treated for a period of 1 hour following removal of the light fractions indicated in the following table. Runs 1, 2 and 3 of Example 1 are included for comparative purposes. There was no discernible production of either noncondensible hydrocarbons or elemental carbon. The results are set forth below.

Temperature, °F	650			700							750
Run	1	4	5	6	7	2	8	9	10	11	3
Light Fraction Removed, Percent by Volume	70	50	30	90	80	70	65	60	50	30	70
Residue Pour Point Before Soaking	80	95	90	----	----	80	90	95	95	90	80
Residue Pour Point After Soaking	80	90	90	----	95	80	85	95	----	85	80
Final Oil Pour Point, °F	50	70	75	55	45	15	20	15	35	50	5

Treatment of Raw Oil with HCl

C.N. Kimberlin, Jr. and H.G. Ellert; U.S. Patent 2,966,450; December 27, 1960; assigned to Esso Research and Engineering Company have found that shale oil is effectively upgraded by treatment with an acidic gas in the presence of a hydrocarbon solvent under certain conditions. There is obtained a sludge or precipitate representing substantially all of the more undesirable coke-forming nitrogen-containing constituents; this sludge nonetheless being extremely small in amount. There is further obtained as an extract a maximum amount of high quality cracking stock.

Referring to Figure 2.6 raw shale oil obtained by retorting oil shale at 800° to 1200°F for 0.1 to 2 hours is fed by means of line **2** to precipitation zone **4**. Though the shale oil may be given a preliminary thermal treatment or subjected to visbreaking conditions, this is normally not necessary. In fact, it is one of the advantages of the process that the visbreaking operation may be dispensed with.

Zone **4** may be an upright vessel provided with means for providing intimate contact of counterflowing streams, such as agitators, packing or perforated plates. Alternately, vessel **4** may be a mixer-settler of conventional design. The shale oil is treated in the vessel with from 0.5 to 10 volumes of a hydrocarbon fraction boiling in the range of from -45° to 250°F. Pentane, hexane, propane, butane or naphtha fractions give good results. This hydrocarbon is introduced through line **5**. The temperature is in the range of 0° to 300°F, preferably 50° to 150°F, and pressure sufficient to keep material in the liquid phase.

FIGURE 2.6: SHALE REFINING PROCESS EMPLOYING HCl TREATMENT

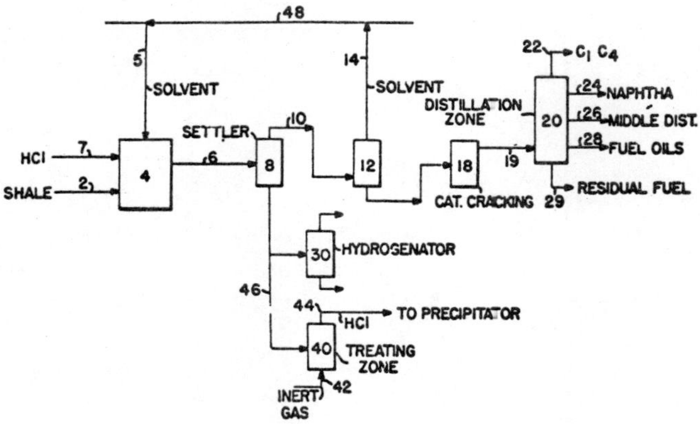

Source: C.N. Kimberlin, Jr. and H.G. Ellert; U.S. Patent 2,966,450; December 27, 1960

An anhydrous acidic gas, such as HCl, HBr or BF_3, but preferably HCl, is passed into vessel **4** through line **7** in amounts sufficient to saturate the oil. 0.5 to 4.0, but preferably 0.9 to 1.5, equivalents of acid per equivalent of nitrogen in the oil may be used. The oil-solvent-gaseous acid mixture is maintained in a state of agitation, preferably for a residence period of from 1 to 60 minutes, and then passes to settling zone **8**.

The extract, containing the oil and solvent, is removed from zone **8** through line **10** to flash distillation zone **12** where the low boiling hydrocarbon solvent is distilled overhead by flash or fractionation and recycled via line **14** to precipitation zone **4**. The shale oil fraction may then advantageously be passed to a catalytic cracking zone **18** for conversion by conventional means into valuable fuels and lubricants.

Returning to settler **8**, the lower layer is a sludge-like material comprising in part high molecular weight hydrogen-deficient material known as Conradson carbon, and also the salt-like reaction product of the HCl and the nitrogen compounds. About 7 to 15 lb of HCl per barrel of oil are required in this process. This low quality sludge, amounting to some 25 to 35% of the total shale oil, may also be utilized to increase the overall yield of motor fuel and lubricants. In one example of the process, it is withdrawn as a lower layer from vessel **8** and passed to a hydrogenation zone **30**. By operating at about 3,000 psig in a conventional manner and at 650° to 1000°F, preferably 700° to 850°F, the nitrogen content of the sludge is reduced from 5 to 0.5% or less. Alternately, this sludge may be partially hydrogenated at less severe conditions and the product recycled through the precipitation zone.

Example 1: A sample of shale oil was treated in accordance with the process described in the presence of n-hexane. Four types of acidic treating agents were employed, with the following result as shown in the following table.

		Solvent (3 volumes n-hexane)			
		Acid			
Treatment	Raw Shale	Anhydrous HCl	60% H_2SO_4*	SO_2 (Gas)	BF_3
Yield, volume percent	-	70	53	86	51

(continued)

Treatment	Raw Shale	Solvent (3 volumes n-hexane) Acid: Anhydrous HCl	60% H_2SO_4*	SO_2 (Gas)	BF_3
Product Quality:					
Gravity, °API	20.6	28.2	25.6	21.6	26.6
Nitrogen, wt. percent	1.9	0.4	0.3	1.5	0.5
Sulfur, wt. percent	0.7	0.5	0.8	0.7	0.7
Con. carbon, wt. percent	4.3	0.9	1.2	4.3	1.7

*4 mols/mol N_2 in oil.

These data demonstrate the excellent results obtained by the process in upgrading shale oil to a catalytic cracking feed stock. It is to be noted that aqueous acid provides a substantially lower oil yield without quality improvement.

Example 2: The necessity of maintaining a solvent in the mixture being acid treated with gaseous HCl is shown by the following data.

Treatment	None	HCl* at 400°F.	Hexane + HCl
Yield, vol. percent	-	67	70
Product quality:			
Gravity, °API	20.6	27.3	28.2
Nitrogen, wt. percent	1.9	1.2	0.4
Con. Carbon, wt. percent	4.3	3.9	0.9

*Added to saturation.

TAR SANDS SEPARATION PROCESSES

The history and recent development of recovery of heavy petroleum from bituminous sands is of considerable interest. By way of definition, bituminous sands are bulk sands impregnated with a heavy petroleum. Early observers named discovered bituminous sands as "tar sands" because of the asphaltic appearance of the sand deposits. Geologists later referred to these sands as "oil sands" since their interest lay in the bituminous sands as a possible source of oil. Through long usage, the term tar sand will probably persist in the art as the proper term for these sands.

Bituminous sands are found in most areas of the world where petroleum is located. It is noted that when oil wells in petroleum fields cease to produce, many countries return to underground workings at the oil field to recover the petroleum from the sands surrounding the oil fields.

The largest known deposit of bituminous sands are the Athabasca bituminous sands which lie in the northern part of the province of Alberta in Canada. These outcrop along the Athabasca River north of McMurray. In the United States large deposits of bituminous sands are found in California and Utah as well as certain other states.

The Athabasca bituminous sand deposit is known geologically as the McMurray Foundation and is of the Lower Cretaceous age. It lies uniformly on Devonian limestone and shale of the Upper Devonian age, and is overlain by the Clear Water formation, a formation of marine shales. The McMurray formation comprises bituminous and fine grained sands, is characterized by erosional disconformity, and includes some clay and shales.

In the United States principal deposits are the Edna deposit near San Luis Obispo, Calif., the Vernal deposits which are located near Vernal, Utah, and are outcroppings along the Green River, and the Sunnyside, Utah deposits.

Various types of recovery treatment and apparatus have been used in connection with the Athabasca bituminous sands and predate experimental methods of recovery in the United States. Most of the procedures used in Canada have been employed on somewhat an experimental basis in the United States.

In Athabasca bituminous sands soluble salts in significant quantity are absent. Where such salts are present as in certain California bituminous sands, the salts have to be leached out before separation can be effected. Extensive investigation with the Athabasca bituminous sands has revealed a thin film of water separates the oil from the sand grain surfaces. Hence, the primary problem is really not one of separation of the petroleum from the sands but rather of concentration.

One technique which has been used in connection with the recovery of petroleum from Athabasca bituminous sands is the so-called Hot-Water Washing Method. Any procedure which treats bituminous sand with warm or hot water results in the oil drawing away from much of the sand and forming oil clots. These oil clots will frequently enmesh sand particles. A practical problem, hence, is to devise a procedure and structure which will result in a withdrawal and separation of oil from the sand, and in the process, collecting a minimum of mineral matter. One must recall that the specific gravity of the heavy petroleum in the sands is greater than that of water.

Another technique that has been employed in connection with recovery of petroleum from bituminous sands is a Cold-Water Washing Method. The principles involved in the cold-water treatment are straightforward. A distillate such as kerosene is mixed with the bituminous sand to increase the fluidity of the oil content, so it will flow readily at room temperatures, and also to reduce its density below that of water so it will float. Also, an alkalizing agent and also a wetting agent are added to the pulp to assist in the disengagement of oil from the sand. The petroleum sand so treated is agitated in water. The diluted oil floats as in a water-in-oil emulsion, and the mineral matter sinks. The oil is collected and then settled to reduce its content of mineral matter and water.

In the cold-water process a net recovery of 95% of the oil in the bituminous sand treated was obtained. But final diluted oil product contained not only about 2% mineral matter but also 25% water.

HOT WATER PROCESSES

Hot Alkaline Polyphosphate Solution

C.O. Walker; U.S. Patent 3,846,276; November 5, 1974; assigned to Texaco Inc. describes a method for separating bituminous petroleum from tar sand recovered from deposits by mining. In one example, the mined tar sand material is fed into a mixing container and contacted with a hot alkaline polyphosphate solution. Bitumen separates from sand without formation of an emulsion or a froth.

The fluid used in conjunction with this separation technique is an aqueous solution of a polyphosphate wetting agent such as sodium acid pyrophosphate plus an alkalinity agent such as sodium hydroxide, heated to a temperature of from about 100° to 200°F. The portion of bituminous petroleum dispersed in the aqueous phase is separated from the phosphate solution by contacting the fluid with a hydrocarbon fluid such as diesel oil.

An attractive feature of the process is the fact that bitumen is not emulsified and no froth is formed on being contacted with the hot aqueous alkaline polyphosphate solution. Bitumen is removed effectively from the sand but remains in a separate phase which will form a discrete layer distinct from the aqueous solution if agitation is stopped. The bitumen may be pumped to the refining process unit, the polyphosphate solution being recycled through the separation equipment.

Example: A tar sand deposit having an overburden varying from 10 to 40 feet in thickness, the deposit having an average thickness of 65 feet, is exploited by means of open pit strip mining. After stripping away the overburden, tar sand is dug from the deposit using scrapers and transported by truck to the separation process equipment. The equipment is theoretically capable of removing the tar sand at a rate of 1,500 tons per hour, although actual production rate is 1,000 tons per hour. The bitumen content of the tar sand is 14% by weight.

Separation is accomplished by using a number of units, each with a capacity of 200 tons per hour. The tar sand is fed by screw conveyor into a 5,000 gallon mixing tank. An aqueous solution of 0.30% by weight sodium acid pyrophosphate and 0.30% by weight sodium hydroxide is formulated in a 1,000 gallon tank and fed into the 5,000 gallon tank at a rate of 1,000 gallons per hour. The polyphosphate solution is heated to a temperature of 160°F by passing through a gas fired heater prior to being added to the 5,000 gallon tank.

Two separation tanks are provided for each mixing tank, so the mixing tank output is passed first to one separation tank and then to the other. Each separation tank has a volume of 3,000 gallons. A screw conveyor removes sand from the bottom of the separation tank. Bituminous petroleum accumulates in a layer immediately above the sand layer, and an aqueous layer having a small amount of bitumen dispersed therein accumulates on top. Diesel oil is added to the separation tank and mixed. Bitumen from the layer immediately above the sand as well as that dispersed in the aqueous phase, dissolves in the diesel oil. After mixing is stopped, a layer of diesel oil and bitumen forms on top, which is decanted and sent to the refinery.

The aqueous sodium pyrophosphate solution is passed through a sand pack filter and then back to the pyrophosphate make-up tank for recycling through the unit. Sand from the settling tank is placed in excavations formed in earlier stages of strip mining operations. The separation technique effectively removes approximately 90% of the bitumen from the tar sand.

In order to verify the operability of the process and further to determine the optimum values for the controllable parameters, the following experimental work was performed. One hundred gram samples of tar sand obtained from the Athabasca area in Alberta, Canada were compressed into cylindrical cakes at 6,000 pounds per square inch pressure. The caked samples were placed in an open container filled with the various test solutions. A motor driven mixer blade was positioned in the solution approximately one and one-half inches from the face of the tar sand cake and rotated. The extent of disintegration of the tar sand at the end of 15 minutes was observed and this observation is reported below. In all cases the temperature was maintained at 125° to 130°F.

Fluid	Cake Condition at End of 15 Minutes
Tap water	Unchanged
Water plus 0.3% sodium acid pyrophosphate	Unchanged
Water plus 0.3% caustic soda	50% disintegrated
Water plus 0.3% caustic soda plus 0.3% sodium acid pyrophosphate	100% disintegrated
Water plus 0.6% caustic soda plus 0.6% sodium acid pyrophosphate	100% disintegrated
Water plus 0.3% sodium acid pyrophosphate NH_4OH to yield pH of 10.8	Slight disintegration

After disintegration of the tar sand samples, the mixtures were allowed to settle. Clean sand settled to the bottom and distinct layers of bitumen and polyphosphate solution formed. Only a slight amount of hydrocarbon was suspended in the aqueous polyphosphate solution. Addition of diesel oil to the solution solubilized the dark material, which floated to the surface and was easily separated. No emulsion or froth was formed.

Treatment with Inorganic Phosphates and Alkylene Oxide Surfactants

A process described by *G.P. Canevari and R.J. Fiocco; U.S. Patent 3,331,765; July 18, 1967; assigned to Esso Research and Engineering Company* involves the recovery of oil from tar sands where in a first or initial stage a substantial amount of the sands is removed by treatment of the sands with water and an oil froth is produced. This oil froth in essence comprises an emulsion of oil and water containing an appreciable amount of solids. This froth is treated in a secondary phase or stage with additional water to remove more solids from the emulsion of oil and water. In a tertiary stage the emulsion, containing some solids, is treated with a chemical mixture to break the emulsion so as to secure an oil phase free of water and solids. A desirable chemical mixture comprises a transfer agent, a chelating agent and a demulsifying agent used in combination with a hydrocarbon solvent. The additive mixture preferably contains from 0.1 to 2.0%, preferably 1% by weight based on the froth, of a transfer agent such as tetrasodium pyrophosphate, or sodium tripolyphosphate. Other equivalent chemicals which may be used are, for example, potassium pyrophosphate, sodium hexametaphosphate and sodium silicate. These transfer agents promote the transfer of the solid fines from the oil-water interface into the aqueous phase by establishing a uniform electropotential over the entire surface of the particle. In general, inorganic builders and condensed phosphates act as transfer agents.

A second constituent of the chemical mixture is a demulsifying additive. These demulsifying additives are nonionic surface active compounds such as, for example, a polyethoxyalkylene compound known as Nalco D-1645 produced by the Nalco Chemical Company. Another preferred demulsifying solution comprises a mixture of (1) one part of the reaction product of diethyl ethanol amine with premixed propylene oxide and ethylene oxide (propylene oxide-ethylene oxide can range from 40/60 to 60/40, preferably 55/45) and (2) approximately 3 parts of a palmitic acids ester of the reaction product of an alkylphenol-formaldehyde resin with ethylene oxide.

$$\frac{\text{alkylphenol-formaldehyde}}{\text{ethylene oxide}} = \frac{1}{1.5}$$

The mixture may be varied in the range from 1 part of the reaction product to 2 parts of the acid ester to 1 part of the reaction product to 4 parts of the palmitic acid ester. Other satisfactory demulsifiers are Pluronic polyols. These Pluronic compounds are polyoxypropylene glycols having a chemical structure as follows:

$$HO(CH_2{-}CH_2{-}O)_a(\underset{\displaystyle CH_3}{\underset{|}{C}}H{-}CH_2{-}O)_b(CH_2{-}CH_2{-}O)_c\,H$$

where a = molecular weight range of 50 to 7,000, b = molecular weight range of 900 to 4,000 and c = molecular weight range of 50 to 7,000. These compounds have molecular weights ranging from 1,000 to over 16,000. The amount of demulsifier used is in the range from 0.01 to 0.5%, for example, 0.2% by weight based on the weight of the total emulsion. In instances where the froth may contain heavy minerals such as zircon, rutile, ilmenite, tourmaline, apatite, staurolite, garnet, etc., it may be desirable to employ chelating agents in addition to a demulsifier and transfer agent. For example, chelating agents which may be used as part of the chemical mixture are ethylenediamine-tetraacetic acid, sodium gluconate, gluconic acid, sodium oxalate and diethylene glycol.

The chemical mixture may contain a hydrocarbon solvent boiling in the range from 250° to 600°F, preferably in the range from 250° to 450°F. Preferred solvents are xylene and naphtha boiling in the range from 250° to 450°F. The amount of solvent used is in the range from 10 to 100% by weight of the froth, preferably 50%.

In related work, *S. Ross, G.P. Canevari and R.J. Fiocco; U.S. Patent 3,296,117; January 3, 1967; assigned to Esso Research and Engineering Company* describe a process for the recovery of oil from tar sands where the oil is removed from the sands in an initial stage or phase comprising a treatment with water to produce a froth containing an appreciable amount of water and solids (sand/clay) particles. This froth is subsequently treated in a secondary stage with water containing a selected chemical transfer agent. In a third stage the substantially solids-free oil and water phase are separated by treatment with a selected chemical demulsifier.

Tetrasodium pyrophosphate is one such transfer agent; other examples are trisodium polyphosphate and sodium silicate. The amount of tetrasodium pyrophosphate added to the water is in the range from 0.1 to 2.0%, preferably 1.0%. The demulsifying solution comprises a mixture of (1) one part of the reaction product of diethylethanolamine with premixed propylene oxide and ethylene oxide (the propylene oxide/ethylene oxide can range from 40/60 to 60/40, preferably 55/45) and (2) approximately three parts of a palmitic acid ester of the reaction product of an alkyl phenol formaldehyde resin with ethylene oxide.

$$\frac{\text{alkylphenol-formaldehyde}}{\text{ethylene oxide}} = \frac{1}{1.5}$$

The mixture may be varied in the range from 1 part of the reaction product to 2 parts of the acid ester to 1 part of the reaction product to 4 parts of the palmitic acid ester. A preferred mixture is 1 part by weight of the reaction product to 3 parts by weight of the palmitic acid ester. In order to illustrate the process a froth comprising 46.6% by weight of bitumen, 42.6% of water and 10.8% sand was contacted in a number of operations, namely, A, B, C, D and E. The results of these operations are illustrated in the following table.

Effects of Transfer Agents and Demulsifying Agents on Unwashed Froth

Operations	Water, Weight Percent	Solids, Percent	Bitumen
A. Untreated sample (control)	42.6	10.8	46.6
B. Water washing with 1% TSPP solution*	36.0	1.79	62.21
C. Water washing plus addition 0.4% demulsifying mixture**	18.0	- - - -	- - - -
D. Water washing with 1% Na_2SiO_3 plus addition 0.4% demulsifying mixture**	7.5	- - - -	- - - -
E. Water washing with 1% TSPP solution* plus addition 0.4% demulsifying mixture**	0.3	0.1	99.6

*Tetrasodium pyrophosphate
**Mixture of the process

From the above, it is apparent that far superior results are obtained if the froth is contacted initially with water containing a transfer agent to remove substantially all of the solid and then contacted in the subsequent stage with a demulsifying agent. In Operation B a substantial amount of solids was removed but the bitumen contained an appreciable amount of water. The same situation occurred in Operation C, as well as in Operation D.

However, when employing the technique of the process by washing the froth in an initial phase containing tetrasodium pyrophosphate and then removing the sand and treating the same in the secondary phase, the oil phase was 99.6% oil, containing only 0.1% by weight of solids and 0.3% by weight of water.

The palmitic acid ester is preferably secured from a phenol-formaldehyde resin where the alkyl group contains from 3 to 4 carbon atoms and is secured from isopropyl alcohol or alkylbutyl alcohol. The molecular weight is in the range from 1,200 to 1,600, preferably 1,400.

Recovery of Oil from Middlings Layer

A process described by *P.H. Floyd, R.C. Schenk, H.L. Erskine, Jr. and J.V.D. Fear; U.S. Patent 3,401,110; September 10, 1968; assigned to Great Canadian Oil Sands Limited, Canada* comprises the combination of a hot water separation step for primary recovery of oil from the sands together with a scavenger step for further treatment of the middlings layer obtained from the primary separation step to recover an additional amount of oil.

The process also involves regulating the amount of water introduced to the primary separation step and the rate of transfer of middlings layer to the scavenger step so as to maintain the density and/or viscosity of the middlings layer. The procedure involves the following.

(1) Forming a pulp of the bituminous sands with a minor amount of water in a pulping zone while heating the mixture with steam;

(2) Removing pulp and mixing it with hot water and hereinafter specified recycle steam in a dilution zone, the amount of the hot water plus the water added to the bituminous sands in forming the pulp being 0.2 to 3.0 lb/lb of the bituminous sands;

(3) Flushing the mixture from the dilution zone into a separation zone;

(4) Settling the mixture in the separation zone at a temperature in the range of 130° to 210°F, more preferably 170° to 210°F, to form an upper oil froth layer, a middlings layer comprising water, clay and oil, and a sand tails layer;

(5) Separately removing the oil froth layer and the sand tailings layer;

(6) Removing a stream of middlings layer from the separation zone and passing it to the dilution zone as the recycle stream;

(7) Passing a second stream of middlings layer to a scavenger zone and subjecting it to air flotation to recover an additional amount of oil froth;

(8) Regulating the amount of water incorporated with the bituminous sands in Step (1) and the rate of passage of the second stream to the scavenger zone in Step (7) to regulate and maintain the density of the middlings layer in the range of 1.03 to 1.50 and/or the viscosity thereof in the range of 0.5 to 10 cp;

(9) Removing from the scavenger zone and discarding from the system middlings material of depleted oil content comprising clay dispersed in water.

Referring to Figure 3.1 the bituminous oil sand is fed to the system through line **10** where it first passes to a conditioning drum or muller **11**. Water is fed to the muller via line **12** and steam is introduced through line **13**. The total water so introduced in liquid and vapor forms is a minor amount based on the weight of the tar sands processed and generally is in the range of 10 to 45% by weight of the mulled mixture.

The conditioning drum is provided with suitable kneading or mixing means to give the desired mulling action. Enough steam is introduced through line **13** to raise the temperature in the conditioning drum to within the range of 130° to 210°F and preferably to above 170°F. Mulling of the tar sands produces a pulp which then passes from the conditioning drum as indicated by line **14** to a screen indicated at **15**. The purpose of this screen is to remove from the tar sands pulp any debris, rocks or oversized lumps as indicated generally at **16**.

The conditioned tar sands pass from screen **15** to a pulp box **17** which serves as a zone for diluting the pulp with additional water before passage to primary separation zone **18**. Hot water from heater **27** is passed through line **19** to the pulp box and additional steam is fed through line **20** if necessary to maintain the temperature in the range of 130° to 210°F and preferably above 170°F. Also a middlings stream which is withdrawn from the primary separator **18** is recycled through lines **21** and **19** to the pulp box. This recycle stream serves to provide sufficient liquid to flush the tar sands from the pulp box and effect transfer of the pulp to the separator.

Another important function of the recycle stream is to cause dispersion of the pulped material as it is fed into separation zone **18**. However, such recycling of middlings is not essential in all cases, particularly when the clay content of the tar sands is high. In this event a relatively high rate of fresh feed water introduction through heater **27** can be employed to compensate for the high clay content while a correspondingly high rate of transfer of middlings layer through line **26** can be maintained. Under these circumstances recycling of the other stream of middlings through lines **21** and **19** to pulp box **17** is not required.

Modifications that may be made in the process as above described include sending a minor portion of the middlings recycle stream from line **21** through a suitable line to muller **11** to supply all or a part of the water needed therein other than that supplied through condensation of the stream which is consumed. Also, if desired, a stream of the middlings recycle can be introduced onto the screen **15** to flush the pulp through and into pulp box **17**.

FIGURE 3.1: RECOVERY PROCESS EMPLOYING PRIMARY HOT WATER SEPARATION AND SCAVENGER ZONE

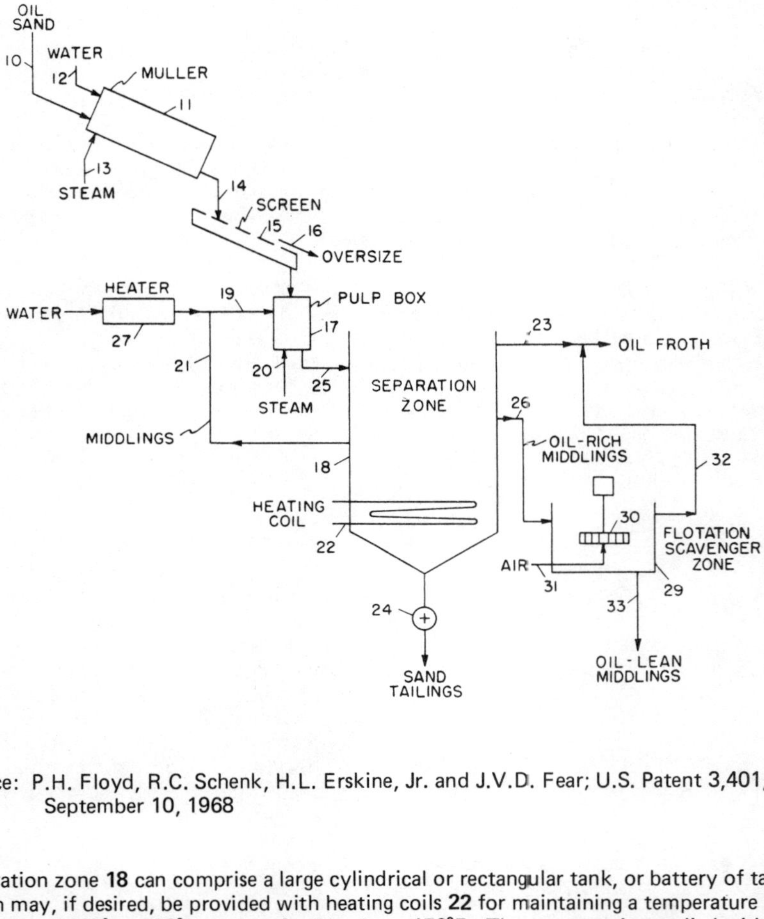

Source: P.H. Floyd, R.C. Schenk, H.L. Erskine, Jr. and J.V.D. Fear; U.S. Patent 3,401,110; September 10, 1968

Separation zone **18** can comprise a large cylindrical or rectangular tank, or battery of tanks, which may, if desired, be provided with heating coils **22** for maintaining a temperature in the range of 130° to 210°F, and preferably above 170°F. The separator is supplied with an oil froth withdrawal line **23** adjacent the top and a sand tailings removal line having star valve **24** or any other suitable control discharge means at the bottom. Separator **18** also has an intermediate withdrawal line **26** through which a stream of middlings layer is removed in addition to that recycled through line **21**. The following example illustrates the process.

Example: The process is utilized to recover the oil from Athabasca tar sands containing by weight 10% bituminous matter and 89% mineral matter. 25% of the latter consists of particles having diameters less than 44 microns. On an hourly basis 100 pounds of the tar sands, 0.1 pound of concentrated caustic soda solution, 30 pounds of water and steam are fed into a conditioning drum and the mixture is heated to 180°F while being mulled.

The resulting pulp is passed through a screen and then to pulp box **17** as shown in Figure 3.1. Hot water at 190°F in an amount of 74 pounds and a middlings recycle stream in an amount of 200 pounds are passed through line **19** into the pulp box and the mixture continuously flushes into separator **18** where the temperature is maintained at 190°F. The pH of the middlings is held at 8.4 by the addition of caustic soda at the above specified rate. From the separator sand tailings are removed at a rate of 78 pounds per hour from the bottom and froth is removed from the top at a rate of 10 pounds per hour. The tailings are composed of 74% mineral matter, 25% water and 1% oil by weight. The froth is composed by weight of approximately 50% oil, 10% mineral matter and 40% water. The oil content corresponds to a recovery of 50% of oil in the original tar sands.

A stream of oil rich middlings in an amount of 112 pounds per hour and composed of 4 pounds of oil, 27 pounds of mineral matter and 81 pounds of water is withdrawn from separator **18** and is transferred to a scavenger zone where it is subjected to air flotation in a subaeration type air flotation cell **29**. This rate of transfer of the oil rich middlings from the separator, in combination with the previously specified rates, maintains the middlings at a viscosity of 1.5 cp and a density of 1.14. From the flotation cell additional oil froth is obtained in an amount of 7 pounds per hour and 104 pounds per hour of oil lean middlings are withdrawn and discarded. Use of the flotation cell increases the recovery of oil from the tar sands to an over-all value of 85%.

In related work, *E.W. Dobson; U.S. Patent 3,594,306; July 20, 1971; assigned to Great Canadian Oil Sands Limited, Canada* describes a process which involves the hot water process for extracting bitumen from tar sands products. A secondary recovery is conducted on the middlings from the separation cell to produce additional froth which usually is combined with the primary froth and treated. It has been found that the froth from the secondary recovery operation can be upgraded in bitumen content by gravity settling. By the process, secondary recovery froth is settled to produce an upper layer upgraded in bitumen content over the secondary froth. This upgraded bitumen layer is then added to the primary froth for further processing.

Coker Naphtha Diluent

E. Nagy; U.S. Patent 3,607,721; September 21, 1971; assigned to Cities Service Athabasca, Inc., Imperial Oil Limited, Atlantic Richfield Corporation and Royalite Oil Company Limited describes a method for the removal of solids and emulsified water from a bituminous froth resulting from the hot water treatment of Athabasca tar sands. The froth is diluted with a light hydrocarbon diluent such as coker naphtha and fed to a settling vessel where the solids and water droplets settle out in the form of a sludge constituting a concentrated stable emulsion containing most of the impurities. The substantially purified bituminous froth is withdrawn as overflow. The settled sludge is removed from the bottom of the vessel at rates of between 4 and 5% by volume of the diluted bituminous froth feed rate and centrifuged to recover its oil content. The following examples illustrate the process.

Example 1: Bituminous froth was mixed with coker naphtha in the ratio of 1 part froth to 0.807 part naphtha by weight and fed into the settling vessel which was maintained at room temperature (about 75°F) where it was retained for about 1 hour of residence time. The diluted feed to the settling vessel comprised 77.46% weight oil, (bitumen and naphtha), 18.31% weight water, and 4.23% solids. Product withdrawn as overflow from the settling vessel comprised 91.02% weight oil (bitumen and naphtha), 7.49% weight water, and 1.49% weight solids. This represents a recovery efficiency of 95.57% of the oil. Sludge comprised 12.24% weight oil, 78.62% weight water, and 9.14% weight solids.

Example 2: Bituminous froth was mixed with coker naphtha to obtain a diluted froth having a naphtha to bitumen ratio of 1.078 and fed to the settling vessel. The froth feed comprised 73.87% weight oil (naphtha and bitumen), 22.12% water and 4.01% weight solids. After similar treatment to that described above in Example 1, the product withdrawn as overflow comprised 92.68% weight oil, 5.77% weight water, and 1.55% weight solids. This represented a 96.13% recovery of available oil (naphtha and bitumen) in the feed. Sludge comprised 12.07% weight oil, 75.96% weight water and 11.97% weight solids.

Example 3: A sample of sludge processed according to Examples 1 and 2 was subjected to centrifuging at 1,000 rpm for 10 minutes. The resulting liquid product comprised about 98.7% weight oil, 0.7% weight water and 0.5% solids. The effective secondary recovery was about 67% weight of the oil in the sludge and results in an over-all combined effective recovery of at least 98%.

Gaseous Hydrocarbons as Flotation Aids

According to a process described by *C.W. Bowman and D.S. Montgomery; U.S. Patent 3,573,195; March 30, 1971; assigned to Cities Service, Athabasca, Inc., Imperial Oil Limited, Atlantic Richfield Corporation and Royalite Oil Company, Limited* bitumen is recovered from bituminous sand by forming a fluid slurry of bituminous sand, water and liquid hydrocarbon diluent having normally gaseous hydrocarbon material dissolved therein at a temperature less than about 110°F, preferably between 80° and 110°F.

The slurry is then introduced into a body of water maintained in a separation zone at a temperature greater than 150°F, usually between 150° and 200°F, so that the gaseous hydrocarbon comes out of the solution and aids in flotation of the bitumen to the top of the water while sand settles to the bottom for removal. Normally gaseous hydrocarbons including ethane, methane and propane may be used and are preferably used in amounts between about 0.5 and 2.0 standard cubic feet of gas per standard cubic foot of bitumen. The liquid hydrocarbon diluent preferably boils between 200° and 500°F and is used in amounts between about 1 and 10 weight percent based upon bitumen.

Referring to Figure 3.2, bituminous sand entering through conduit **12**, water entering through conduit **13** and hydrocarbon diluent containing dissolved normally gaseous hydrocarbons entering through conduit **14** are mixed in mixer **15** to form a slurry. The aqueous slurry thus formed may be passed from the mixer through conduit **16** to flood cell **17** in which additional flooding water entering as through conduit **18** may be mixed with the slurry to increase the water content thereof if desired.

From the flood cell, the flooded slurry is passed through conduit **19** to separating tank **21** containing a body of water into which the slurry is introduced. Bitumen rises to the top of the body of water in the separating tank and may be removed from the upper portion thereof in the form of a bituminous emulsion as through conduit **22** while sand settles to the bottom and may be removed as through conduit **23**.

As is conventional in such processes, water may be introduced through conduit **24** to wash sand prior to its removal and a side stream of water may be removed through conduit **26** and may, if desired, be recycled to the mixing step or treated for recovery of bitumen or solids therefrom.

Prior to introduction into the separating tank, the slurry is maintained at a temperature between about 80° and 110°F while the body of water in the separating tank is maintained at a temperature between about 150° and 100°F. Due to the higher temperature in the separating tank, the normally gaseous hydrocarbon material originally dissolved in the liquid hydrocarbon diluent introduced through conduit **14** is evolved as gas within the separating tank and aids in flotation of bitumen.

This gas may then be collected and removed from the separating tank as through conduit **27** and recycled together with makeup gas introduced as through conduit **28** to absorber **29** where it is absorbed into liquid hydrocarbon diluent introduced as through conduit **41**. The diluent containing absorbed normally gaseous hydrocarbon material may then be utilized in the process as described below.

Liquid hydrocarbon diluent rises together with the bitumen in the separating tank and may be recovered through conduit **22** along with the bituminous emulsion. This diluent may either be passed from the process as product along with the bitumen or may be recovered from the bitumen as by conventional fractionation and recycled to absorber **29** for reuse in the process.

FIGURE 3.2: ADDITION OF LIGHT HYDROCARBON GAS TO TAR SAND

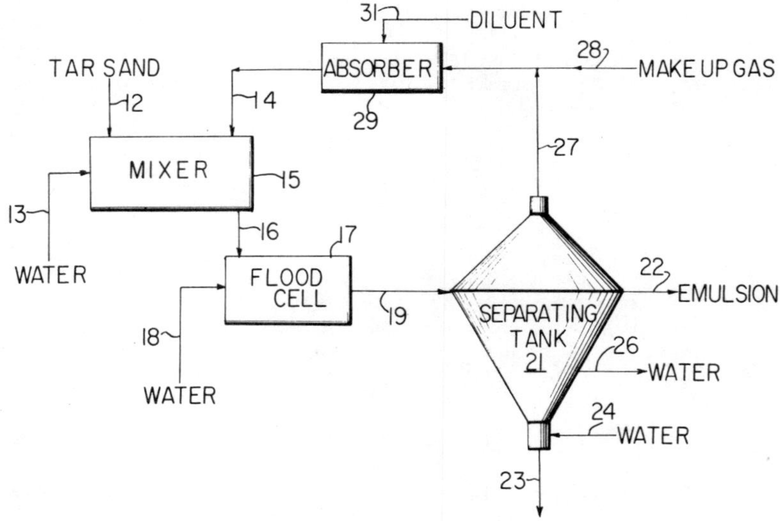

Source: C.W. Bowman and D.S. Montgomery; U.S. Patent 3,573,195; March 30, 1971

Example: In this example, the equipment and process described above are used in the recovery of bitumen from tar sand originally containing 11.75 weight percent bitumen, 3.40 weight percent water and 84.85 weight percent solids. This tar sand is mixed in the mixer at the rate of 120 pounds per hour with water and liquid hydrocarbon diluent. The liquid hydrocarbon diluent is naphtha having a boiling range between 200° and 400°F and which has dissolved in it one standard cubic foot of methane per standard cubic foot of bitumen in the tar sand. Sufficient water is mixed with the tar sand and diluent to form a slurry containing 48 weight percent solids, 45 weight percent water, 6.7 weight percent bitumen and 0.3 weight percent hydrocarbon diluent.

The slurry is then passed to the flood cell where additional water is added to raise the water content of the slurry to 60 weight percent and the flooded slurry is then transported to the separation tank. During its formation and until introduction into the separation tank, the slurry is maintained at a temperature of about 110°F while the body of water in the separation tank is maintained at a temperature of 160°F thereby resulting in release of methane gas within the separation tank.

The methane and air entrained and dissolved in the slurry during its formation act to float bitumen and liquid hydrocarbon diluent to the top of the body of water in the separation tank and a mixture containing 3 weight percent diluent, 72 weight percent bitumen, 5 weight percent solids and 20 weight percent water is withdrawn therefrom through the conduit **22**. Methane is recovered from the separation tank through the conduit **27** and is recycled to the absorber **29** along with fresh makeup methane. In the absorber, methane is absorbed by the naphtha introduced through the conduit **31** so that the naphtha introduced into the mixer through the conduit **14** contains one standard cubic foot of absorbed methane per standard cubic foot of bitumen in the tar sand introduced through the conduit **12**.

Sand tailings withdrawn through conduit **23** from the separation zone contain 75 weight percent solids, 0.3 weight percent bitumen and approximately 25 weight percent water. Sufficient water is removed through conduit **26** to maintain a steady emulsion-water interface at the top of the separating tank.

In related work *A.W. Hyndman; U.S. Patent 3,574,086; April 6, 1971; assigned to Cities Service Athabasca, Inc., Imperial Oil Limited, Atlantic Richfield Corporation and Royalite Oil Company, Limited* describes another process in which bitumen is recovered from bituminous sand containing the same by forming a slurry of bituminous sand and water in a gaseous hydrocarbon atmosphere.

The aqueous slurry, which contains entrained and dissolved gaseous hydrocarbons, is then introduced into a body of hot water so that bitumen rises to the top in the form of a bituminous emulsion while sand settles to the bottom. The gaseous hydrocarbon atmosphere is preferably maintained throughout the process and between about 0.05 and 5 weight percent hydrocarbon gas based upon bitumen content of the slurry is preferably contained in the slurry.

Controlled Amounts of Fine Sand in Intermediate Layer

C.W. Bowman and V.P. Kaminsky; U.S. Patent 3,619,406; November 9, 1971; assigned to Canada-Cities Service, Ltd., Imperial Oil Limited, Atlantic Richfield Corporation and Royalite Oil Company Limited describe an improvement in water separation processes for bitumen recovery from bituminous sand. A fluid slurry of bituminous sand is introduced into a separation zone containing water and bitumen is floated to the top as a froth for recovery, while sand is allowed to settle to the bottom for removal.

In this process controlled amounts of fine solids are maintained in an intermediate layer of water in the separation zone. Between 10 and 17 weight percent fines are preferably maintained in the intermediate water layer and such fines are mostly less than 20 microns in size. The desired concentration of fines in the water layer is maintained by recycling water from the water layer back to the separation zone or to an earlier stage of the process with removal of fines from the recycle stream in response to measurements of the amounts of fines in the water layer to maintain the desired concentration of fines in the water layer. Fresh water addition to the process is minimized by removing from the process only that water which is contained in the bituminous froth and the minimum amount of water needed to withdraw solids from the process.

Referring to Figure 3.3 bituminous sand entering through a conduit **12** may be mixed with water and steam in a mixer **14**. Water may enter through a conduit **16** and steam through a conduit **18**. In the mixer the bituminous sand is thoroughly moistened and mixed with the water to form a thick slurry. The mixer may take any suitable form such as a conventional rotating cylindrical tumbler having perforations for slurry to pass through and provision for rejecting oversized rocks and lumps of sand which cannot be broken down as through a conduit **20**.

Slurry from the mixer preferably containing between about 20 and 40 parts of water per 100 parts of tar sand passes through a conduit **22** where it may be further mixed with additional quantities of recycled water introduced through a conduit **24** and passed through the conduit **22** to a separation vessel **26**. Fresh makeup water, in addition to that added to the slurrying unit, may be required to maintain the level of the froth-water interface **49**.

The level of the interface is detected by a level control device **53** such as a Foxboro d/p cell which activates valve **54** and admits fresh water through conduit **56** and thence to conduit **22** where it is admixed to the slurry. In the separation vessel **26**, the slurry is allowed to settle in a body of water preferably maintained between about 130° and 170°F to form an upper froth layer **28**, an intermediate water layer **30** and a lower sand layer **32**. Bituminous froth is removed from the froth layer through a conduit **34** as a product of the process for recovery of bitumen therefrom while sand tailings are withdrawn through a conduit **36** for disposal.

FIGURE 3.3: CONTROL OF SOLIDS IN PROCESSING BITUMINOUS SAND

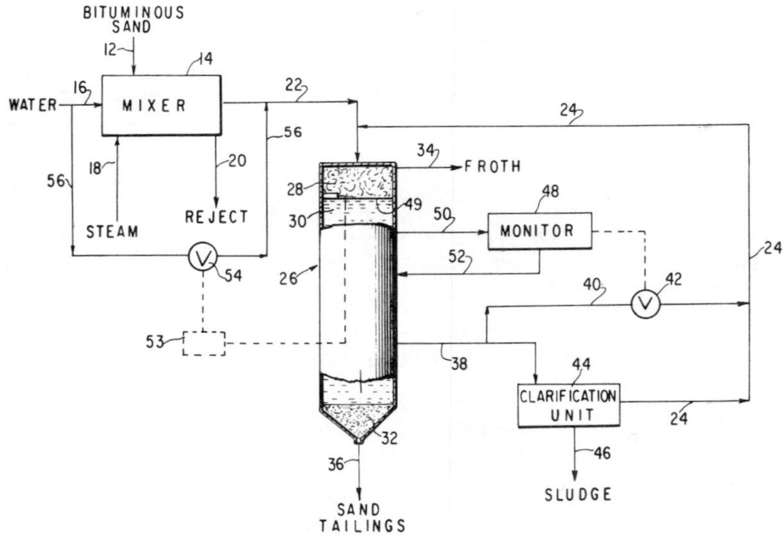

Source: C.W. Bowman and V.P. Kaminsky; U.S. Patent 3,619,406; November 9, 1971

Froth removed through the conduit **34** typically contains between about 65 and 75 weight percent bitumen, between about 15 and 32 weight percent water and between about 3 and 10 weight percent solids, mostly in the form of fines of less than 44 in micron size. Sand tailings withdrawn through the conduit **36** preferably contain between about 70 and 80 weight percent solids, between about 20 and 30 weight percent water and not more than about 1 weight percent bitumen. The bitumen is for the most part associated with the water and by reducing the water content of the tailings stream as much as possible, the amount of bitumen withdrawn with the tailings is considerably reduced from normal practice. The amount of water withdrawn with the tailings is preferably just enough to fill the voids in the sand.

In operating the process to minimize loss of bitumen from the process with wastewater and sand tailings, the amount of fresh water added in the slurrying step is only sufficient to provide the minimum needed for losses with froth sludge and tailings. Since the bituminous sand frequently contains between about 2 and 8 weight percent water to start with, it is preferred that no more than between about 20 and 40 parts of water per 100 parts of bituminous sand be introduced into the process. This water is used in forming the initial slurry, although some may be added at a later stage in the process, if desired or required for maintaining the level of the froth-water interface.

For best separation in the separation zone **26**, it is preferred that the slurry transported to the separation zone contain more than the relatively small amount of water preferably used in forming the slurry. Rather than use fresh water for this purpose, and thereby introduce unwanted additional water into the process, it is preferred that additional water be added as through the conduit **24** and that such additional water be obtained by recycling water from the intermediate water layer. Such additional water is preferably recycled in sufficient quantities so that the slurry as introduced into the separation vessel comprises between about 40 and 55 weight percent water.

In the separation zone, as normally operated, most of the solid particles larger than about 20 microns in size settle to the bottom and are removed with the sand tailings. Particles smaller than about 20 microns in size, however, do not settle as readily and tend to accumulate in the intermediate water layer. Typically, the solids in the intermediate water layer are about 90% smaller than 20 microns in size.

In normal operation, the bitumen particles rise to the surface of the separation zone where they coalesce to form a froth layer. The formation of the froth layer is in essence an inversion of an oil in water emulsion to a water in oil emulsion.

It is apparent that accumulation of large quantities of fines in the intermediate water layer would eventually result in such large concentration of fines in the separation zone that increasing amounts of fines would eventually be carried off with the froth product through the conduit **34**, thereby reducing the effectiveness of separation of bitumen from solids.

It has been found that as the concentration of fines in the intermediate water layer increases, the quality of the froth progressively deteriorates, and above a critical fines concentration emulsion inversion does not occur and no froth is formed. For practical purposes this renders the process inoperative.

To prevent undesirably large buildups of fines in the separation zone, while still restricting the amount of fresh water used in the process and wastewater removed from the process to a bare minimum, the intermediate water layer in the separation zone is treated for removal of controlled amounts of fines in order to maintain the fines content of the water layer at a desired level. This may be done either by direct treatment of water from the water layer for removal of fines followed by return of such water to the water layer, or in a preferred embodiment by recycling a stream of water from the water layer to an earlier stage in the process with treatment of at least a portion of the recycled water for removal of appropriate amounts of fines therefrom in response to indications of fines content of the water layer.

As shown in Figure 3.3, a stream of recycled water may be withdrawn from the water layer as through a conduit **38**. Some or all of this water may be recycled directly to the process as through a conduit **40**, valve **42** and the conduit **24** referred to above while the remainder, if any, may be treated in a clarification zone, such as a clarification unit **44** for removal of some or all of the fines therefrom before being returned to the process through the conduit **24**. Fines removed from the recycle stream in the clarification unit are removed in the form of sludge as through a conduit **46**.

Control of the amount of fines removed to maintain the desired fines content in the water layer may be accomplished in any suitable manner. As shown in Figure 3.3, a monitor **48** is used to monitor the fines content of a stream of water withdrawn from the water layer through a conduit **50** and returned thereto through a conduit **52**. The monitor may take any suitable form. For instance, a conventional density measuring device, such as a Dynatrol density cell (Automation Products, Inc.) may be used for this purpose.

The output signal from the monitor is then preferably used as indicated in the figure to control the setting of the valve **42** to vary the amount of water passed through the valve in response to the fines content of the water layer. Thus, if the fines content of the water layer rises above the desired level, the valve is at least partially closed to thereby divert more of the recycle stream through the clarification unit for removal of fines. Likewise, as the fines content of the water layer falls below the desired level, the valve may be opened to allow additional quantities of the recycle stream to bypass the clarification unit.

Fines removed from the recycle stream by the clarification unit are preferably those in the size range of between about 5 and 20 microns. If centrifuging rather than cycloning is used for this purpose, the size range of fines removed may be increased to cover those particles larger than about 1 or 2 microns, whereas if settling vessels, with or without flocculating agents, are used for this purpose, the smallest size particles removed are usually in the order of 8 microns. Whichever method of fines removal is used, the residence time of the water being treated in the clarification unit should be kept to a minimum, preferably less than 1 minute, and certainly less than 1 day. Excessive residence times result in de-aeration of bitumen so that bitumen settles with solids and is lost from the process with the sludge rather than being recycled with the recycle stream.

It has been found that the desirable amount of fines in the intermediate water layer is usually between 10 and 17 weight percent. The presence of more than 17 weight percent fines in the water layer substantially reduces the effectiveness of the process and increases the amount of fines recovered with froth while less than 10 weight percent fines in the water results in somewhat reduced recoveries of bitumen.

It has been suggested previously that excess water be used in the process to keep fines from accumulating in the water layer in such amounts as to render the process inoperable. This is usually accomplished by adding additional water to the process prior to the separation step, withdrawing water from the intermediate water layer, recovering additional bituminous froth therefrom in a secondary recovery process and then discarding the water from the secondary recovery process.

Unfortunately, such secondary recovery processes yield a very poor secondary froth of comparatively low bitumen content and high water and solids content. This, of course, lowers the over-all quality of the bituminous froth recovered as a product of the process and increases the costs of treating such froth for removal of water and solids therefrom. In addition, such secondary recovery processes usually discard treated water as waste material. This wastewater contains substantial quantities of fines and disposal is difficult because of pollution of the rivers or lakes into which it may be discharged.

When practicing the process, a high quality of bituminous froth is maintained because no secondary froth is involved. Waste material is kept to a minimum by limiting the use of fresh water and recycling water from which fines have been removed to the process. Water removed with the sand tailings through the conduit **36** and with the sludge from the clarification unit through the conduit **46** is kept to the absolute minimum necessary. Preferably, the total water content of these waste streams is just sufficient to fill the voids in the solids contained therein and is usually restricted to between about 20 and 30 weight percent of the combined sand tailings and sludge streams.

Calcium Carbonate as pH Controller

C.W. Bowman; U.S. Patent 3,623,971; November 30, 1971; assigned to Cities Service, Ltd., Canada, Imperial Oil Limited, Atlantic Richfield Corporation and Royalite Oil Company, Limited has found that addition of calcium carbonate to bituminous sand is extremely beneficial in controlling the pH of the sand both before and during treatment for recovery of bitumen and also in increasing the processability of the bituminous sand. The use of calcium carbonate is particularly beneficial in preventing deterioration of bituminous sands exposed to the atmosphere before processing for recovery of bitumen.

In the process calcium carbonate is usually added to aid in extraction processing of bituminous sands in amounts between about 0.01 to 1.0 weight percent based on total weight of bituminous sand with the use of amounts in the range between about 0.02 and 0.1 weight percent being preferred. The use of calcium carbonate, preferably in powder form to accelerate the reaction, will normally maintain or adjust the pH of the bituminous sand to within the range of between about 7 to 8.5. If slightly higher pH values are desired, conventional pH controlling additives such as sodium hydroxide may be added during processing of the bituminous sand.

Example 1: Overburden was removed from portions of a typical deposit of bituminous sand suitable for feed to a commercial extraction process and calcium carbonate in the form of crushed limestone was spread over a portion of the freshly uncovered deposit.

Samples were taken of material at a depth of 5 feet to obtain information concerning the pH of the deposit immediately following removal of overburden and after various periods of time. It was found that an untreated portion of the deposit deteriorated from an initial pH of 7.25 to a pH of 6.0 after only 4 months of exposure to the atmosphere and to a pH of 4.75 after 2 years. By contrast, the portion of the deposit treated with calcium carbonate had a pH of 7.77 after 1 year. In this test, calcium carbonate was applied at a rate of approximately 0.1 pound per square foot of surface area of exposed bituminous sand.

Example 2: To illustrate the effect of addition of calcium carbonate on the pH of bituminous sand, varying amounts of calcium carbonate were added to samples of bituminous sand which had been exposed to the atmosphere for long periods of time and had a pH of only 3.4. Addition of varying amounts of calcium carbonate increased the pH of the bituminous sand as follows.

Weight Percent $CaCO_3$ Added	pH
0	3.4
0.05	6.6
0.07	6.9
0.50	7.6

Example 3: To illustrate the advantages of adding calcium carbonate to slurries of bituminous sand being treated by the hot water process for recovery of bitumen, two different samples of a bituminous sand comprising 11.36 weight percent bitumen, 4.43 weight percent water and 84.21 weight percent solids were treated by a bench scale hot water type process for recovery of bitumen. The bituminous sand had an initial pH of 4.5.

In these tests, about 500 grams of each sample of bituminous sand was slurried with 150 milliliters of water at 180°F for about 10 minutes and simultaneously aerated. Next the slurry was flooded with water at a temperature of about 180°F and stirred, without aeration, for about 10 minutes to separate the bituminous froth formed from the water-sand slurry. This froth was then removed and analyzed.

For the first run, no additives were used. Results of this run included a recovery in the froth of 41% of the bitumen present in the original bituminous sand. The froth contained 69.2 pounds of solids per 100 pounds of bitumen and 56.3 pounds of water per 100 pounds of bitumen. In the second run, 0.68 weight percent calcium carbonate was added to the slurry, raising the pH of the slurry from 4.5 to 7.2. In this run, 78.9% of the bitumen was recovered in the froth and the froth contained only 29.8 pounds of solids per 100 pounds of bitumen and 48.6 pounds of water per 100 pounds of bitumen.

Example 4: To further illustrate the advantages of adding calcium carbonate in hot water type processes, two additional runs were made using the process described in Example 3. For these runs, the bituminous sand used comprised 11.53 weight percent bitumen, 3.37 weight percent water and 84.75 weight percent solids and had a pH of 8.25.

In the first run, no additives were used and 66.4% of the bitumen was recovered in the froth. The froth also contained 14.3 pounds of solids per 100 pounds of bitumen and 36.8 pounds of water per 100 pounds of bitumen.

On the second run, 0.07 weight percent calcium carbonate was added to the slurry and bitumen recovery increased to 72.7% with the froth containing only 10.7 pounds of solids per 100 pounds of bitumen and 34.2 pounds of water per 100 pounds of bitumen.

From the above examples it can be seen that the addition of calcium carbonate increased recovery of oil even when the pH was at the reasonably good level of 8.25 to start with. When the initial pH was low, as in Example 3, addition of calcium carbonate resulted in recovering almost twice as much bitumen than where the calcium carbonate was not used.

Reduction of Solids Using Secondary Froth Wash Technique

V.P. Kaminsky; U.S. Patent 3,738,930; June 12, 1973; assigned to Canada-Cities Service, Ltd., Imperial Oil Limited, Atlantic Richfield Canada, Ltd., and Gulf Oil Canada Limited describes a process for recovering bitumen from a secondary-separator froth stream produced from the aeration of a primary separator middlings stream during the processing of Athabasca tar sands. The process comprises washing the secondary separator froth stream with fresh hot water.

The secondary separator froth is intimately contacted with hot water to cause deaeration of the froth and separation of the entrained water and solids from the secondary separator froth. The washed bitumen is separated from the water and solids as product and may be recycled to combine with the primary separator froth for the further recovery of bitumen. It is preferred, when utilizing the process that the washing step utilize fresh hot water rates between about 1 and 15 pounds of water per pound of solid in the secondary separator froth. It is further preferred that there be intimate contacting of the fresh hot water with the secondary separator froth. This is best accomplished by stirring. Provisions for the addition of surfactant to the fresh hot water stream for further separational improvement may also be included.

The process obviates the utilization of a cyclone or centrifuge froth cleanup operation in order to reduce the amount of solids in a secondary froth to a concentration which becomes economically feasible. In the process a separation of water and solids from the bitumen in a secondary separator froth is accomplished by froth water washing equipment. The specific operation of the process may be more readily understood by referral to Figure 3.4 in which a primary-secondary washing circuit is described where raw tar sand **18** is introduced with steam **20**, optional chemical addition **22** and slurry water **24** into a tumbler **28**. Flood water **26** is added to the slurry **30** produced in the tumbler. The flooded slurry is passed through a screen **34** and enters the separation cell **42** through conduit **44** and a dispersing device **40**.

The water-bitumen-sand mixture introduced into the separation cell separates into a bitumen enriched primary-separator froth **14** at the top of the cell **42** and the solid materials **32** accumulate through settling at the bottom of the separation cell and are removed as sand tailings **46** by pump **48**. In order to minimize the bitumen losses to the sand tailings, the amount of water removed with the sand tailings must be minimized.

Since the aggregate amount of water removed from the separation cell by way of the primary-separator froth and sand tailings is less than the total amount introduced into the separation cell as slurry, the excess water must be removed through some other port. This is accomplished by withdrawing a middlings stream **50** by pump **72** and reintroducing the middlings stream into secondary separation cell **82** via conduit **54**.

A dispersing device **56** located in the secondary separation cell causes bitumen from the middlings to contact self-induced or pressured air **66** and form a buoyant voluminous secondary froth **90**. Most of the solids settle to the bottom of the secondary cell and are removed by the effluent pump **58** through conduit **59**. Because of excessive aeration in the secondary cell, the secondary separator froth contains large amounts of water and solid matter which must be removed. These contaminants are removed from the secondary separator froth by overflowing the froth into a froth washer **126**.

The secondary separator froth is contacted by a hot water spray **104** which may contain chemicals, for example, a surfactant comprising a sulfonated petroleum fraction. The spray of fresh hot water deaerates the secondary separator froth and becomes a carrier for the collapsed froth that is introduced within the froth washer.

The collapsed froth and spray water mixture is passed into the baffle mixing chamber **105** of the froth chamber to allow intimate contacting between the bitumen and the fresh hot water so as to accomplish a separation of the bitumen from the solids-water mixture.

FIGURE 3.4: SECONDARY FROTH WASH TO REDUCE SOLIDS

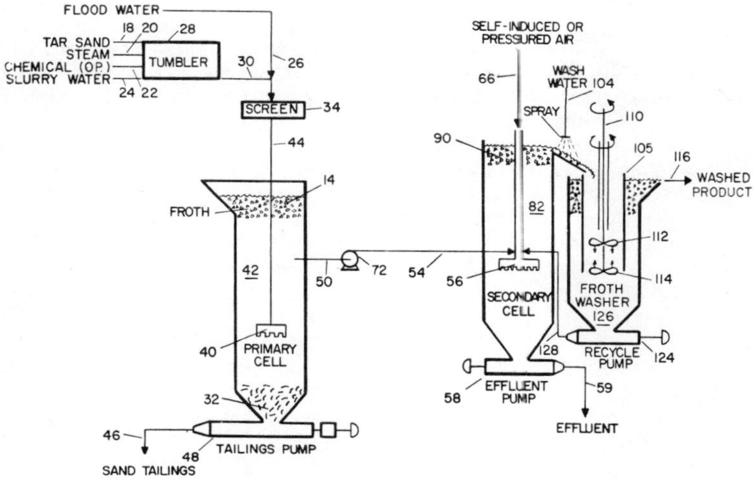

Source: V.P. Kaminsky; U.S. Patent 3,738,930; June 12, 1973

This intimate contacting may be enhanced by the imposition of two opposite throw marine type propellers. The upper propeller **112** forces the bitumen water mixture downward where it meets an upward moving stream produced by the lower up-throw type marine propeller **114**. Propellers **112** and **114** may be rotated by shaft **110** powered by an auxiliary motor, not depicted herein. The colliding streams impart a washing action which releases solids entrained in and with the bitumen flakes. The down-throw vector of the upper propeller **112** is designed to be larger than the lower up-throw propeller, thereby assuring a positive feed through-flow from the agitated zone and into the quiescent settling zone outside the mixing zone.

The function of the draft tube **105**, contained within the froth water, is to insure a positive directional flow past both propellers and prevent short circuiting of the bitumen into the settling zone. The froth-hot water mixture is confined for maximum bitumen-water contact and provides a quiescent settling zone below the propellers and outside the draft tube for the inversion of bitumen to reform into a more concentrated froth.

The washer effluent **128** may be recycled through recycle pump **124** to the secondary cell **82** for scavenging any bitumen losses or the washer effluent may be returned to the primary extraction cell as flood water. If chemicals such as sodium hydroxide or sodium tripolyphosphate and other surfactants are added to the washwater **104**, the effluent may be recycled to the tumbler as slurry water for the residual chemicals to have a beneficial effect on the primary extraction. Washed bitumen product **116** may be combined with the primary separator froth for further treatment.

Example 1: A washer similar to that shown in Figure 3.4 consisting of a 7-inch by 7-inch by 16-inch high vessel with a 60° pyramidal cone at the bottom was utilized as a froth washer. Secondary froth was introduced with the aid of a hot water spray to a 4-inch diameter, 14-inch long draft tube within the froth washer. Within the draft tube were positioned two opposed 3-inch flat vane marine type propellers. The vanes of the upper (down-throw) propeller were set at 18° from the horizontal and the lower (up-throw) propeller was set at 15° from the horizontal. A 3-inch separation between the propellers was accomplished with the lower propeller being positioned approximately 2 inches above the lower end of the draft tube. The draft tube confined the washing action and separated the agitated zone and the quiescent settling area below the propellers.

Based on an average of 25 pounds of solids per 100 pounds of bitumen in the secondary froth, the effect of wash water upon the deaeration and separation thereof is shown in the following table. It can be readily seen that the contaminants in the secondary froth may be reduced to less than 10 pounds of solids and approximately 50 pounds of water per 100 pounds of bitumen by the washing process with the use of less than 15 pounds of fresh water per pound of solids in the secondary froth. Losses to the recycle stream are reclaimed in a scavenging step.

Effects of Wash Water Rates on Bitumen Loss

Water rate (lbs./lb. of solid in froth)	Washed froth quality at— 100 s/o	100 w/o	Bitumen lost to recycle stream (percent of total)
3.6	9.7	53.8	7.4
7.4	8.7	42.9	8.1
14.0	7.3	46.7	15.9

Based on the results of the above table, an experiment was conducted using 7.4 pounds of fresh water per pound of solids in a secondary froth washing scheme. The table below depicts the results of this experiment at various temperatures and shows that higher washing temperatures produce the best quality froths although the bitumen losses to the recycle stream increased with an increase in washing temperature. Washing temperatures of approximately 180°F are preferred as a compromise between good froth quality and low bitumen losses to the recycle stream.

Effect of Temperature in Froth Quality and Bitumen Losses

Temperature (° F.)	Froth quality at— 100 s/o	100 w/o	Bitumen lost to recycle stream (percent of total)
100	13.4	69.6	7.4
150	11.6	62.4	7.1
180	8.7	42.9	8.1
200	5.0	37.5	11.6

Example 2: In order to determine the degree of agitation and mixing required with the equipment identified in Example 1 at the optimum temperature of 180°F wash temperature and 7.4 pounds of water per pound of solids in the secondary froth, various impeller velocities were tested in order to determine an optimum recovery of bitumen from the froth and an optimum loss of bitumen to the recycle stream. The table below depicts the quantities shown. Therefore, it can be seen that higher impeller speeds produce better quality froths and increase bitumen losses to the recycle stream. The bitumen losses to the recycle stream are, however, recovered by reaeration after recycle to the secondary separator cell.

Effect of Agitation on Bitumen Loss

Impeller velocity (ft./min.)	Froth quality at— 100 s/o	100 w/o	Bitumen lost to recycle stream (percent of total)
235	13.8	152.0	4.8
470	12.8	126.3	6.5
705	9.6	54.7	7.4
940	8.0	50.1	7.3

Generally, from the experiments, it can be seen that higher impeller speeds, high wash water rates and higher wash water temperatures yield the best quality froths. However, these same conditions also produce a very finely dispersed bitumen in the water emulsion and a potential for high bitumen losses to the recycle streams if the washer effluent is discarded. In practice, it has been preferred that the washer effluent be recycled to the secondary vessel which then also serves as a scavenger cell for recovery of the potential bitumen losses.

Addition of Tar Sand Oil to Slurry

S. Andrassy; U.S. Patent 3,208,930; September 28, 1965 describes a process for the separation of oil from tar sands which comprises forming a mixture of tar sands, oil of the same consistency as the oil in the tar sands and water in an amount sufficient to form a water layer on top of the tar sands, needling the water and oil-tar sand mixture and heating the water-oil-tar sand mixture to the boiling point of water thereby causing oil trapped in the tar sands to separate from the tar sands and float to the top of the water layer, maintaining the temperature at 190° to 200°F so as to allow the separated oil to accumulate on the top of the water layer, and recovering the oil from the water layer.

One of the essential features in the process is the addition to the tar sands of oil of the same constituency as the oil trapped in the tar sands. It has been found that the addition of this oil, preferably oil previously extracted from the tar sand itself has a profound effect upon the oils present in the tar sands and allows for virtually complete stripping of the oil trapped in the tar sands so as to leave clean white tailings of sand behind. The best results are obtained when 15 to 25% of oil based on the weight of the tar sands is added to the tar sands prior to heating the mixture. When less than 15% oil is added, the process takes longer, and when more than 25% oil is added, no particular advantages were found.

It is believed that one of the primary reasons for the efficacy of this process is that the added oil has an affinity for the entrapped oil thereby helping to strip the entrapped oil from the sands. The increased volume of the oil, which changes the proportions of the basic material, may also be a contributing factor. During the heating process, the specific gravity of the crude oil trapped in the sand becomes less than that of the water itself and consequently floats to the surface of the water where it may be easily removed.

Water at 60°F has a specific gravity of 0.9990 and at 212°F it has a specific gravity of 0.9584. The crude oil trapped in the sands generally has a specific gravity at 60°F of 1.002 to 1.02 and has a specific gravity at 212°F of from 0.926 to 0.944. Thus, the addition of excess oil from a previous extraction plus the increase in temperature to near the boiling point of water causes the specific gravity of the oil trapped in the sands to decrease below that of water and consequently permits it to rise to the surface. The following examples illustrate the process.

Example 1: One hundred parts by weight of Athabasca oil sands were mixed with 20 parts by weight of oil previously extracted from oil sands and with 100 parts by weight of water. The mixture was heated to the boiling point of water and held at 212°F for 2 minutes. The heating was stopped and the oil which had floated to the surface of the water was skimmed off. 46 parts by weight of the oil and water were skimmed off. Upon drying, 37 parts by weight were obtained. This yielded 17 parts by weight of the recovered oil on a dry basis. The Athabasca tar sand deposits have an oil content of 10 to 17% (D.S. Pasternak, *Chem. Eng. Progress,* Volume 56, April 1960). On the basis of the tar sands tested, the recovery appears practically theoretical.

Example 2: Twenty-five parts of oil recovered from a previous batch of tar sands were mixed with 300 parts of water and heated until the oil floated to the surface. The mixture was then cooled to slightly below 190°F and 100 parts of tar sands were added to the mixture. The mixture was then needled and heated to 212°F until the oil began to separate from the tar sands and rise to the surface. The temperature was then decreased to between 190° and 200°F and maintained until the oil accumulated on the surface of the water. The separation of the oil from the tar sands took somewhat longer than usual and was not as effective since some oil still remained in the tar sands. The temperature was decreased and the free oil permitted to settle to the bottom of the reaction chamber. The mixture was then reheated with needling in the manner previously described. A clean separation of the oil from the tar sands was obtained.

Water Washing and the Use of Fluid Bed Combustion Chamber

A process described by *C.E. Hemminger; U.S. Patent 2,940,919; June 14, 1960; assigned to Esso Research and Engineering Company* involves recovering oil from oil-bearing solids, particularly tar sands, by the use of hot liquid extraction procedures. Experimental work has indicated that when the tar sands are subjected to a hot aqueous wash, e.g., 150° to 200°F, oil is liberated from the oil sands and passes into the aqueous phase and it is ultimately separated and recovered as product.

In the process, spent washed tar sands are passed from the washing zone preferably after a drying step, to a fluidized solids combustion zone, e.g., a fluid bed burner. The tar sands as mined generally range from 10 to 200 mesh in size and thus are naturally suited to fluidization. By utilizing a fluidized combustion zone as opposed to other types of furnaces or burners, it is possible to effectively burn the less than 2 weight percent residual bitumen in the sand, thus liberating thermal energy.

Such fluidized combustion zones have been found to burn materials having carbonaceous contents of as low as 0.1 to 0.7 weight percent. The heat liberated from the fluidized combustion zone is then utilized to supply the thermal requirements of the over-all system, e.g., heating the wash medium, effecting drying of spent sands, preheating fresh sand feed, etc. Additionally, it can be converted into mechanical or electrical energy for the mining and transportation of the sand as well as providing power for the washing system itself.

In a process described by *L. Clark; U.S. Patent 3,052,621; September 4, 1962; assigned to Cities Service Research and Development Company* crude hydrocarbon oil is separated from bituminous sand containing the same by the process which comprises mixing the sand with sufficient water to form a slurry, introducing the slurry into the lower portion of a mixing tower, and adding additional water to the slurry in the lower portion of the mixing tower. The slurry, together with the added water, is then flowed upwardly through the mixing tower and hydrocarbon diluent is introduced at an intermediate point in the mixing tower.

The mixture of crude oil, sand, hydrocarbon diluent and water then passes to the upper portion of a sand separation tower in which it is allowed to flow downwardly over a series of downwardly slanted baffles. Due to its lower specific gravity, oil tends to collect under the baffles and an oil-rich mixture of oil and water is drawn off under the upper portions of the baffles.

This oil-rich mixture may be treated by conventional methods such as flotation for separation of oil and water. The remainder of the mixture in the sand separation tower preferably continues downwardly through the tower and is contacted in the lower portion of the tower with a countercurrent flow of water introduced into the lower portion of the tower. This water serves to strip the sand and stripped sand may be withdrawn from the bottom of the tower. Slanted baffles similar to those in the upper portion of the sand separation tower are preferably contained also in the lower and intermediate portions of the tower and stripping water containing small amounts of oil may be withdrawn from under the baffles at the lower part of the tower while water containing slightly greater quantities of oil may be withdrawn from under baffles in the intermediate part of the tower.

The oil-rich mixture of oil and water from the upper portion of the sand separation tower is preferably passed to the upper portion of a surge tower located at a slightly lower elevation than the separation tower. The mixtures of oil and water withdrawn from the intermediate and lower portions of the separation tower may then be passed to an intermediate portion of the surge tower. The use of a surge tower in this manner serves to minimize the effects of variations in flow rates, etc., and to insure a uniform pressure drop in removing liquid streams from the separation tower. By maintaining a uniform, relatively low pressure drop across the valves or outlets, the possibility of forming undesirable emulsions is substantially reduced.

The lower portion of the surge tower preferably contains slanting baffles similar to those used in the sand separation tower and oil tends to become trapped under these baffles as the oil-water mixtures from the intermediate and lower portions of the sand separation tower move towards the bottom of the surge tower. Such oil may be withdrawn from under the upper portions of these baffles and allowed to pass through a bypass conduit to the upper portion of the surge tower. A product mixture of oil and water may then be withdrawn from the upper portion of the surge tower for recovery of oil by suitable processes such as flotation while a stream of water is withdrawn from the lower portion of the surge tower. Part or all of this water is preferably recycled for forming the original slurry with the bituminous sand as described above.

Thermal Dehydration of Emulsion

R.A. Given and M.P. du Plessis; U.S. Patent 3,338,814; August 29, 1967; assigned to Imperial Oil Limited, Canada, Atlantic Richfield Company and Royalite Oil Company Limited, Canada describe a process where the separation of oil from the emulsion is effected by thermally dehydrating the emulsion to remove water and applying specific gravity separation to the dehydrated oil to divide the oil into portions having a high and a low solids concentration to permit processing of each portion separately.

The process comprises thermally dehydrating the bituminous emulsion under elevated temperature and pressure, dividing the dehydrated emulsion by specific gravity separation into a first fraction having a concentration of mineral solids substantially less than the first fraction, and separately treating each fraction to recover a hydrocarbon product substantially free of mineral solids and residue. The reaction conditions, the hydrocarbon products, or the residue differ for each of the separate streams which are treated.

The dehydration temperature and pressure can vary over wide ranges such as temperatures of from 225° to 550°F and pressures from 4 to 1,000 psig. Preferably, the temperature should be from 350° to 450°F and the pressure from 100 to 300 psig. The dehydrated emulsion can contain from 0 to 5% water and preferably from 0 to 2% water. Additionally, the dehydrated emulsion contains from 5 to 40% solids, these values generally being from 5 to 25% solids and from 79 to 94% bitumen.

The dehydrator can be any suitable vessel for separating the vaporized water from the emulsion under the specified temperatures and pressures. The heat removed from the dehydrator in the vapor overhead is utilized by direct use of the steam in the extraction process, or to produce clean steam by using heat exchangers. In this regard the dehydrator conditions of pressure and temperature can be set, within the limits specified above, so as to produce steam of high enough temperature and pressure to be suitable for use elsewhere.

After dehydration, the mixture of bitumen, mineral solids and small quantities of water is subjected to specific gravity separation to divide the mixture into a portion having a high solids concentration and a portion having a solids concentration substantially less than the first portion. Since the mineral solids contain large amounts of fines, it is extremely difficult to make a good specific gravity separation between oil and solids. Preferably, the concentration of mineral solids of the separated streams or portions of the dehydrated emulsion is at least 10% higher in the high solids portion as compared to the low solids portion and it is particularly desirable that the difference in concentrations of mineral solids in the different portions be at least 20%.

Thus, with a minimum differential of 10% solids, one portion of the dehydrated mixture can contain from 35 to 70% solids, 29 to 64% bitumen, and no more than 4% water; whereas the second can contain from 1 to 10% solids, 97 to 89% bitumen and no more than 4% water. In the specific gravity separation a larger concentration of water will be found in the portion having the lower solids concentration. The specific gravity separation of the dehydrated mixture is preferably effected at elevated temperatures such as those of from 300° to 700°F and preferably from 400° to 700°F, so that the bitumen viscosity is greatly reduced.

Cycloning is a preferred method for achieving specific gravity separation in the dehydrated mixture of bitumen and mineral solids in order to effect the desired separation. In cycloning the dehydrated mixture, conventional hydrocyclones and procedures can be employed. Two stage cycloning can be employed when it is desirable to do so, such as when the mineral solids in the dehydrated mixture encompass a large range of particle size. Two stage cycloning is accomplished by first feeding the dehydrated mixture into a large diameter cyclone or cyclones operating in parallel. The ensuing centrifugal force applied to the mixture in the cyclone(s) causes a small portion thereof, containing a relatively high concentration of mineral solids, to underflow from the cyclone(s).

The mineral solids in the underflow are primarily from the larger sized mineral particles in the feed. The remainder of the mixture, comprising the bulk of the fluid in the mixture and a relatively low concentration of mineral solids, overflows from the cyclone(s). The overflow from the large diameter cyclone(s) is fed into smaller diameter cyclones operating in parallel. This feed is separated into overflow and underflow streams which exit from the cyclones.

The underflow stream comprises a relatively small portion of the feed, but it contains a relatively high concentration of mineral solids. The overflow from the small diameter cyclones contains the bulk of the hydrocarbon in the feed stream to the large diameter cyclones, but much of the mineral solids contained in the feed stream have been removed in the underflow streams effluing from the large and small diameter cyclones. The streams underflowing the smaller diameter cyclones and larger diameter cyclones are combined and processed to recover the hydrocarbon value.

Dehydration Using Heated Dry Bitumen

A process described by *S. Lupul; U.S. Patent 3,796,652; March 12, 1974; assigned to Canada-Cities Service, Ltd., Imperial Oil Limited, Atlantic Richfield Canada, Ltd., and Gulf Oil Canada Limited* is concerned with that type of thermal dehydration process where the bitumen froth is heated, to vaporize contained water, and is then flashed to separate the water vapor from the bitumen. The process has been developed with the objects of: (1) avoiding the use of heat exchangers; (2) minimizing foaming and keeping it within predictable limits; and (3) keeping equipment costs low.

In the process, dry bitumen is used as the vehicle for heating the froth. The bitumen is preheated to at least 300°F, preferably 600° to 700°F. It is mixed with the froth in a cocurrent flow mixer. The bitumen temperature, the amount of bitumen used, and the mixer residence time are selected having regard to the requirement that the desired amount of vaporization must be completed within the mixer. The mixer product is flashed in a flash zone of sufficient volume and reduced pressure to produce water as overhead product and dry or substantially dry bitumen as the bottom product. Part of this bottom product may be recycled to the preheating step.

The mixer itself may be a downflow vessel having staggered, vertically spaced plates which blank off the vessel cross section to restrict the flow area to about 20%. Alternatively, the mixer may be of the cyclone type.

The process is operated so that substantially all water vaporization is carried out in the mixer. This is achieved by maintaining the temperature (Tm) of the mixer product stream about the same as the temperature (Tf) of the flash vessel bitumen stream. Preferably the temperature difference is maintained at less than 10°F. One way of doing this involves feeding the hot dry bitumen to the mixer at a constant rate and temperature, monitoring

Tm and Tf, and manipulating the rate at which froth is admitted to the mixer to maintain Tm equal to Tf.

The process is characterized by several advantages. By heating dry bitumen instead of froth, erosion of furnace tubes is reduced because of lower fluid velocities. The capital cost of the mixer and flash vessel is less than that of the equipment used in prior art processes. By carrying out vaporization in the mixer and maintaining Tm and Tf equal, it is found that the amount of foam generated in the flash vessel is reduced and its behavior within the vessel is reasonably predictable; this latter characteristic permits one to select the smallest possible flash vessel for a given feed stock and operating conditions. Finally, the bitumen product contains only a small amount of water and this content is reasonably consistent over a prolonged period of operation. The process is illustrated by the following example.

Example: An Athabasca tar sand bitumen froth feedstock, comprising 30% by weight water, 10% solids and 60% bitumen, was fed, together with a hot, dry bitumen stream comprising ½% by weight water, 15% solids and the balance bitumen, into a mixer. The relevant flow rates, temperatures and analyses were as follows for a run of 20 hours.

Flow rates (lbs./hr.)			Temperature, ° F.					Analyses (percent water)		
Froth	Hot, dry bitumen	Water	Froth	Hot, dry bitumen	Mixture	Bitumen product	Water	Froth	Bitumen product	Water
1,400	3,500	490	150	610	305-310	300	290	33	0.5	95
640	2,000	220	175	600	300-310	300	287	33	0.5	95

The mixer comprised a series of 6 cyclone mixers. Each cyclone had a cylindrical top section, provided with a tangential inlet, and a conical bottom section. The top section had a diameter of 4 inches and length of 6 inches. The bottom section reduced over a length of 4 inches to an outlet diameter of 1 inch. The cyclones were connected by a 1-inch diameter pipe. The inlet pressure to the first cyclone was 55 psi and the outlet pressure from the last cyclone was 15 psi. The temperature of the product stream was maintained at about 300°F.

The mixture was fed directly into a flash drum. This vessel was cylindrical in shape and had an interior diameter of 8 inches and a length of 38 feet. Its feed inlet was located 10 feet from its top. The drum was operated to maintain a fluid depth of about 20 feet, a foam disengagement zone about 8 feet in length, and a pressure of 4 psi. The flow rates, temperatures and compositions of the product streams from the drum are given in the above table.

As shown in the table, the process was operated to maintain the mixer product stream at the same temperature as the flash drum bottom product. This was done by manipulating the rate of froth feed to the mixer. It was found that the levels of the various layers within the drum remained substantially constant under these conditions.

This work has shown that the process must be operated with a recycle oil temperature between 300° and 700°F to vaporize the water at an acceptable rate. Preferably, it should be operated between 600° and 700°F. Below 600°F the material recirculation rate becomes excessive; above 700°F cracking of the bitumen begins to occur.

Micellar Dispersions for In Situ Treatment

J.T. Kelly and F.H. Poettman; U.S. Patent 3,648,771; March 14, 1972; assigned to Marathon Oil Company have found that oil from subsurface tar sands having an injection means in fluid communication with a production means is recovered by injecting an oil-external micellar dispersion at a temperature above 100°F into the tar sands and displacing it toward the production means to recover hydrocarbon.

The micellar dispersion can be preceded by a slug of hot water, (e.g., above 100°F) and the water can have a pH greater than 7. Also, the water within the micellar dispersion can have a pH of 7 to 14, and preferably the dispersion is at a temperature greater than 150°F. The oil-external micellar dispersion is comprised of hydrocarbon, surfactant, and aqueous medium. Optionally, alcohol and/or electrolyte can be incorporated. Examples of volume amounts include 4 to 50% hydrocarbon, 10 to 90% aqueous medium, at least 4% surfactant, 0.01 to 20% cosurfactant, and 0.001 to 5% by weight of electrolyte. In addition, the dispersion can contain other additives such as corrosion inhibiting agents, bactericides, sequestering agents, etc.

Example: This example is presented to show specific examples of micellar dispersions useful with this process. These examples are prepared at room temperature with minimal agitation. The compositions of the micellar dispersions are indicated below.

	Surfactant		Hydrocarbon		Aqueous medium		Cosurfactant	
Sample	Percent	Type	Percent	Type	Percent	Type	Ml. 100 ml.	Type
A	5.2	Ammonium petroleum sulfonate (average EqW=440, 81% active).	24.75	Crude oil	70	60% Henry plant water.[3] 40% Palestine water.[4]	0.08 0.25	n-Amyl alcohol. Isopropanol.
B	10	"Pyronate 50"[1]	20	do	70	60% Henry plant water. 40% Palestine water.	3.25	n-Hexanol.
C	10	"Petronate L"[2]	20	Straight-run gasoline.	70	60% Henry plant water. 40% Palestine water.	4.25	n-Amyl alcohol.
D	20	"Duponol WAQE"	10	do	70	Distilled water	14	i-Amyl alcohol.
E	10	"Energetic W-100"	20	do	70	do	8.5	Do.
F	10	"Triton X-100"	20	do	70	do	5.5	Do.
G	20	"Arquad 12-50"	10	do	70	do	17.0	Do.
H	16.6	Sodium petroleum sulfonate (avg. EqW=465, 62% active).	16.6	do	66.6	do	1.3	Isopropanol.
I	10	Ammonium petroleum sulfonate (avg. EqW=440, 81% active).	5	Crude oil	85	60% Palestine water. 40% Henry plant water.	2.5	n-Amyl alcohol.
J	7.4	Sulfonate defined in Example H	49.6	Straight-run gasoline.	43	Water containing 500 p.p.m. dissolved solids.		
K	11.6	do	60.6	do	24	Water defined in Example J.		3.8% isopropanol.
L	12 6	do	51.0	Pentane	33	do		3.4% isopropanol.

[1] Pyronate 50, a sodium petroleum sulfonate, average equivalent weight 350. Sold by Sonneborn Chemical Co., 300 Park Avenue South, New York, New York, 10010.

[2] Petronate L, a sodium petroleum sulfonate, average equivalent weight 422. Sold by Sonneborn Chemical Co., 300 Park Avenue South, New York, New York, 10010.

[3] Henry plant water is obtained from the Henry lease in Illinois; contains about 18,000 p.p.m. of dissolved salts.

[4] Palestine water is obtained from the Palestine water reservoir in Palestine, Illinois; contains about 420 p.p.m. of dissolved salts.

NOTE.—The amount of cosurfactant is based on ml. of cosurfactant per 100 ml. of liquid containing surfactant, hydrocarbon and aqueous medium—except in examples K and L wherein the amount of alcohol is based on percent of the total volume of dispersions.

In related work *J.T. Kelly and F.H. Poettmann; U.S. Patent 3,637,018; January 25, 1972; assigned to Marathon Oil Company* have found that oil from subsurface tar sand having an injection means in fluid communication with a production means is recovered by injecting a water-external micellar dispersion at a temperature above 100°F into the tar sands, displacing it toward the production means and recovering the oil.

The micellar dispersion can be preceded by a slug of hot water which can optionally have a pH greater than 7. Also, the micellar dispersion can have a pH of 7 to 14 and preferably a temperature greater than 150°F. The micellar dispersion contains hydrocarbon, surfactant, aqueous medium, and optionally cosurfactant and/or electrolyte.

Water-External Micellar Dispersions

J.T. Kelly and F.H. Poettmann; U.S. Patent 3,644,194; February 22, 1972; assigned to Marathon Oil Company have found that oil within tar sands can be recovered by contacting the tar sands, preferably comminuted, with a water-external micellar dispersion in amounts sufficient to solubilize economically feasible portions of the oil from the tar sands. Thereafter, the micellar dispersion containing the solubilized oil is separated from the spent tar sand and subsequently the oil is recovered. The water-external micellar dispersion is preferably at a pH of 12 and also can be at temperatures in excess of 100°F, and preferably 150°F or more. Volume amounts of from 0.05 to 30 volumes of the micellar dispersion per volume of tar sand are useful.

The water-external micellar dispersion useful in this process is comprised of hydrocarbon, surfactant, aqueous medium, and optionally cosurfactant and/or electrolyte. Examples of volume amounts include 1 to 50% hydrocarbon, 40 to 95% aqueous medium, at least 4%

surfactant, 0.01 to 20% or more of cosurfactant, and 0.001 to 5% by weight of electrolyte. In addition, the dispersion can contain other additives such as corrosion inhibiting agents, bactericides, sequestering agents, etc. The surfactants useful with the micellar dispersion can be nonionic, cationic, and anionic surfactants. Specific examples of useful surfactants include those found in U.S. Patent 3,254,714.

Specific examples of useful cosurfactant include alcohols, amino compounds, esters, aldehydes, and ketones. Useful examples include isopropanol, n- and isobutanol, amyl alcohols such as n-amyl alcohol, 1- and 2-hexanol, 1-octanol, decyl alcohols, alkaryl alcohols such as p-nonyl phenol, alcoholic liquors such as fusel oils, etc. Preferably the electrolyte is an inorganic base or inorganic salt, e.g., sodium hydroxide, sodium chloride, sodium sulfate, sodium nitrate, ammonium chloride, ammonium hydroxide, potassium chloride, etc. Examples of specific electrolytes useful with the process include those found in U.S. Patent 3,330,343. Specific examples of useful water-external micellar dispersions are found in U.S. Patents 3,506,070 and 3,506,071.

Oil-External Micellar Dispersions

J.T. Kelly and F.H. Poettmann; U.S. Patent 3,660,268; May 2, 1972; assigned to Marathon Oil Company describe a process where oil from tar sands is extracted by contacting the tar sands with an oil-external micellar dispersion (containing 55 to 90% water). Then the sands are separated from the micellar solution containing solubilized oil, and the oil is recovered from the micellar dispersion. The micellar dispersion can be at a temperature in excess of 100°F and the pH of the water within the dispersion can be about 7 to 14. Volume amounts of 0.5 to 30 volumes of micellar dispersion per volume of tar sand are useful with the process.

The micellar dispersion also contains hydrocarbon, surfactant, and optionally cosurfactant and/or electrolyte. Examples of volume amounts include 4 to 40% hydrocarbon, 55 to 90% aqueous medium, at least 4% surfactant, 0.01 to 20% cosurfactant, and 0.001 to 5% by weight electrolyte. The micellar dispersion can contain other additives such as corrosion inhibiting agents, sequestering agents, etc.

Centrifugal Separation

G.R. Coulson; U.S. Patent 2,968,603; January 17, 1961; assigned to Can-Amera Oil Sands Development Ltd., Canada describes a process which involves the clarification of the water of the quiescent bath, and of the separation of the froth or overflow product of quiescent bath into relatively pure constituents, by the application of centrifugal force rather than by settling tanks or thickeners. In the process, the froth or overflow product from the quiescent bath and oil sludge from plant water are first mixed with a suitable diluent, i.e., a hydrocarbon oil capable of dissolving the oil constituent of the froth or overflow product, and of substantially lowering its specific gravity.

While relatively low boiling hydrocarbons such as benzene, xylene, toluene, gasoline, naphtha or the like may be used, it is preferred to use diluents having an initial boiling point of 350° to 400°F or higher such as kerosene, furnace distillates or diesel fuels, etc. to prevent substantial losses by evaporation. The diluent should be thoroughly mixed with the froth or overflow product and oil recovered from plant water to ensure uniform distribution therethrough. The diluted product is then subjected to the action of centrifugal force.

It is essential, during the process of separation through the application of centrifugal force, that sufficient water be present to form a continuous layer or barrier between the oil constituent and the solids constituent as well as completely saturating the voids in the solids component. It has been found that if the solids and oil are allowed to contact each other, they will, in many cases, recombine and require additional centrifuging to reseparate them. If the froth or overflow product contains sufficient water to ensure the maintenance of such a barrier and saturation of sand void space, it may be introduced without further addition, but if the water content is not sufficiently high, it will be necessary to add further water to the diluted product prior to the application of centrifugal force.

Under operating conditions in the centrifuge, a water layer or barrier of at least one inch should be continuously maintained between the oil and the solids. If water is to be added, the lower limit of water addition is the least amount at which the solids removed are virtually oil free, and the oil is virtually solids free. This point may be readily determined when the process is in operation. The only upper limit on the amount of water to be added is determined by the efficiency of the process, for when sufficient water is added to form the water barrier, no additional amount of water will influence the recovery, but a large excess of water would be uneconomical to handle in the centrifuge.

The application of centrifugal force will result in a three-way separation of the constituents by densities, namely, an outer layer of solids, an intermediate layer of water and an inner layer of diluted oil. The oil recovered is of a high grade, substantially water-free, virtually solids-free and may be further processed in a standard refinery without the necessity for settling, thickeners or evaporators. The water recovered will be found to be substantially solids-free and may be recycled into the process for recovery of the heat content thereof.

Continuous Extraction in Centrifugal Separator

According to a process described by *R.C. Terry and R.E. Lane; U.S. Patent 3,738,929; June 12, 1973; assigned to Allied Chemical Corporation* hydrocarbon bearing materials are introduced as a slurry into a centrifugal separator where the material is substantially concentrated.

The concentrated material (separator underflow) discharges into a vessel containing hot water at a temperature in a controllable range of about 200° to 300°F. The difficulty of obtaining the 200°F temperature at atmospheric conditions, because of the blending with cold materials, results in the temperature range of above 212° to about 300°F with pressure conditions for the hot water being preferable. This temperature is controlled by direct steam injection into the vessel in such a fashion that it also aids in agitating the influent materials. As the material passes through the hot water zone, the hydrocarbons are freed from the gangue material and float to the surface of the water. Brine may be used in the slurry to aid in floating the hydrocarbons. The hydrocarbons are removed in one embodiment through a float controlled automatic valve at some upper level.

The application of heat reduces viscosity of the hydrocarbons and their specific gravities. Because the water preferentially wets the sand particles and excludes oil particles, the oil particles are, therefore, freed from the sand. The lowered viscosity and hence lower surface tension allows for the action of the water as described. Because the oil is lighter than the water, a gravity separation takes place. The cleansed material settles to the bottom of the vessel where it is automatically flushed as tailings.

Referring to Figure 3.5, hydrocarbon bearing material such as tar sands is introduced as a water-tar sand slurry into centrifugal separator **1**. The slurry is concentrated from approximately 30% tar sands to approximately 75% tar sands. This concentration is necessary in order to minimize heat requirements. The underflow or concentrated slurry discharges into the extraction unit, as at **1A**. A fluid level is maintained in the extraction unit as indicated in section **A** by flooding section **D** with cold water, the volume of which is regulated by control valve **3** which is actuated by level control mechanism **2**.

As the bituminous materials move downward by gravity through section A they must pass through steam sparge ring **5** as a result of direction by cone **14**. Steam entering at this point provides heat to section A and also provides positive agitation of the materials, a necessary function in cleansing the sand particles. The amount of steam is regulated by control valve **7** which is actuated by the temperature sensing transmitter device **6**. Section **A** is defined in this case by (1) the upper level **L** of the fluid surface, (2) the cone **14** and (3) the sides of the unit.

FIGURE 3.5: CONTINUOUS EXTRACTION IN CENTRIFUGAL SEPARATOR

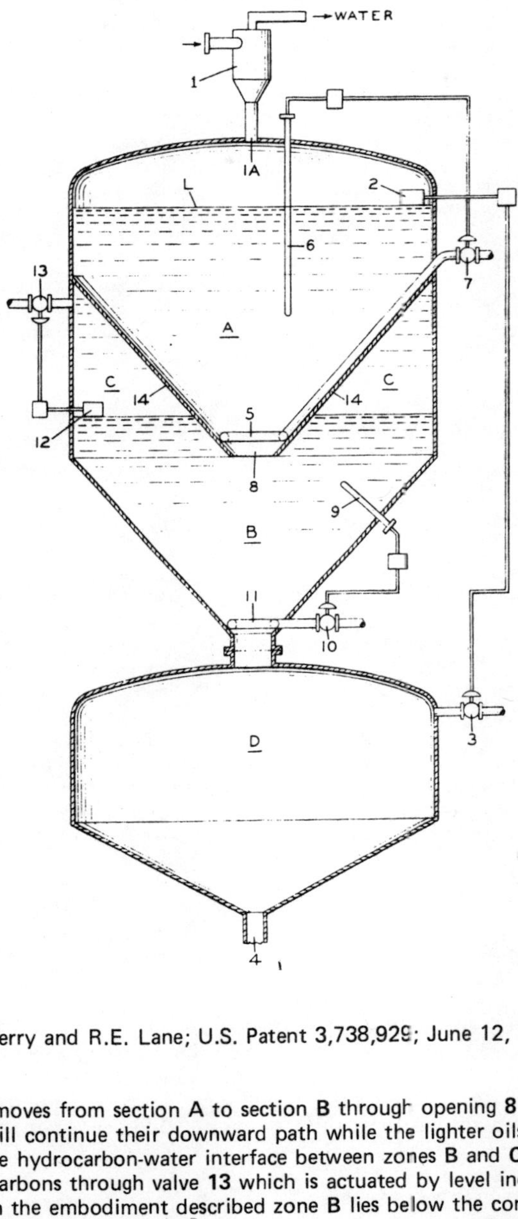

Source: R.C. Terry and R.E. Lane; U.S. Patent 3,738,929; June 12, 1973

As the material moves from section **A** to section **B** through opening **8** in the cone **14** the cleansed sands will continue their downward path while the lighter oils will tend to float into zone **C**. The hydrocarbon-water interface between zones **B** and **C** is controlled by removing hydrocarbons through valve **13** which is actuated by level indicator-transmitter **12**. As shown in the embodiment described zone **B** lies below the cone **14** and zone **C** is on top of zone **B** but also limited in its upward extent by the cone **14**.

Heat is applied to section B by the introduction of steam through sparge ring **11**, the volume of which is regulated by valve **10**, which in turn is actuated by temperature sensing-transmitting device **9**. The introduction of steam at this point further aids in cleansing the material and maintains temperatures at described levels.

After passing through sparge ring **11** the cleansed sand enters section **D** where it is preferably continuously flushed out fixed exit orifice **4**, although a valving arrangement could be provided. The fixed orifice is sized such as to provide enough restriction to allow maintenance of pressures above the orifice at desired levels by the addition, for example, of cold water into section **D** through, for example, valve **3**.

Preferably the pressure drop through the orifice remains substantially constant no matter what the varying characteristics of the slurry, for example, fluctuation in the viscosity of the material going through the orifice. This is accomplished as stated by introducing water into the slurry at a point before it goes through the orifice and maintaining the flow of the liquid introduced such that the conditions set forth above occur.

In a specific example, slurry made up of tar sands and water is introduced into the separator to obtain a rate of about 40 tons per hour of tar sand, with a resultant hydrocarbons output exiting from valve **13** at a rate of 5 to 15 barrels per hour, depending upon the hydrocarbon content in the ore. In the range of about 150 to 300 gallons per minute, for example, of cold water is introduced into the cold water section in this specific example. Steam is introduced at a rate to maintain temperatures within the unit in a range from about 212° to 300°F.

Cyclone Settling Zone Prior to Centrifuging

According to a process described by *R.H. Smith; U.S. Patent 3,808,120; April 30, 1974; assigned to Atlantic Richfield Company* part of the centrifuges previously used in the treatment of tar sands froth are eliminated providing a substantial savings in capital expense and operating expense.

This is accomplished by subjecting the froth, upstream of any centrifuging zone, to at least one cyclone settling zone in which the froth is tangentially injected into the cyclone zone so that it moves in a circular manner as it also moves downwardly under the force of gravity thereby using a centrifugal force, which is not mechanically generated as with centrifuges, to amplify the settling rate of solids and water from the froth.

In this way, substantial amounts of the more easily separated water and solid particles, particularly coarse particles, are removed by cyclone settling prior to treatment in centrifuging zones thereby decreasing the volume of material which the centrifuging zones must operate on. This allows the use of smaller numbers and/or smaller sizes of centrifuges in each centrifuging zone.

This process also provides a method for reducing bitumen losses by the use of countercurrent washing of cyclone underflows with diluent which is lower boiling than bitumen. Bitumen is thus replaced by the diluent which can be recovered by distillation.

Figure 3.6 shows a feed bitumen froth passing by way of pipe **1** into cyclone settling zone **2**. The froth contains bitumen, water, coarse and fine mineral particles, diluent if different from water, (e.g., a hydrocarbon diluent such as naphtha). Diluent is added by way of pipe **3**. Cyclone zone **2** separates out from the froth some particles, particularly coarse particles, and some water to produce a first underflow in pipe **4**. The underflow is removed from the froth thereby creating a bitumen enriched material which remains and is removed by way of pipe **5** as a first overhead. Diluent may or may not be added to this first overhead by way of pipe **6**.

There can be one or more cyclone zones **2** and each zone can contain one or more cyclone devices as desired. Thus, a substantial amount of relatively easily removable solids and water is settled from the froth feed prior to passing that froth into any centrifuge zone. This way the bitumen enriched material in pipe **5** is already reduced in materials to be separated so that first centrifuge zone **7** which feeds on material in pipe **5** can be sized to more efficiently remove a certain class of materials such as remaining coarse particles and some water.

FIGURE 3.6: FROTH TREATMENT EMPLOYING CYCLONE SETTLING ZONE

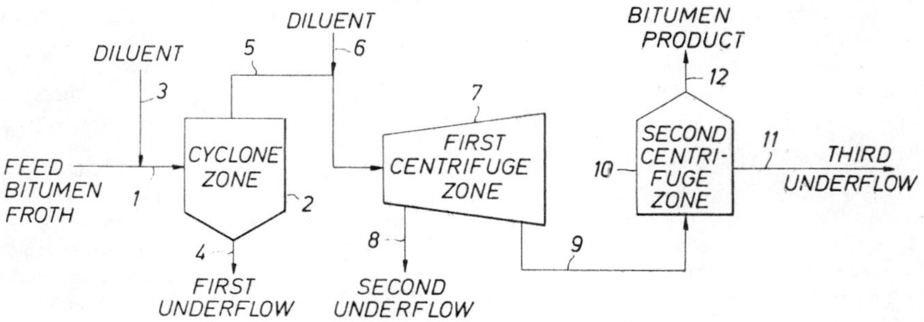

Source: R.H. Smith; U.S. Patent 3,808,120; April 30, 1974

The particles and water, and any diluent if present, are removed in a second underflow by way of pipe **8** thereby forming a bitumen-rich overhead which is removed as second overhead in pipe **9** and passes as feed to a second centrifuge zone **10**.

There can be one or more first centrifuge zones **7** and each can contain one or more centrifuge machines. Preferably, first zone **7** removes most of the remaining coarse particles and is a centrifuge of the solid-bowl or scroll type.

Second centrifuge zone **10** is preferably designed to remove a substantial amount of the remaining fine particles and a substantial amount of water and therefore can employ centrifuges of the disc-nozzle type. As with zone **7**, there can be one or more zones **10** and each zone can contain one or more centrifuges. Zone **10** removes a substantial amount of water and fine particles, and possibly some residual coarse particles and diluent if any, from the process by way of pipe **11** thereby leaving a bitumen product as the overhead from zone **10**. This product is removed separately by way of pipe **12** from the third underflow in pipe **11** for further processing as desired.

Example: Bitumen froth feed composed of 63 weight percent bitumen, 9 weight percent solid particles (coarse and fine), 28 weight percent water, all weight percents being based on the total weight of the feed, is treated in the system shown in Figure 3.6 where naphtha diluent is added to the froth and passed first through a cyclone settling zone **2** to produce a first underflow **4** containing, based on the first underflow, 51 weight percent coarse particles, 6 weight percent bitumen, 3 weight percent naphtha and the remainder water. The bitumen enriched overhead in pipe **5** passes into a scroll-type centrifuge which constitutes first centrifuge zone **7** and which produces a second underflow **8** which contains, based on the second underflow, 63 weight percent coarse solids, 4 weight percent bitumen, 2 weight percent naphtha, and the remainder water.

The bitumen-rich overhead from zone **7** is then passed to second centrifuge zone **10** which contains a disc-nozzle centrifuge and which produces a third underflow **9** containing, based on the third underflow, 6 weight percent fine particles, 2 weight percent bitumen, 1 weight percent naphtha and the remainder water. The bitumen product in the overhead from second centrifuge zone **10** contains, based on the bitumen product, 60 weight percent bitumen, 36 weight percent naphtha, 0.6 weight percent solid mineral particles (coarse and fine) and 3 weight percent water, all weight percents being based on the total weight of the overhead in pipe **12**. By use of cyclone settling zone **2**, a single centrifuge machine of modest size may then be employed in each of zones **7** and **10** and the desired degree of removal of water and solids from the bitumen froth will still be obtained.

Two-Stage Separation System

L.M.O. Cymbalisty; U.S. Patent 3,847,789; November 12, 1974; assigned to Canada-Cities Service Ltd., Imperial Oil Limited, Atlantic Richfield Canada Ltd., and Gulf Oil Canada Limited describes an improvement on the hot water process for recovering bitumen from tar sand. The aqueous slurry of tar sand is introduced into a first vessel, termed the sand separation cell, containing a body of hot water. The coarse sand settles out and is discharged as tailings.

The top product, comprising bitumen, water and fine sand, is transferred to a second vessel, termed the froth formation cell, containing a second body of hot water. In this latter cell, the bitumen forms froth and is recovered, and the fine solids and some water are recycled to the lower end of the sand separation cell. Because the coarse sand has previously been removed in the sand separation cell, good distribution of the feed across the cross-sectional area of the froth formation cell is achieved. This leads to good recoveries and froth quality. By recycling the fines from the froth formation cell to the vicinity of the tailings outlet of the sand separation cell, the fines can be eliminated from the system without the need for a middlings dragstream, as is required in the prior art.

By way of exemplification, the apparatus is described with reference to the bench scale system shown in Figures 3.7a and 3.7b. The front end of the system, up to the sand cell, is conventional. It comprises a pug mill **1** having an internal steam sparger means. The tar sand, water and steam are fed into the pug mill and are mixed for sufficient time to heat and disintegrate the tar sand. At the same time, air bubbles are entrained in the slurry which is formed. The product slurry leaves the pug mill and drops through the screen **2** into the slurry hopper **3**. On leaving the hopper, the slurry is blended with a stream of hot water. The diluted mixture is then delivered by auger **4** into the sand cell **5**. The sand cell is an upright vessel having a cylindrical middle section and conical top and bottom sections **6** and **7**.

A submerged cylinder **8** is mounted coaxially within the sand cell. This cylinder is disposed in spaced relation to the cell's side wall and cooperates therewith to form a vertical annular passage **9**. Transverse baffles **10** extend across the top and bottom openings **11**, **12** of the cylinder **8**, to protect the zones above and below from the turbulence within it. As shown, the auger **4** delivers the slurry into the cylinder. The auger thus provides means associated with the sand cell intermediate its ends for introducing the slurry thereinto. A pump or like means could be used for the same purpose. An impeller **13** rotates within the cylinder and mixes its contents to further disperse the sand and bitumen particles.

The conical top section **6** defines a zone **14** which is in communication with the top opening **11** of the cylinder **8** and the annular passage **9**. The bitumen-rich product of the cell converges into this zone and passes out the cell's top outlet **15**. This outlet provides means at the upper end of the cell for removing therefrom a first product stream, comprising water, bitumen, fine sand particles and a minor amount of coarse sand particles.

A cone **16** is positioned immediately beneath the cylinder bottom opening **12**. This cone has a pair of upwardly extending conduits or tubes **17** which connect the underside of the cone with the annular passage. The recycle line **18**, from the tailings outlet **19** of the froth cell **20**, extends through the wall of the sand cell and terminates beneath the cone **16**, for discharging a recycle stream beneath the cone.

The conical bottom section **7** of the vessel defines a sand concentration zone **32** and includes a bottom outlet **33**. The sand tailings leave the vessel through this outlet and are pumped by pump **34** to a tailings pit. The outlet and pump therefore provide means associated with the sand cell vessel at its lower end for removing therefrom a second stream.

The sand cell's top outlet **15** is connected through a line **21** to the froth cell **20**. The froth cell is also an upright vessel, having a cylindrical side wall and conical bottom. Its interior cross-sectional area is greater than that of the sand cell **5**.

FIGURE 3.7: TWO-STAGE SEPARATION SYSTEM

a.

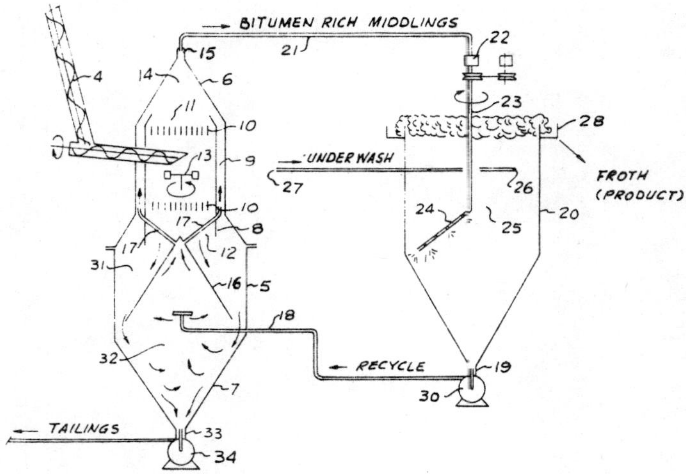

Process Overview

b.

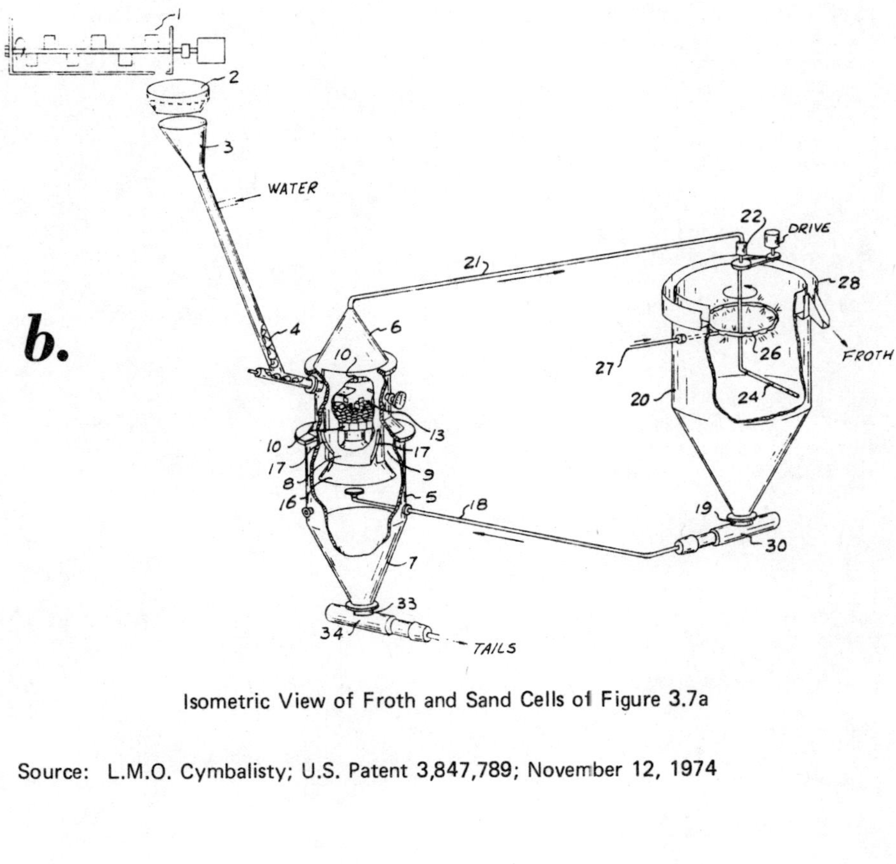

Isometric View of Froth and Sand Cells of Figure 3.7a

Source: L.M.O. Cymbalisty; U.S. Patent 3,847,789; November 12, 1974

The line **21** from the sand cell is connected through a swivel joint **22** to a rotatable vertical feed pipe **23** extending into the froth cell. The feed pipe has a horizontal leg **24** at its lower end; this leg is perforated at intervals along its length. When the feed pipe is rotated, its leg is swept through the interior of the second body of water and evenly distributes the stream from the sand cell across the froth formation zone **25**. The feed pipe **23** provides means, associated with the froth cell vessel intermediate its ends, for distributing the first stream thereinto uniformly across its cross-sectional area. The line **21** functions to connect the sand cell outlet **15** with the feed pipe for transferring the first stream therebetween.

A circular, perforated underwash tube **26** is positioned in the froth cell **20** above the feed leg **24** and below the expected level of the froth-water interface. A source **27** is connected to the underwash tube for supplying it with low solids water, which is to be introduced into the froth cell. A launder **28** runs around the top rim of the froth cell for recovering a third stream, namely, the bitumen froth.

The froth cell has a tailings outlet **19** at its bottom end. This outlet provides means for removing from the froth cell a fourth stream, comprising water, fine sand particles, and minor amounts of coarse sand and bitumen. It is connected by a pump **30** and recycle line **18** with the sand cell, as stated above.

In operation, the slurry is introduced into the sand cell through the auger. As the slurry swirls around within the cylinder **8**, the sand and bitumen particles are further dispersed. Bitumen particles, together with some solids, float up through the cylinder and out its upper opening **11** into the zone **14**. Most of the coarse sand, together with some fines and bitumen, settle out through the cylinder's bottom opening **12** into the sand separation zone **31**. Here they are distributed outwardly by the cone **16**.

As they move through the sand separation zone, bitumen particles and some fines begin to rise and float up through the annular passage **9** into the zone **14**. The coarse sand particles, together with some bitumen and fines, drop into the conical sand concentration zone **32**. Here they move downwardly, toward the tailings outlet **33**, and are simultaneously packed together. Some bitumen particles trapped in the sand are squeezed out; these particles rise back through the cell **5** or collect above the tailings outlet. The coarse sand particles are removed through the tailings outlet as a second product.

A stream of water, fines and minor amounts of bitumen and coarse sand is recycled from the froth cell tailings outlet **33** into the lower end of the sand cell. This stream establishes an upward flow through the tubes **17**, annular passage **9** and zone **14**. The constant circulation functions to keep the fines in the sand cell from gelling and rendering the cell inoperative. It also functions to carry nonbuoyant bitumen from the lower end of the sand cell up to the zone **14**.

While it is preferred to recycle the froth cell tailings to the sand cell, one may alternatively supply water to the recycle line from another source, such as the tailings pond, and treat the froth cell tailings in a secondary oil recovery unit, if desired. In fact, one may even operate the system without introducing water at the lower end of the sand cell, although many advantages of the process will be lost in doing so.

The first product from the zone **14** is transferred to the line **21** into the froth cell **20**. It enters the body of hot water contained therein through the perforations in leg **24** and is distributed evenly across the cross-sectional area of the body. At the same time, hot water is fed into the froth cell through the underwash tube **26**. The bitumen floats upwardly and forms a froth which is recovered in the launder. The water and fines move downwardly to the tailings outlet **19** and exits as a fourth product.

Example: A system in accordance with Figure 3.7b was operated continuously for 4 hours as follows. 240 pounds per hour of tar sand, comprising 11.33% by weight bitumen, 4.33% water, and 84.04% solids, including 17.9% fines, were fed to a pug mill. The tar sand had a pH of 7.8. Sufficient fresh hot water, preheated to 180°F, was added to the tar sand to produce a slurry containing 24% by weight water. The pug mill was externally heated with a steam jacket. The product slurry was observed to have a temperature of 154° to 155°F. The residence time in the pug mill was approximately 2½ minutes.

The slurry was fed by a screw auger from the pug mill into a sand cell. Water, at 180°F, was added to the slurry at the screw auger to raise the water content to approximately 28%. The sand and froth cells were both filled with clear water at 180°F at the time the test run was initiated. The underflow from the froth cell was recycled to the underside of the cone at a rate of 6.6 pounds per minute. Clear water, at a temperature of 180°F, was fed into the froth cell through the underwash ring at a rate of 7 to 10 pounds per 100 pounds of tar sand being introduced to the system. The following table lists the compositions, in weight percent, of the sand cell tailings, recycle, and product froth streams.

Stream	Percent Oil	Percent Water	Percent Solids
Sand cell tailings	0.56	35.48	63.96
Recycle	1.70	80.28	18.03
Product froth	81.4	13.7	4.9

The oil or bitumen recovery from the system was 93.5%. The total water used was 40.3 pounds per 100 pounds of tar sand processed.

Integration of Extraction Processing Streams

A process described by *R.A. Given; U.S. Patent 3,726,786; April 10, 1973; assigned to Canada-Cities Service, Ltd., Imperial Oil Limited, Atlantic Richfield Canada, Ltd., and Gulf Oil Canada Limited* involves the integration of extraction processing streams obtained when processing tar sands from areas exhibiting low fines concentrations, for example, from upper tar sands beds with extraction processing streams obtained when processing tar sands from areas of high fines, for example, from lower tar sands beds, while efficiently maintaining the extraction cell operating conditions at a minimal expense. The maximum amount of suspended solids in the extraction processing stream which an extraction cell can tolerate and efficiently operate the cell is an easily determinable variable.

Generally, a fines concentration of from 13 to 17 volume percent is preferred with the concentration, preferably held at 15 volume percent. In the process for recovery of bitumen from tar sands in a water extraction cell, the method comprises measuring the suspended solids level in the extraction processing stream of the several extraction cells. In accordance with the determination of an unsaturated concentration of solids in comparison with the maximum tolerable solids concentration for efficient bitumen extraction, the extraction processing streams are cross-blended with other extraction processing streams to provide the maximum tolerable saturation solids for efficient bitumen extraction.

The operation of the process is described with reference to Figure 3.8. A first extraction cell **10** is depicted having a draft tube **36**. A rotation arm **38** is connected to a stirrer **40** and extended within the draft tube. Tar sands **26** and extraction water **24** are introduced into the first extraction cell through pump **28** and line **30** feeding mixer **32**. The effluent from the mixer is directly fed to the first extraction cell through line **22**.

The arrangement allows maximum concentration of tar sands introduced into the first extraction cell with the extraction processing stream present in the first extraction cell. The separated solids **44** are removed through exit **42** located in the lower portion of the first extraction cell. The removal of froth **70**, containing the bitumen, from the first extraction cell, is provided by decanting.

The solids density within the extraction cell may be measured through density sensing probe **46** electrically connected by line **48** to a density meter **50** so as to activate simultaneously through electrical connection **52** the pump **92** and valve control apparatus **58**. The density meter activates extraction processing stream removal pump **92** removing extraction processing stream **62** from the extraction cell **10**.

FIGURE 3.8: PROCESS FOR INTEGRATION OF PROCESSING STREAMS

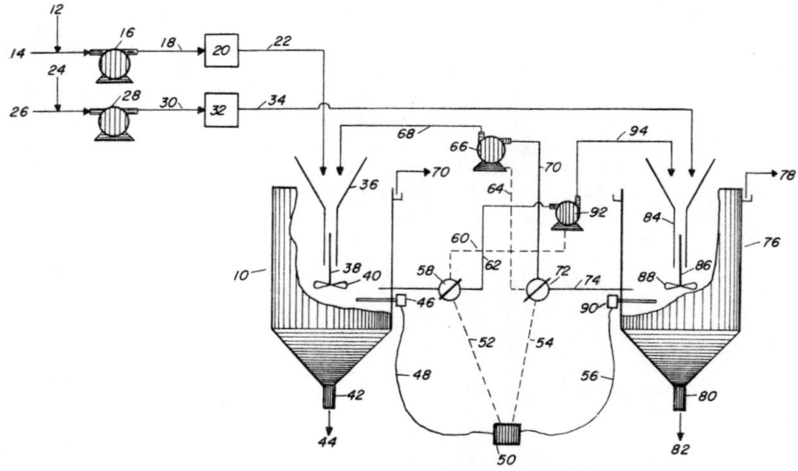

Source: R.A. Given; U.S. Patent 3,726,786; April 10, 1973

The particular density sensed by density sensing probe **46** determines what quantities of extraction processing stream the control valve **58** will allow through pump **92**, depending upon whether or not the density of the extraction processing stream is at the maximum tolerable level for efficient operation of the first extraction cell.

The extraction processing stream **94** is introduced into a second extraction cell **76** through a draft tube **84** having a rotation arm **86** and stirrer **88**. Solids **82** are removed from this extraction cell through exit **80** located in the lower portion. A density sensing probe **90** determines the solids concentration within the extraction processing stream if it is less or greater than the tolerable solids concentration through electrical line **56** to density meter **50**.

The extraction processing stream may be passed to the first extraction cell for immediate use as makeup water within the first extraction cell through fluid line **74** controlled by valve **72** and pump **66** connected by lines **70** and **68** to draft tube **36**. Froth **78** is decanted from the upper portion of the second extraction cell. Therefore, simultaneously with the removal of the extraction processing stream **74** from extraction cell **76**, the control apparatus activates removal of extraction processing stream **62** from extraction cell **10** as determined by particular levels and concentration of the solids within cells **10** and **76**. Makeup waters **12** and **14** may be required to adjust the solids concentrations of the extraction processing streams to the desired solids volume concentration of about 15%.

Therefore, the maximum amount of suspended solids in the extraction processing stream which the extraction zone can tolerate and still operate efficiently is determined in accordance with the process. Portions of the extraction processing streams are withdrawn from one or more extraction cells in processing tar sands from areas which yield multivariant concentrations of suspended solids concentration and are cross-blended to obtain the desired concentration levels for operation.

When processing tar sands from a lower bed which produces an extraction processing stream containing suspended solids in excess of a tolerable quantity, a portion of the extraction processing stream is commingled with a portion of an extraction processing stream from an upper tar sands bed having a low solids concentration in order to balance the solids concentrations within the extraction processing streams. When the supply of the two or more streams has been cross-blended, the process can be integrated and made self-sustaining. Extraction processing streams of low solids concentrations are returned in appropriate volumes to those of the high solids content removed from other extraction cells.

During those periods when the extraction processing streams are all low in solids content, the cross flow is curtailed and straight run processing is accomplished. By utilizing a continuous solids concentration sensing device such as a density meter depicted in the accompanying Figure 3.8 to control the various pumps and valves, the process may be automatically controlled at predetermined ranges of solids contents in the extraction processing streams and extraction cells.

Therefore, the process provides an improved system for optimum use of extraction processing streams water within extractions cells for the processing of tar sands. The process provides a means wherein tar sands mined in areas in which the tar sands fines concentration is radically changed may be used to the advantage of the tar sands processing in order to provide an efficient and operable process for the efficient utilization of aid of waters in the tar sands bitumen recovery project.

Introduction of Water Immediately Below Emulsion Layer

G.R. Gray and L.M.O. Cymbalisty; U.S. Patent 3,567,621; March 2, 1971; assigned to Cities Service Athabasca, Inc., Imperial Oil Limited, Atlantic Richfield Corporation and Royalite Oil Company Limited describe a process for the recovery of bitumen from tar sand in which a fluid slurry of tar sand is introduced into a body of hot water with bitumen rising to the top in the form of a layer of bituminous emulsion and sand settling to the bottom.

Water is introduced immediately below the layer of bituminous emulsion to displace solids, froth particles associated with coarse solids, and prevent entrainment of fines in the emulsion. The additional water is preferably introduced into a quiescent zone in which a very slow downward velocity of water is maintained and water may be withdrawn from the lower portion of the quiescent zone to maintain a suitable downward velocity of water within the quiescent zone.

By introducing additional water into the body of water in the separation cell immediately under the layer of froth or emulsion which forms at the top of the water, it is possible to displace a considerable amount of the solids which are normally carried up and entrapped in the froth layer. This results in improved froth quality, i.e., lower solids content, and in a reduction of the total amount of water needed for the recovery process. Water may be added in any suitable quantities such as between 0.1 and 5 pounds of water per pound of bitumen in recovered emulsion.

While merely adding water to the separation cell immediately below the froth layer as described above is beneficial in displacing solids and preventing entrainment of solids in froth, it is preferred that a disengaging zone be established immediately below the froth layer in which a small downward movement of water is maintained in order to carry solids away from the froth layer, thereby further reducing chance of entrainment of solids in the froth layer. This may be accomplished in any suitable manner such as by the use of baffles to create a quiescent zone.

Addition of water immediately beneath the froth layer is preferably controlled so that the downward movement of water in the disengagement zone is between about 0.05 and 2.5 feet per minute. Other movement of mixtures of water, bitumen and solids within the disengagement zone is also preferably maintained less than about 2.5 feet per minute in order to reduce agitation and further reduce entrainment of solids in froth.

Water may be withdrawn from a point below the introduction of additional water but above the point of introduction of the slurry into the body of water in the separation cell to minimize downward velocities in the lower portion of the separation cell and thereby minimize loss of bitumen to sand tailings. Water withdrawn in this manner may be recycled to earlier stages of the recovery process and if desired may be treated for secondary recovery of bitumen or removal of solids.

Multicell Processing Apparatus

R.S. Brimhall; U.S. Patent 3,605,975; September 20, 1971 describes an apparatus for processing bituminous sands to recover petroleum. The apparatus is preferably a multicell structure, principal cells of which include bottom stirrer zones and upper quiet zones separated from the stirrer zones by baffles. Continuous through-plant feed is provided for the plant structure by partially processed sands being urged into subsequent cells for continued processing.

An oil slick forms on the upper surfaces of the fluid levels in each cell and the so-formed oil slick gradually spills over into a collection trough. Special impeller or stirrer means are provided to provide the requisite, slight agitation at the bottom of each cell for oil separation and yet retain the stirred mixture at the bottom of the cell to the bottom zone of each cell affected. Adjustable gate means are provided to accommodate regulation of feed-through through the plant structure.

Referring to Figure 3.9a, tank **10** is supported by legs **11** and includes partitions **12** for dividing the tank into three cells, **A**, **B** and **C**. For convenience of reference, cells **A** and **B** shall be referred to as primary recovery cells. Cell **C** may be considered as a settling cell or tank in which some secondary recovery of oil may occur. As to cells **A** and **B**, a plurality of baffle plates **14** are secured to the inner side walls of the tank, preferably in a stacked configured fashion as illustrated in Figure **3.9**b. See also Figure 3.9c.

Disposed over the top of the tank and preferably welded to the upper side surfaces of partitions **12** is a bearing support angle **15**. The latter mounts plural journal members **16** which receive stirrer shafts **17**. The journal members may comprise bushings, pillow-block bearings, or other suitable type of journaling means. The stirrer shafts mount at their lower ends revolvable stirrers **18**.

The stirrers comprise individual units, each having a central boss **19** pinned or otherwise secured to the lowermost extremity of its shaft **17**. Affixed to the respective bosses are stirrer paddles **20** through **23**. Paddles **20, 21** and **22** are preferably curved or arcuately formed as indicated to offset centrifugal action and optimize particle distributions in the water. Additionally, it is preferable that they be inclined forwardly in connection with their direction of travel. See Figures **3.9**a and **3.9**f.

When so formed, they tend to stir or agitate slowly particles contained within the tank at the respective cells **A** and **B** and, in addition, to keep such stirred portions of the liquid mixture downwardly within the respective cells. One of the paddles **23** may be straight, if desired, for successively urging accumulations at the bottom of the cell toward and through gate opening **24**. See Figure **3.9**c. Auger, interior conveyor belt, or other conventional means might be used equally as well in conveying sand droppings from one cell to a next adjacent cell, then to a third cell, and so on.

Each of the partitions **12** includes a respective gate opening **24** over which is disposable movable gate **25**. The latter may be secured to a gate actuating rod **26** having ring or handle **27**. The gate, hence, may be positioned in accordance with the pinning of shaft **26**, at a respective aperture **28** thereof, to aperture **29**, a positioning aperture located in each respective partition. This may be accomplished by a pin or removable key **30**, the latter preferably being secured by flexible chain **31'** to rivet or screw **32** affixed to the respective partition.

FIGURE 3.9: MULTICELL APPARATUS FOR PROCESSING BITUMINOUS SANDS

a.

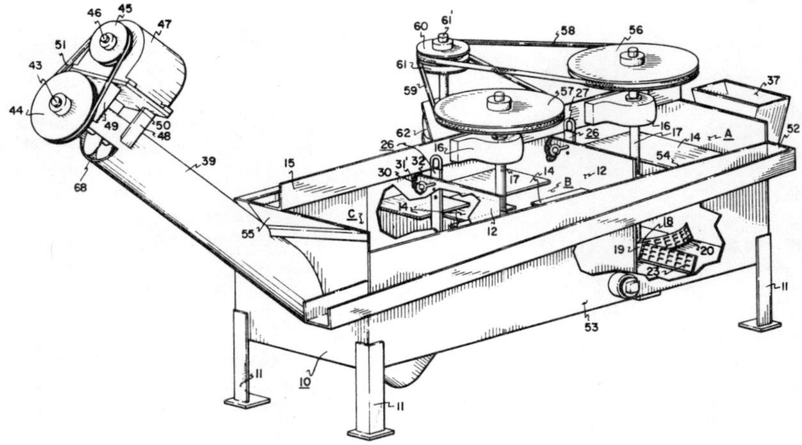

Perspective View of a Petroleum Sand Treatment Plant

b.

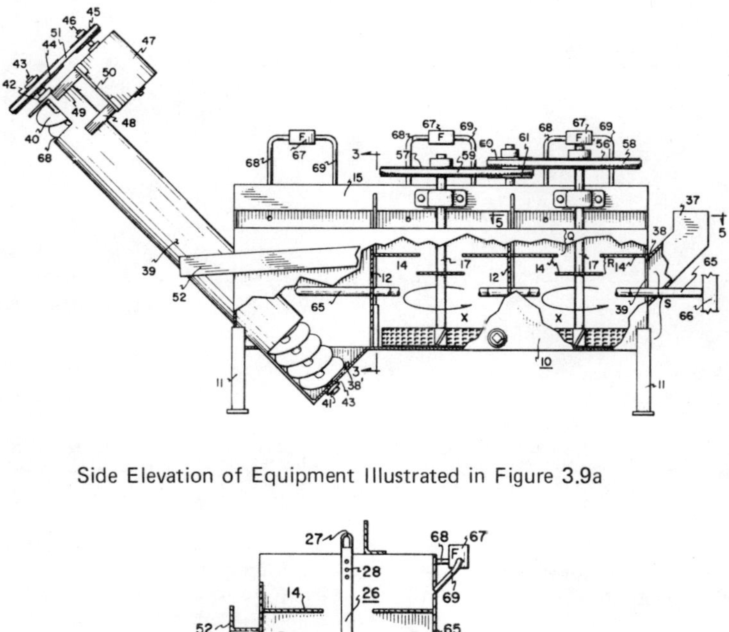

Side Elevation of Equipment Illustrated in Figure 3.9a

c.

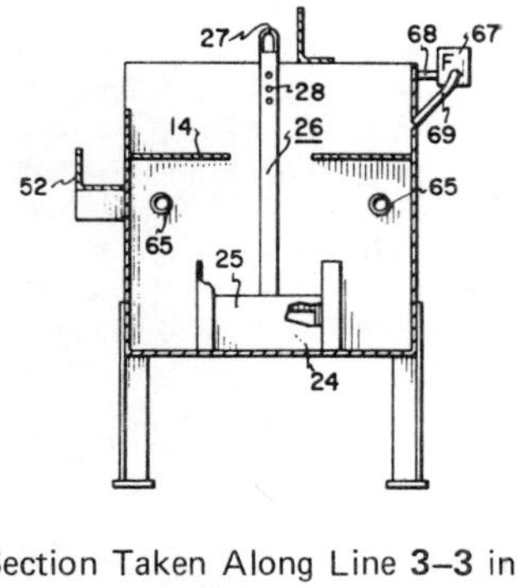

Transverse Section Taken Along Line **3–3** in Figure 3.9b (continued)

FIGURE 3.9: (continued)

d.

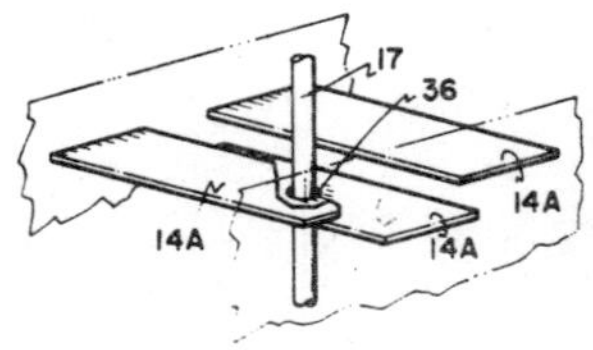

Baffle System

e.

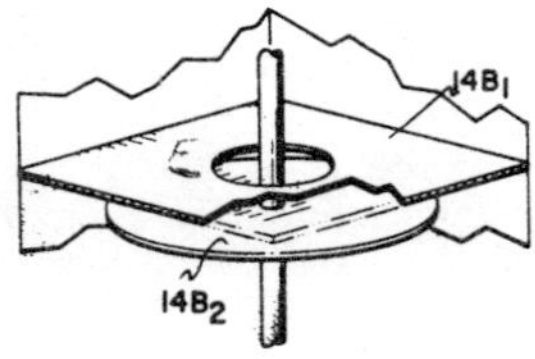

Alternate Baffle Arrangement

f.

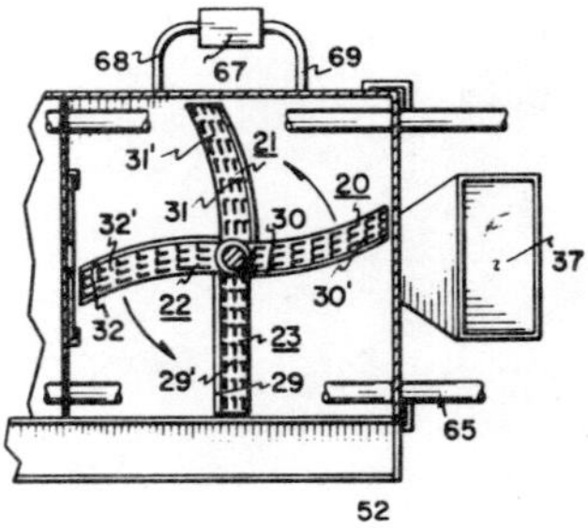

Fragmentary View Along Line **5–5** in Figure 3.9b

Source: R.S. Brimhall; U.S. Patent 3,605,975; September 20, 1971

Accordingly, the gate **25** may be selectively raised and lowered to new positions so as to uncover a selected portion of the gate **24** opening. This in turn will predetermine the feed rate of material accumulations proximate the lower portion of cell **A** and its advancement through the gate to cell **B**, then to gate **C**, and so on.

In returning to a consideration of the several stirrer paddles associated with each shaft **17**, it is seen that the same preferably are perforate, forming a perforate grate or wire mesh which may be supported, for example, by a radially extending bar **29**, of straight configuration relative to stirrer paddle **23**, and arcuate support bars **30, 31** and **32** of blades or paddles **20, 21** and **22**. Each of the stirrer paddles includes the grates made of wire mesh or bar mesh as at **29', 30', 31'** and **32'** so as to supply respective perforate paddles.

The reason for inclusion of perforate paddles is to provide for maximum turbulence per incremental volume areas of the mixture at the bottom of cells **A** and **B** without producing undue turbulence of the liquid level. While solid stirrer paddles may be employed, it is preferable in operation of the system that the stirrer paddles be perforate. This is to provide the slight internal agitation needed to keep the oil flecks suspended and available for rising without causing marked undulations in the surface level of the cell. The gratings may be made of sufficient thickness as to individual elements to provide sufficient forward movement of accumulations at the bottom of each cell **A** and **B** forwardly through their respective gates.

Figures 3.9a through 3.9f illustrate various baffle arrangements which may be used in the structure. This is for the purpose of dividing the cells into quiet zones **Q**, baffle areas **R**, and stirring areas S. Each cell A and B may be partially or fully baffled. The term "fully baffled" refers to the situation wherein all upward movements of oil particles must assume circuitous paths in order to advance to the upper surface of the liquid contained with tank **10**. In such event, the baffles may be overlapping and staggered as illustrated in Figure 3.9d. Note baffles **14A** in this regard and the vertical projections of which overlap, with the central baffle including an aperture **36** for receiving a respective stirrer shaft **17**. Alternatively, baffles $\mathbf{14B_1}$ and $\mathbf{14B_2}$ may be employed in each of the cells A and B to provide complete circuitous movement of rising oil particles or droplets as these advance upwardly through the liquid to the upper surface thereof.

Inlet hopper **37** is secured to the structure as by welding at **38** and communicates with opening **39** leading into the initial oil recovery cell **A**. Disposed to the left of the leftward partition **12** in Figure 3.9b is the settling tank or settling cell **C**. The latter may include a depending floor area **38'** forming an outwardly extending auger housing **39**. The latter includes auger **40** which is journaled to the structure at **41** and **42** in preferably included liquid seal glands **43** of conventional design.

The auger includes a drive shaft **43** upon which driven pulley **44** is mounted. Drive pulley **45** is keyed to shaft **46**, the latter being an integral part of motor **47**. The motor is bracketed in place by brackets **48** and **49**. These latter may support a mounting plate **50** to which the motor is bolted or otherwise secured. The V-belt **51** intercouples drive pulley **45** with drive pulley **46**.

Collecting trough **52** is formed either as a separate part or integral with the outermost side **53** of the processing tank **10**. The collecting trough is shown to slope downwardly for optimum running condition, and the level of liquid is adjusted to within the entire tank such that the oil slick formed on the quiet upper surface of the included water will automatically spill over edge **54** into the collecting trough.

Suitable bracing **55** may be provided for the auger housing **39** as shown. Shafts **17** each include respective pulleys **56** and **57** which are pinned thereto. V-belts **58** and **59** intercouple these pulleys with drive pulleys **60** and **61** as shown. These outer pulleys are keyed to shaft **61'**, the latter of which may be an extension of the motor shaft or motor **62**. The motor may be electrically driven or may comprise an internal combustion engine or fluid motor, as desired.

Optimum results are obtained when the fluid mixture in the tank of Figure 3.9a is heated. Such heating can be effectuated through the provision of a heating coil, or simply a U-configured steam line, as seen in Figure 3.9b to take the form of a hairpin-configured line **65** proceeding through all of the cells of **A**, **B** and **C** and being connected to a steam heat generating source **66**. Other forms of heating, of course, are possible. The steam line is ideally placed in Figure 3.9b where the same is shown to be disposed beneath the baffle areas. Thus, the area of agitation is that area in which the oil particles or droplets are heated so as to assist the latter to rise to the top of the liquid contained in the cells.

For each of the cells there may be provided a fluid circuit including a filter **67** and lines **68** and **69** connected thereto and communicating with the interior of the tank immediately beneath the oil slick area of the tank. The filter may include a pump so as to withdraw colloidal or coagulant accumulations immediately beneath the slick, filter the same out by conventional appropriate means, and return the liquid phase back into the tank. If desired, each cell of the tank may be provided with the same, in the manner as shown in Figure 3.9b. It is to be noted that the liquid phase is returned to a substantial area beneath the portion of the liquid both at the oil slide and immediately beneath it where the colloidals will form.

The apparatus operates as follows. Introduced in inlet hopper **37** is bituminous sand feed stock. This feed stock comprises bituminous sand mixed with a diluent such as kerosene. The mixture feed stock will contain from approximately 6 to 20% diluent, depending upon the character of the specific bituminous sand to be processed as well as the petroleum cut selected as the diluent. Steam should not be introduced directly in the feed stock since this tends to produce too-early separation of the water film that separates bitumen from each sand particle, thereby causing the bitumen to stick to the sand particles. The alkalizing agent should not be introduced into the pulp or slurry but into the processing tank directly, thereby avoiding bitumen reaction and preserving proper pH in the aqueous mixture. Preferably, the bituminous sand will have a maximum particle bulk size less than 2 inches in diameter.

Steam is applied via source **66** so that the aqueous mixture is heated to the boiling point of water for the altitude at which the apparatus is being used. As the bituminous sand, diluent oil and alkalizing agent are deposited into initial cell A, they are mixed by the combined agitating action of the stirrer paddles as they are rotated about the vertical axis of shaft **17**.

With the mixture heated by circuitous steam lines **65**, the oil sand will tend to separate into the constituents of oil, water and sand. Sand will drop to the bottom of cells A and B whereas the additional water will tend to vaporize to rise to the top of the liquid within tank **10**. The diluent will aid the oil particles within the mixture also to rise since the diluent serves to decrease the specific gravity of the combined oil within the mixture. The sand drops to the bottom and is urged through the gate of the right-hand partition **12** into the second cell B. At this point the process is continued to provide for a secondary recovery of oil from the sand. As the oil rises and forms an oil slick on the upper surface of the liquid, the oil slick tends to spill over the edges **54** to collecting trough **52**.

The oil so collected moves downwardly along the trough and into an optionally included drying stage, not shown. The stirrer paddles of the individual stirrers **X** keep the oil sand in suspension in the cell's liquid mixture a sufficient time in order that the heat from the steam line **65** may coact with the bulk water present thereat in releasing through the oil from the sand, permitting the oil to rise upwardly. The two stages A and B, hence, are substantially identical. The third cell **C** is a settling cell permitting any residual oil to rise to the top and permitting the sand advanced through the system to drop down into the cavity in which the lower portion of auger **40** is disposed. Rotation of the auger by motor **47** causes the sand to be advanced and out of the discharge opening **68**. It has been found through experiment that clean sand drops from **68**.

The baffle system is selected so that the upper portions of each cell form quiet zones so that the upper surface of the water is not disturbed. It is this water which supports the film of oil which forms on the upper surface. Oil recovery is had where the surface is not disturbed and where the oil flows smoothly over the edge **54** in Figure 3.9a. This process does not rely upon the formation of froth or an oil emulsion as in flotation type equipment.

Filters **F** with their associated pumps (not shown) allow the discharge of colloidals or coagulants from immediately underneath the surface of the oil slick and a return of the liquid phase to the tank. This prevents the bottom of the oil slick from becoming gummed with forming colloidals. The maintenance of pH at about 8.2 within the tank will deter most colloidals from forming. Those which do form can be pumped out, filtered and discharged. Several conventional filters **F** including their respective pumps **P** may include conventional self-cleaning apparatus.

The process relies upon heating an aqueous mixture containing the oil sands and allowing the oil separated thereby to form at the top of a tank as a quiet slick or collection pool. Stirring at the base of the respective cells is prevented, by suitable baffles, from interfering with the upper, quiet regions of the respective cells. The agitation thus produces a forced suspension of oil sand for heat processing and oil separation. This is done in a manner such that the uppermost column portion of oil liquid is maintained in a quiescent state so that a quiet, nonundulating oil slick may be formed on the upper surface of the water. The slick thus formed is then allowed to deposit into collecting trough **52**.

HOT WATER-CLARIFICATION PROCESSES

Clarification of Middlings Water

R. Schutte; U.S. Patent 3,816,305; June 11, 1974; assigned to Gulf Oil Canada Limited, Canada-Cities Service, Ltd., Imperial Oil, Limited and Atlantic Richfield Canada, Ltd., Canada describes a process for the clarification of tailings and middlings waters produced in conjunction with the hot water treatment of tar sands. In general, the process comprises the introduction of sulfuric acid into the water to reduce the pH to within the range 5.5 to 7.0 to cause flocculation and coagulation of the clays and small solids contained therein. The product is allowed to stand for several days to allow the pH to recover to the substantially neutral condition and the solids from the middlings water to settle.

The process may further comprise stirring the treated tailings or middlings water until a pH equal to about 7.0 is obtained wherein the water may be returned to the native water source subsequent to clarification and neutralization. The following examples illustrate the process.

Example 1: To investigate the time dependence of pH, two sets of examples of middlings, and a tailings water treatment were studied. Various amounts of sulfuric acid were added to study the clarification rate. The initial set was allowed to settle naturally with the second set (the even numbered samples in Table 1) being stirred after each pH measurement. Table 1 relates the amount of acid added and the pH of the supernatant fluid as a function of time for each sample. After seven days settling, the pH of all the samples was found to have a pH over the neutral value.

TABLE 1: VARIATION OF pH WITH TIME

Sample No.	H_2SO_4 ml	g/g Solids $\times 10^{-4}$	Hours after addition of acid: 2	5	12	23	29	47	76	167
	(0.13N)	Middlings from a non-chemical run. 5.6 wt. % Solids. 500 ml Samples								
M1	1		6.90	6.90	7.30	7.14	7.00	7.10	7.35	7.44
M2	1	1.92	6.93	6.65	7.11	7.17	7.07	7.20	7.23	7.76
M3	2		6.67	6.52	6.92	6.93	6.9[illegible]	6.82	7.08	7.52
M4	2	3.91	6.78	6.60	6.97	6.99	7.00	7.18	7.10	7.80
M5	4		6.53	6.51	6.79	6.68	6.70	6.81	7.00	7.18
M6	4	7.48	6.56	6.50	6.79	6.90	6.80	6.98	6.92	7.30
M7	8		6.10	6.28	6.43	6.23	6.4[illegible]	6.40	6.75	6.97
M8	8	15.0	6.14	6.22	6.60	6.63	6.29	6.58	6.85	7.11
Blank	—	—	7.30	7.10	7.48	7.40	7.32	7.42	7.44	7.81

Table 2 represents middlings samples obtained from a semicontinuous unit run in which 0.03 weight percent sodium hydroxide, an often used processing chemical, was first added in the process. It can be seen that after five days' settling the pHs of the clarified portion of all samples were also returned to a neutral value. Table 3 represents a set of tailings water samples containing about 40% solids obtained from a nonchemical run in which 1.0 to 5.82 x 10^{-4} grams of sulfuric acid per gram of solids was utilized in order to depict a time dependent rise of pH to the basic side of the pH scale.

TABLE 2: VARIATION OF pH WITH TIME

Sample No.	H_2SO_4 ml	g/g Solids $\times 10^{-4}$	pH: Hours after addition of acid: 1	9	19	25	43	49	121	140	547
	(0.382N)		Middlings contained .03 wt % NaOH. 12.5 wt % Solids								
1	2		8.05	8.0	8.0	7.87	8.23	8.05	8.07	8.05	8.00
2	2	5.64	8.18	8.0	8.09	7.72	8.20	7.88	8.10	8.10	
3	4		7 66	7.52	7.70	7.83	8.04	7.77	7.20	8.07	8.00

(continued)

TABLE 2: (continued)

Sample No.	H_2SO_4 ml	g/g Solids × 10^{-4}	pH: Hours after addition of acid 1	9	19	25	43	49	121	140	547
4	4	11.19	7.72	7.61	7.52	7.57	8.07	7.90	7.50	8.10	
5	6		7.09	7.1	7.60	7.59	7.70	7.70	7.09	7.70	7.80
6	6	16.94	7.10	7.0	6.94	7.0	7.80	7.70	7.17	7.80	
7	8		6.60	7.13	7.17	7.20	6.83	6.80	7.48	7.30	7.60
8	8	22.04	6.66	6.52	6.65	6.65	7.35	7.00	7.30	7.32	
Blank			8.79	8.70	8.49	8.38	8.17	8.02	8.26	8.20	

TABLE 3: VARIATIONS OF pH WITH TIME

Sample No.	H_2SO_4 ml	g/g Solids × 10^{-4}	pH: Hours after addition of acid 2	5	12	23	29	47	76	167	318
			Tailings from a non-chemical run, 40 wt % Solids, 500 ml Samples								
	(1. ON)										
T1	1		6.27	6.10	6.49	6.68	6.97	6.88	6.80	7.38	7.70
T2	1	1.56	6.33	6.18	6.48	6.54	6.81	6.64	6.58	7.47	7.80
T3	2		5.80	5.81	6.03	6.01	6.18	6.13	6.41	6.94	7.24
T4	2	3.08	5.90	5.98	6.17	6.20	6.30	6.35	6.42	7.19	7.54
T5	4		5.52	5.57	5.40	5.21	5.39	4.90	5.11	5.50	6.20
T6	4	5.74	5.53	5.57	5.54	5.52	5.71	5.19	5.40	5.81	6.56
T7	8		5.13	5.16	5.08	4.70	4.80	4.31	4.53	4.42	3.98
T8	8	11.81	5.14	5.22	5.30	5.01	5.21	4.58	4.64	4.43	4.30
Blank			6.88	7.10	7.33	7.20	7.13	7.23	6.97	7.50	7.62

A more rapid return of the pH to a basic value was found in the stirred rather than the nonstirred samples. The stirred samples also showed that the amount of change of the middlings and tailings water brought about with the addition of acid was permanent, even when the pH was initially basic excellent clarification was experienced.

Example 2: The clarified portions of the samples of Table 1 were centrifuged after 170 hours. The specific conductance, ultraviolet (UV) spectrum, surface tension and pH were measured. The results are given in Table 4. Color of the samples was light yellow (Sample M1) to clear (Samples M7 and M8). An increase in acid gave a small increase (20%) in conductance, a decrease in UV absorbance by a factor of 2.5 and an increase in surface tension of about 6%.

TABLE 4: PROPERTIES OF CENTRIFUGED CLARIFIED PORTIONS

Sample	Specific Conductance X 10^{-4}	UV Absorbance at 250 nm	Surface Tension (relative values)	pH
M1	3.98	0.79	68.1	7.55
M2	4.32	0.49	68.3	7.58
M3	4.35	0.55	69.7	7.52
M4	4.41	0.40	70.9	7.50
M5	4.75	0.42	72.2	7.44
M6	4.88	0.36	70.5	7.56
M7	5.12	0.33	70.7	7.45
M8	5.44	0.29	72.7	7.60
Blank	4.27	1.69	68.7	7.58
Distilled Water	1.66 × 10^{-6}	—	74.4	—

The supernatant liquid from several sludge settling experiments was combined to give 6 liters of clarified water of a yellowish appearance and a pH of 5.5. This water was used in small batch extraction tests to show that the acid-clarified water did not adversely affect the oil recovery and the solids content of the froth, but increased the water content of the froth. The latter was found to be corrected by addition of 0.02 wt % NaOH to the slurry.

Example 3: To determine the commercial significance of the process, the clarification of middlings water from a nonchemical run and chemical run involving 0.03 weight percent sodium hydroxide were studied. Costs per gallon of clarified water present were analyzed in order to determine the significance of the process, utilizing the acid and the natural tendency of the middlings water and solids contained therein in order to reneutralize the solution after acid addition and subsequent coagulation and flocculation have occurred. Table 5 relates the rate of clarification of the middlings from a nonchemical run showing the cost of about one per thousand gallons clarified water utilizing sulfuric acid where the stream is initially at a neutral pH level.

Table 6 depicts a cost of from about 9 to 18 cents per thousand gallons for clarification of middlings from a chemical run in which a slightly basic solution is subjected to clarification. Therefore, the utilization of the process with other chemical treatments is possible, but may be a deterrent in providing an economic means for clarifying the waters to an essentially neutral solution with the precipitated sludge removed therefrom.

TABLE 5: CLARIFICATION OF MIDDLINGS FROM A NONCHEMICAL RUN*

Sample	Wt % Solids	0.1N ml	H_2SO_4 mg	Water Clarified ml	Water Clarified %	Residue after evaporation of supernatant liquid (wt. %)	c/1000 gallons (Imperial) clarified water
1	3.14	2.1	10.3	156	32	0.303	1.06
2	4.76	3.1	15.2	190	39	0.164	1.28
3	6.34	4.1	20.1	245	50	0.114	1.31
4	7.94	5.0	24.5	265	55	0.096	1.48

*500 ml samples, 18 hr.

TABLE 6: CLARIFICATION OF MIDDLINGS FROM A CHEMICAL RUN*

Sample	Wt % Solids	0.382N ml	H_2SO_4 mg	Water Clarified ml	Water Clarified %	Residue after evaporation of supernatant liquid (wt %)	c/1000 gallons (Imperial) clarified water
1	12.5	2	37.44	69	14	.078	8.70
2	do.	4	74.87	78	16	.090	15.40
3	do.	6	112.31	102	21	.095	17.70
4	do.	8	149.74	188	40	.109	12.80

*0.03 wt % NaOH, 18 hr.

As indicated in the data, water clarified in accordance with the process usually requires only 3 to 7 days to recover pH to essentially neutral condition. This compares with approximately 3 months required for untreated material to settle to the point where clarified water may be returned to a river. The process is considered most useful in connection with treatment of primary separator middlings water and secondary separator tailings water.

Freeze-Thaw Separation Technique

O.M. Elliot; U.S. Patent 3,751,358; August 7, 1973; assigned to Great Canadian Oil Sands, Limited, Canada describes a process for treatment of water discharged from a hot water process for separating bitumen from tar sands. The treatment comprises agglomerating clay dispersed in the water, alternately freezing and thawing the water into a sludge layer and a clarified water layer. The clarified water layer is substantially reduced in clay content compared to the water discharged from the hot water process and is suitable for recycle or discard.

The agglomeration step on the pondwater can be carried out by adding a conventional agglomerating reagent to the water with gentle agitation. Among the various reagents useful for agglomerating clay are sodium and calcium chlorides, aluminum sulfate (alum), polyalkylene oxides such as polyethylene oxide, compounds of calcium such as calcium hydroxide, calcium oxide, calcium chloride, calcium nitrate, calcium acid phosphate, calcium sulfate, calcium tartrate, calcium citrate, calcium sulfonate, calcium lactate, the calcium salt of ethylene diamine tetraacetate and similar organic sequestering agents. Also suitable are guar flour or a high molecular weight acrylamide polymer such as polyacrylamide or copolymer of acrylamide and a copolymerizable carboxylic acid such as acrylic acid.

Preferably agglomeration is accomplished by changing the pH of the water. Variation of pH causes a change in the charge on the edges of the clay particles, thus allowing agglomeration. The effluent discharge from the hot water process has a pH ranging from about 7.5 to 9.0 typically about 8.3. In this range, the contained mineral material is not agglomerated. Raising the pH above about 9.0 or lowering it below about 7.5 causes agglomeration. More preferably, agglomeration can be accomplished by adding sulfuric acid (to the middlings or effluent discharge portion as the case may be) to about pH 5.

The freezing and thawing steps can be conducted by any means known in the art. Preferably however, freezing and thawing of the pondwater is conducted in the pond zone by dividing the zone into two areas. The first area is allowed to freeze by exposure to ambient temperatures below 32°F. Water being discharged from the distribution pipes through the sand zone collects in a second pond area. During periods when the ambient temperatures are greater than 32°F, the clear clarified water from the first area is discarded or recycled. The water in the second area is diverted to the first along with current runoff from the distributed pipes. This pondwater is collected until again average ambient temperature falls below 32°F.

Another approach is to alternatively add warm water at various ends of a frozen pond at certain intervals over a period when the ambient temperature is below 32°F. This approach results in repeated freezing and thawing-settling cycles. Another method of conducting the freezing-thawing-settling technique is to circulate water from beneath frozen pondwater to the surface of the ice accompanied by addition of additional runoff water either beneath or onto the surface of the ice.

Several freeze-thaw runs on pondwater from a hot water process for separating bitumen from tar sands are reported in Tables 1 and 2. In Table 1 pondwater feeds of various wt % clays were treated with various agglomerating agents, then frozen, thawed and settled and compared to an untreated sample. The results in Table 1 indicate that various agglomerating agents can be used and that the process is applicable to the clarification of pondwaters of various clay compositions.

TABLE 1

Reagent	Feed, wt. percent clay	Sludge, wt. percent clay	Reagent dosage,[1] pounds/1,000 gallons: Clear	Cloudy	Note
$Al_2(SO_4)_3$	6.5	19.7	13.8	5.5	
	12.0	15.8	13.8	5.5	
$FeCl_3$	6.5	13.0	7.14	3.5	
$CaCl_2$	6.5	10.3	9.17	4.5	
	12.0	18.1	9.17	4.5	
H_2SO_4	6.5	26.6	5.05	3.33	pH=4.5
	10.0	23.2		[2] 5.05	pH=5.0
None	6.5	6.5			

[1] Intermediate dosages not tested. "Clear" indicates sample settled within 1 hour to produce clear supernatant. "Cloudy" supernatants did not settle further within 1 week. "Cloudy" supernatants contained >200 p.p.m. suspended solids (by visual inspection).
[2] Supernatant settled clear within 48 hours.

TABLE 2

Freeze-thaw cycles completed	Weight percent solids in sludge: Test 1, 5.05 pounds H_2SO_4/1,000 gallons	Test 2, 5.05 pounds H_2SO_4/1,000 gallons	Test 3, 13.8 pounds $Al_2(SO_4)_3$/1,000 gallons
1	9	21.6	8.1
2	30.7	29.5	8.1
3		30.5	29.1

The runs in Table 2 were made on pondwater containing 6.5 weight percent clay to determine the extent of sludge compaction for repeated freeze-thaw cycles. These runs show that it is possible to achieve compactions of 30 weight percent solids by using the process.

Treatment of Discharge Water with Gaseous Effluent

J.F. Paulson; U.S. Patent 3,607,720; September 21, 1971; assigned to Great Canadian Oil Sands, Limited, Canada describes an improved process for treating bituminous tar sand comprising forming a mixture of tar sands and water; passing the mixture into a separation zone to form an upper bitumen froth layer, a middlings layer comprising water, finely divided mineral, and bitumen, and a sand tailings layer; removing the bitumen froth layer, middlings layer, and sand tailings layer from the separation zone; resolving the bitumen froth layer to a liquid bitumen product; and coking the bitumen product to produce coke and a coker distillate, combusting the coke to produce a hot flue gas comprising carbon dioxide and sulfur dioxide.

The improvement is describable as passing the hot flue gas product into intimate contact with at least a portion of the middlings to dissolve carbon dioxide and sulfur dioxide in the portion, thereby flocculating the finely divided mineral in the portion, and settling the middlings portion to produce a clarified water substantially reduced in mineral content.

The process is advantageous in two respects. Firstly, the steam formed by contact between the water and hot gaseous effluents can be used as a source of power and heat to provide a more self-sufficient overall hot water process. Secondly, the contacting with hydrogen sulfide, carbon dioxide, or sulfur dioxide containing gaseous effluents causes flocculation of clay in the discharge water. The flocculated clay can then be settled from the water and the treated water reused in the hot water process or discarded.

It is significant that contact with gaseous effluents followed by settling clarifies the discharge water in view of its unique composition, i.e., water containing 0.1 to 1.5 weight percent bitumen and up to 20 weight percent minerals between 80 and 100% of which is fine clay of a size smaller than 2 microns. A substantial portion (about 50% by weight of the mineral) is fines of a size smaller than 0.2 micron. This fine clay has extremely poor settling characteristics. It is this material in combination with other components of the water which makes this water portion difficult to clarify to a degree such that the water is suitable for reuse or discard.

In Figure 3.10, bituminous tar sands are fed into the system through line **1** where they first pass to a conditioning drum or muller **3**. Water and steam are introduced from **2** and mixed with the sands. The total water so introduced is a minor amount based on the weight of the tar sands processed and generally is in the range of 10 to 45% by weight of the total mixture. Enough steam is introduced to raise the temperature in the conditioning drum to within the range of 130° to 210°F and preferably to above 170°F.

Monovalent alkaline reagents can also be added to the conditioning drum, usually in amount of from 0.1 to 3.0 tons per ton of tar sand. The amount of such alkaline reagent preferably is regulated to maintain a pH of the middlings layer in separation zone **12** within the range of 7.5 to 9.0. Best results seem to be obtained at a pH value of 8.0 to 8.5. The amount of the alkaline reagent that needs to be added to maintain a pH value of the middlings layer in separation zone **12** within the range of 7.5 to 9.0 may vary from time to time as the composition of the tar sands as obtained from the mine site varies. The best alkaline reagents to use for this purpose are caustic soda, sodium carbonate or sodium silicate.

Mulling of the tar sands produces a pulp which then passes from the conditioning drum as indicated by line **4** to a screen **5** which is used to remove from the tar sand pulp any debris, rocks or oversized lumps as indicated generally at **6**. The pulp then passes from screen **5** as indicated by **7** to a sump **8** where it is diluted with additional water from **9** and a middlings recycle stream **10**. In the event that the clay content of the tar sands is high, a relatively high rate of fresh or treated feed water introduction through **9** can be employed to compensate for the high clay introduction while a correspondingly high rate of transfer of middlings layer through line **15** can be maintained.

FIGURE 3.10: HOT WATER PROCESS

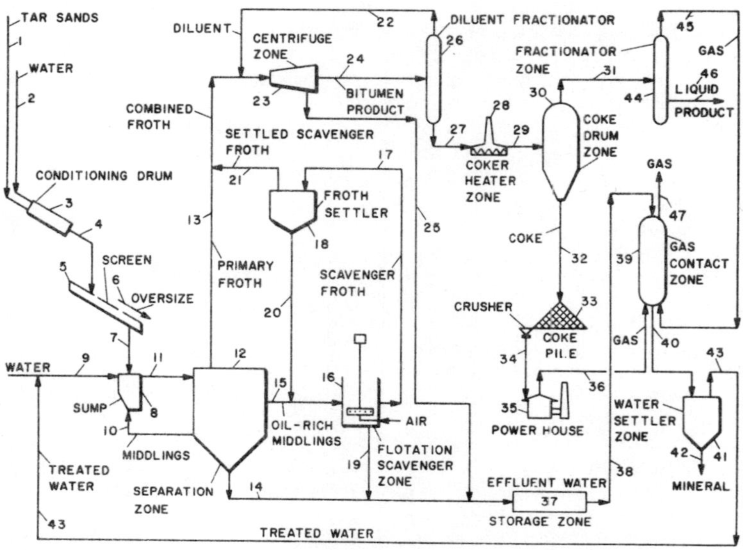

Source: J.F. Paulson; U.S. Patent 3,607,720; September 21, 1971

The pulped and diluted tar sands are pumped from the sump through line **11** into the separation zone **12**. This zone comprises a cell which contains a relatively quiescent body of hot water. In the cell, the diluted pulp forms into a bitumen froth layer which rises to the cell top and is withdrawn via line **13**, and a sand tailings layer which settles to the bottom to be withdrawn through line **14**. An aqueous middlings layer between the froth and tailings layer contains silt and clay and some bitumen which failed to form froth. In order to prevent the buildup of clay in the system it is necessary to continually remove some of the middlings layer and supply enough water in the conditioning operations to compensate for that so removed. The rate at which the middlings need to be removed from the system depends upon the content of clay and silt present in the tar sands feed and this will vary from time to time as the content of these fines varies.

If the clay and silt content is allowed to build up too high in the system, the viscosity of the middlings layer will increase. Concurrently with such increase, an increase in the proportions of both the bitumen and the sand retained by the middlings will occur. If the clay and silt content is allowed to build up too high in the system, effective separation no longer will occur, and the process will become inoperative. This can be avoided by regulating the recycling and withdrawal of middlings and input of fresh water per the process disclosed and claimed in U.S. Patent 3,401,110. However, even when the separation step is operating properly, the middlings layer withdrawn through line **15** will contain a substantial amount of bitumen which did not separate. Hence the middlings layer withdrawn through line **15** is, for purpose of description, referred to as oil-rich or bitumen-rich middlings.

The oil-rich middlings stream withdrawn from separator **12** through line **15** is sent to a scavenger zone **16** where an air flotation operation is conducted to cause the formation of additional bitumen froth. The processing conducted in the scavenger zone **16** involves air flotation by any of the air flotation procedures conventionally utilized in processing of ores. This involves providing a controlled zone of aeration in the flotation cell at a locus where agitation of the middlings is being effected so that air becomes dispersed in the middlings in the form of small bubbles.

Figure 3.10 illustrates a flotation cell of the subaeration type where a motorized rotary agitator is provided and air is fed thereto in controlled amounts. Alternatively, the air can be fed in through the shaft of the rotor. The rotor effects entraining of the air in the middlings. This air causes the formation of additional bitumen froth which passes from the scavenger zone **16** through line **17** to a froth settler zone **18**. A bitumen-lean middlings stream is removed from the bottom of the scavenger zone **16** via line **19**.

In the settler zone **18**, the scavenger froth forms into a lower layer of settler tailings which is withdrawn and recycled via line **20** to be mixed with bitumen-rich middlings for feed to scavenger zone **16** via line **15**. In the settler zone an upper layer of upgraded bitumen froth forms above the tailings and is withdrawn through line **21** and is mixed with primary froth in line **13**. The combined froths are at a temperature of about 160°F. The combined froths are further heated with steam and diluted from line **22** to reduce the viscosity of the bitumen for centrifuging in centrifuge zone **23**. This step is carried out in a battery of centrifuges operating in parallel. The bitumen-naphtha phase is discharged under pressure via line **24** while the water and mineral is discharged through line **25** and is passed to an effluent storage zone **37**, along with the sand tailings from line **14** and the oil-lean middlings from line **19**.

The dilute bitumen product passes to a diluent fractionator zone **26** where the naphtha is removed and recirculated via line **22** to be mixed with froth prior to centrifuging. A resolved bitumen product, preferably containing less than 5 weight percent water and 1 weight percent of mineral, passes from the diluent fractionator zone via line **27**. The bitumen product is heated up to about 900°F in fired heaters **28** and is then charged via line **29** into coke zone **30**. A coker distillate is produced from the bitumen in the coke zone. This distillate passes via **31** onto further processing. Coke is deposited as a solid which is cut from the coke drums in the zone **30** and is removed via line **32** to coke pile zone **33** where it is crushed into a material suitable for combustion in powerhouse zone **35**. The crushed coke is conveyed via **34** to the powerhouse zone **35** where it is combusted in boilers to produce steam, and a flue gas **36** containing carbon dioxide and sulfur dioxide.

The oil-lean middlings in line **19**, the sand tailings from the separation zone **12**, and the water and mineral from the centrifuge zone **23** are combined to form an effluent discharge which is delivered via line **14** to an effluent water storage zone **37**. The effluent contains between 25 and 50 weight percent sand and silt material which is larger than about 2 microns.

Preferably, the effluent storage zone is a pond area surrounded by deposited sand. The water in the effluent discharge percolates down through and over the deposited sand to the pond where it collects as pond water containing up to about 12 weight percent suspended solids, between 80 and 100% of which is fine clay of a size smaller than 2 microns. The pond water also contains between about 0.1 and 1.5 weight percent bitumen. Because of the particular composition of this pond water, and especially because of the extreme fineness of the suspended clay material which has extremely poor settling characteristics, the water cannot be discarded or to any great extent recycled back into the hot water system.

In a preferred form of the process, pond water is withdrawn from storage zone **37** and is fed via **38** into contact zone **39** where it is contacted with flue gas from powerhouse zone **35**. For the purpose of this process, pond water is effluent water discharge from a hot water process which effluent has been settled to give a composition comprising water containing up to about 20% solids, between 80 and 100% of which is fine clay of a size smaller than 2 microns. The effluent discharge from a hot water process comprises middlings material of depleted bitumen content which has undergone final treatment, the sand tailings layer from the process, and other dishcarged water containing fractions which are not the primary products of the hot water process. The discharge is removed from the process areas as a slurry of about 25 to 60, typically 45, percent solids by weight. The effluent contains virtually all of the clay material which was present in the feed. Typically, the amount is 2 to 10 weight percent of the feed. This material is smaller than 2 microns and has extremely poor settling characteristics.

The contacting step conducted in zone **39** can comprise any procedure which will scrub the carbon dioxide and sulfur dioxide from the flue gas into the water. For example, the operation in zone **39** can be vertical countercurrent flow of the gas and the water through packed, plate or spray towers. The contact with the sulfur dioxide and carbon dioxide containing flue gas causes flocculation of the mineral contained in the pond water. Flocculation of mineral occurs through the formation of insoluble salts in the water which causes mineral flocs to form and also through the change in pH of the water below about 7.5. The effluent discharge from a hot water process has a pH ranging from about 7.5 to 9.0 typically about 8.3. In this range, the contained mineral material does not flocculate; however, lowering the pH below 7.5 does cause flocculation.

After contact with the flue gas, the flocculated mineral is settled from the water. In a batch contacting operation, the settling can be conducted by merely discontinuing gas flow to the gas contact zone **39**, permitting the fines to settle and recovering clarified water from the zone top and flocculated mineral from the zone bottom. In a continuous operation the water, treated with sulfur dioxide and carbon dioxide, can be continuously withdrawn from zone **39** via line **40** to settler zone **41** where the water is permitted to settle into flocculated mineral which is removed via **42** and discarded and into a clarified water fraction which can be introduced, via line **43**, back into the system as all or a portion of the water in line **9** to the sump as shown or can be alternatively introduced into the system via lines **2** or **10** or as a screen wash or at any desired point of introduction into the process.

Acrylic Acid Polymers as Flocculants

A process described by *W.J. Lang and D.A. Hentz; U.S. Patent 3,723,310; March 27, 1973; assigned to International Minerals & Chemical Corporation* is based upon the discovery that finely divided clay having hydrocarbons in intimate association therewith which is suspended in an aqueous medium having a pH of at least 6.0, preferably from 6.0 to 10.0, can be flocculated and settled by the addition of acrylic acid or a water-soluble acrylic acid-derived polymer in an amount sufficient to form a floc of the clay.

In the process, a nonionic polymer is utilized as a secondary flocculant. The employment of a cationic polymer or an inorganic salt as the secondary flocculant possesses the disadvantage that it is best to separate the floc produced by the addition of the primary flocculant prior to the addition of the secondary flocculant. However, this is not the case when a nonionic polymer is used as the secondary flocculant. Therefore, the use of a nonionic polymer as the secondary flocculant in the preferred embodiment possesses the advantage of allowing the concomitant addition of two flocculants, if desired.

The process is especially useful for flocculating and settling solids from aqueous suspensions such as those produced by utilizing the hot water process for recovering oil from tar sands. The following example illustrates the process.

Example: This example illustrates the effectiveness of the two stage treatment of this process for flocculating and settling finely-divided material from an aqueous suspension. The suspension which was utilized in this experiment was a sample of the complex oil-clay-sand slimes, containing about 2.0% by weight of suspended solids, obtained from the tailings storage pond from the hot water process employed by the Great Canadian Oil Sands Limited. The primary flocculant was a graft copolymer produced by copolymerizing about

100 parts acrylic acid and about 1 part of carboxymethylcellulose. An aqueous solution containing 0.5% of the graft copolymer was added to a 100 milliliter sample of the slime in a graduated cylinder at the rate of about 1.5 lb of polymer per ton of slime solids. The graduated cylinder was inverted several times to allow mixing. It was noted that there was rapid flocculation of a portion of the suspended solid content of the slime. A 0.05% aqueous solution of a nonionic polymer sold under the trademark of Polyox was then added to the liquid in the cylinder which was again inverted to allow mixing. A sufficient quantity of the aqueous solution was added to add the polymer at the rate of 2 pounds of polymer per ton of original slime solids. It was noted that this combination of polymers resulted in the rapid coagulation and settling of all of the suspended material to leave a clear supernatant liquid.

Tailings Disposal System

A process described by *R. Schutte; U.S. Patent 3,869,384; March 4, 1975; assigned to Canada-Cities Service Ltd, Imperial Oil Limited, Atlantic Richfield Canada Ltd and Gulf Oil Canada Limited, Canada* relates to the treatment of the tailings stream from a hot water extraction plant for recovering bitumen from tar sand. This stream comprises water and coarse and fine solids. The stream is fed into a mined-out cavity to settle the coarse solids. Some of the decanted water from the cavity is transferred to a large settling pond for clarification. The remainder of the decanted water is returned to the plant and added to the tailings stream to fluidize it and render it pumpable.

The process has the following advantages: some of the fines become trapped in the beach of coarse sand formed in the cavity, thereby reducing the volume of sludge which is generated in the settling pond. The use of decanted water to fluidize the tailings issuing from the extraction process reduces the amount of water needed for the extraction process itself, thereby reducing the amount of clarified water required for the plant.

The first step in the process is to provide two settling zones. Preferably this is done by mining the tar sand with dragline equipment, thereby providing a mined-out cavity which can be used as the first settling zone. However, it is to be noted that the first settling zone need not be a cavity; all that is needed is a disposal site for the sand beach, isolated from the second settling zone. A dike can be built to provide the second settling zone, which will usually be of much greater lateral extent than the first zone.

The extraction unit is preferably operated with only the minimum amount of water needed to obtain good bitumen recovery, e.g., 92%. Preferably this may be done by using 50 to 70 lb of water per 100 lb of tar sand. Of this amount, about 20 lb is fresh water and the balance is water recycled from the second settling zone. The tailings from the extraction process are pumped to the first settling zone. Here the coarse sand immediately falls out and forms a beach, trapping about 17 lb of the tailings water per 100 lb of tar sand processed in its void spaces. The tailings water has fines suspended in it.

The water collecting in the first settling zone is then withdrawn and divided into two streams. The first stream is recycled to the extraction plant to fluidize the tailings to render them pumpable to the first settling zone. The second stream is transferred to the second settling zone for prolonged settling of the contained fines. In an alternative version of the process, some of the sludge may be withdrawn from the second settling zone and mixed with the first stream being recycled from the first settling zone to fluidize the tailings. In this manner, the concentration of fines suspended in the tailings water may be increased, to thereby increase the proportion of fines trapped in the beach.

Frothing Treatment of Middlings Stream

A process described by *L.M.O. Cymbalisty; U.S. Patent 3,864,251; February 4, 1975; assigned to Canada-Cities Service, Ltd, Atlantic Richfield Canada, Ltd and Imperial Oil Limited, Canada* relates to treating the middlings stream from the primary separation cell used in the hot water process for extracting bitumen from tar sand. In the process, the middlings stream is diluted with hot water, preferably contacted with minute air bubbles to aerate nonbuoyant bitumen, and introduced into a settler vessel. In the quiescent and dilute environment within this vessel, the buoyant bitumen forms a froth product which is reduced in solids and water content in comparison to the prior art secondary recovery froth obtained by subaeration flotation. Solids, water and traces of bitumen are withdrawn from the vessel through the tailings outlet. Part of this tailings stream may be recycled to the front end of the process for repeated treatment in the separation cell. In this manner, hot water is conserved within the process. The remainder of the tailings stream may be discarded or otherwise processed for recovery of contained bitumen. The froth product from the settler is preferably combined with the primary froth from the separation cell and forwarded on to the froth treatment circuit. The following example illustrates the process.

Example: 1,360 lb/hr of Athabasca tar sand, comprising 11% by weight bitumen, 4% water, and 85% solids, including 12% fines, is fed to a tumbler. Sufficient fresh hot water, approximately 290 lb/hr, is added to the tar sand in the tumbler to produce a slurry containing 22% by weight water. The product slurry is observed to have a temperature of 160°F. The residence time in the tumbler is approximately 3½ minutes.

The slurry is fed from the tumbler into a separation cell. Water at 180°F is added to the slurry at the screen to raise the water content to approximately 42%. The separation cell and settler are both filled with clear water at 180°F at the time that the test run is initiated. Froth, comprising 70% bitumen, 6% solids, 24% water, is produced at the surface of the separation cell at a rate of 170 lb/hr. A tailing stream, comprising 70% solids, 0.8% bitumen and 29.2% water, is produced from the cell at a rate of 1,180 lb/hr.

A middlings stream, comprising 2% bitumen, 15% solids, and 83% water, is withdrawn from the separation cell by a pump at a rate of 900 lb/hr. This stream is mixed with 200 lb/hr of water and approximately 0.1 cfh air in a mixer 7 and is then pumped to the settler The diluted middlings stream is introduced to the central portion of the settler through a rotating, perforated distribution pipe. A froth product is continuously formed and withdrawn at the rate of 19 lb/hr from the settler. This froth product comprises 77% bitumen, 5% solids, and 18% water. A stream of dilute solids and water is continuously withdrawn through the tailings outlet of the settler vessel and is recycled to the front end for extraction. By way of comparison, the same middlings stream, when fed directly to a subaeration secondary recovery cell, gives a froth product comprising 52% bitumen, 13% solids and 35% water.

COLD WATER PROCESSES

Addition of Kerosene and Recycled Oil

In a process described by *G.R. Coulson; U.S. Patent 2,825,677; March 4, 1958* the bituminous sands are mixed with a solvent which must be capable of dissolving substantially all of the bituminous constituents and also of reducing the specific gravity of these constituents so that the oil-solvent mixture has a specific gravity substantially below that of the aqueous phase. While relatively low boiling, normally liquid hydrocarbons such as benzene, xylene, toluene, gasoline, either cracked or straight run, petroleum naphtha, coal tar naphtha or the like may be employed provided precautions are taken to prevent loss by evaporation, it is preferred to employ solvents having an initial boiling point of 350° to 400°F or higher, for example, the higher boiling petroleum naphthas, kerosene, gas-oil in the boiling range of furnace distillates or diesel fuels, etc.

This obviates to a large degree the loss difficulties which might be encountered with the lower boiling materials and the recoveries of crude oil obtained are satisfactory. When lower boiling solvents are used, a vaporproof system should be used to avoid loss of solvent Oils of an aromatic or olefinic character are the most efficacious since certain constituents of the tar-sand oils are more soluble in these than in predominantly paraffinic distillates. Indeed, where the tar-sand oil is of an asphaltic base, some constituents may be almost entirely insoluble in paraffinic solvents, particularly the lower boiling ones, and hence would not be recovered.

It is advantageous to recycle a portion of the recovered oil mixed with the solvent since this, in some instances, appears to increase the ultimate recovery of oil from the sand. This reduces the proportion of solvent used, which, however, must be present in sufficient amount to produce the specific gravity within the desired range of the recovered product. It is preferred to employ the more unsaturated and/or aromatic distillates, either alone or mixed with some of the recovered oil which may be produced during the thermal or catalytic cracking or coking of hydrocarbon oils. The constituents of the tar sand seem to be somewhat more compatible with solvents of this character and hence the degree of recovery is increased.

Where such cracked distillates are available on the premises they may be used in the form of side streams such as that normally employed as recycle stock in the cracking process. Thus, if the recovered oil is being cracked or coked at or near the recovery site, a side stream may be taken from the cracking or coke plant fractionating tower, mixed with the sands, the oil recovered and then the entire mixture returned to the cracking or coking operation, if desired by way of the fractionator. In view of the fact that the crude oil recovered is too viscous for transportation through a pipeline, it is necessary to convert it to a more fluid form. A particularly desirable way of doing this is to subject a portion or all of the crude oil to a coking operation and the crude oil can be reduced to a gravity and viscosity suitable for transporting through a pipeline.

A particularly suitable diluent comprises the overhead distillate from such an operation. The lighter gravity oil thus produced may be used as such or in a mixture with a portion of the oil as recovered from the sands. Instead of using all of the overhead distillate from the coking operation, a lower boiling fraction of the coke distillate, such as that boiling in the range of heavy naphtha, kerosene, or light gas oil, may be employed, alone or in admixture with the recovered crude oil. The solvent performs the function of dissolving the oil and reducing its viscosity and making it more mobile and susceptible to the stripping forces of the aqueous phase. It also reduces the specific gravity to a point below that of the aqueous phase, thus facilitating separation. This has the effect of reducing power and equipment requirements.

The amount of the solvent or diluent which is added will vary with the raw sands, and factors influencing the amount to be added are the porosity of the sand where the voids between the grains are filled to a greater or lesser extent; the degree of saturation, i.e., the amount of bituminous matter present in the sand; the density, fluidity and melting point of the bitumen; the specific gravity of the diluent; and the viscosity of the resulting mixture of diluent and oil. The best separation takes place when the specific gravity of the bitumen is reduced to between 0.79 and 0.95 by the addition of a suitable amount of the lower gravity diluent.

If the specific gravity of the bitumen is 1.025 (a common value for oils in sands found in the Athabasca River area and containing 10 to 20% by weight of oil) and if a kerosene of 0.75 specific gravity is used in equal parts by volume, the mixture of oil and kerosene will have a specific gravity of 0.887, which is in the desired range and has been found to be highly useful in this process.

The following comparative results, run as a batch, show the effect of varying degrees of dilution, and of the addition of a surface active agent and were arrived at through the following procedure. A uniform sample of raw Alberta oil sand was divided into a number of equal portions.

In each case two volumes of water were added after dilution, and the diluted sand and water thoroughly agitated before centrifuging. The first table below shows the results using kerosene (specific gravity = 0.81) as the solvent. The second table shows the results using recycled diluted oil recovered from the process (specific gravity = 0.87) as the solvent.

Kerosene as Solvent, 2 Volumes of Water

Percentage Solvent by Weight of Oil Sands	Specific Gravity of Diluted Oil	Ratio by Volume Diluent: Oil	Net Oil Recovery (Percentage of Oil Content by Weight)	S. A.,[1] Percent
50	0.842	5.25	91	----------
30	0.859	3.15	91	----------
25	0.865	2.63	91	----------
10	0.910	1.05	87	98
7.5	0.923	0.78	63	77
5.0	0.942	0.52	54	71
2.5	0.973	0.26	44	44
0	1.02	0.00	Nil	Nil

[1] Surface active agent Ethomid HT/60, 1 part to aqueous phase 40,000 parts.

Recycled Diluted Oil as Solvent, 2 Volumes of Water

Percentage Solvent by Weight of Oil Sands	Specific Gravity of Diluted Oil	Ratio by Volume Diluent: Oil	Net Oil Recovery (Percentage of Oil Content by Weight)	S. A.,[1] Percent
50	0.892	4.88	91	----------
30	0.905	2.93	87	----------
25	0.912	2.44	96	96
20	0.920	1.95	98	91
15	0.928	1.46	63	86
5.0	0.966	0.49	54	71
2.5	0.988	0.25	34	62
0	1.02	0.00	Nil	Nil

[1] Surface active agent Ethomid HT/60, 1 part of aqueous phase 40,000 parts.

The above data shows that when no solvent is used, the recovery of oil from the sands is nil. However, when the solvent is added in proportions amounting to 0.5 to 5 volumes of diluent per volume of oil contained in the sands, recoveries of from more than 50% of the oil up to more than 90% of the oil can be obtained. It should also be noted that when a surface active agent is added in the areas of lower diluent ratios substantial improvement in oil recovery is obtained. This means that by use of a surface active agent in conjunction with the diluent and water a lower amount of diluent can be added than is otherwise necessary. Similar results are obtainable when using a portion of the recovered diluted oil as the solvent.

In related work *G.R. Coulson and L. Clark; U.S. Patent 2,885,339; May 5, 1959; assigned to Can-Amera Oil Sands Development Ltd., Canada* have found that coarse sand may be separated from the oil, water, clay and fines by treatment in a fluidizing tower subjected to a countercurrent flow of water through a sand column.

Generally, the method consists in feeding downward, into the top of a tower, a slurry of oil bearing sand, diluent and water, maintaining a constant level of sand in the tower by constantly withdrawing sand from the bottom, and forcing a stream of water upward through the sand column, countercurrent to the slurry, and withdrawing the overflow of oil, clay and other fines, and water.

The raw oil-bearing sand may conveniently first be mixed with a diluent oil capable of dissolving the oil content. The use of such a diluent has the effect of reducing the viscosity and the specific gravity of the oil constituent thus permitting separation. In their natural state specific gravity of the oil content varies from 1.000 to 1.025. It is desirable to dilute the oil constituent to such an extent as to reduce the specific gravity of the diluted constituent substantially below 1.000 and preferably in the range of from 0.79 to 0.95.

This slurry is fed into the upper portion of the body of a column in which a constant sand level is maintained. To the bottom of the tower, water is supplied under pressure to create an upward flow at a rate of approximately 0.2 to 1.5 ft/min free water velocity in the column, more or less as determined by the size of the sand grains. In order to ensure even distribution of the water stream over the entire cross-sectional area of the column, the water may conveniently take the form of a plurality of upwardly or downwardly projecting jets located near the bottom of the tower and disposed at regular intervals over the cross section of the tower.

Sand is continuously withdrawn from the bottom of the column to maintain the sand level in the column, while the overflow from the column, consisting of oil, clay and other fines, and water is continuously withdrawn from the top of the column. In an installation of this type, the sand in the column may be considered as settled and initially motionless. As the rate of flow of the rising fluid is increased, the bed particles begin to move due to the lifting effect of the rising fluid. As the flow is increased, the motion of the particles increases. At a certain upward velocity, the sand becomes fluidized to a point where the abrasive action of the sand particles on each other causes rupture of the surrounding oil films, without however being sufficient to carry the solid particles upward in the stream.

The oil thus released is carried upward with the rising water and discharged from the tower as an overflow product together with the clay and other fines of the raw sand. In a continuous column with slurry feed at top and with sand grains descending against the rising water, the rising water velocity required is reduced in part by the rate of descent of the sand in order to maintain good attrition condition. This process is continuous, the oil-sand water slurry being continuously charged into the top, the water countercurrent from the bottom, the sand removed continuously from the bottom and the oil, clay, water and fines removed continuously by way of overflow, all at such rates as to maintain constant operating conditions in the tower.

Alkali Metal Silicate Treatment

A process described by *R.J. Stegemeier and P.W. Fischer; U.S. Patent 2,924,565; February 9, 1960; assigned to Union Oil Company of California* involves a low temperature process using a warm aqueous solution of a special alkali metal silicate, with or without other reagents, and a moderately heavy hydrocarbon diluent to separate the heavy oil from the bituminous sands. In this process special procedures and apparatus are used in handling the effluent from the mixing step in which these materials are heated and agitated with one another to effect the separation of the heavy oil from the sand.

In related work *P.W. Fischer, V. Kenny and J.W. Scheffel; U.S. Patent 2,903,407; September 8, 1959; assigned to Union Oil Company of California* treat and recover hydrocarbon oil from tar sand by pulping it with a mixture of aqueous sodium silicate and a hydrocarbon solvent at slightly elevated temperatures to separate the oil from the sand. Briefly, the process comprises the mining of surface or near surface deposits of tar sand not exceeding 12 inches in average dimension.

The mined material is immediately transferred without excessive exposure to the atmosphere to one or more storage basins into which it is introduced and kept submerged in an aqueous medium. Although water alone prevents many of the disadvantages of storage, preferably this aqueous medium is a dilute solution of sodium silicate. This storage maintains the material out of contact with air of the atmosphere, prevents water evaporation from the tar sand itself, and further prevents oil wetting of the sand grains.

The temperature of this pretreatment is preferably atmospheric or slightly above, such as between 60° and 160°F. At the higher temperatures evaporation losses of water become high, but a partial digestion and pretreatment of the tar sand is also accomplished, a result which is accelerated with increases in temperature. At the lower temperatures, the chunks of tar sand are completely inhibited from coalescing and hardening into a single mass similar to that from which they were mined. The stored and pretreated material is removed from storage and is fed to the treatment process.

J.W. Scheffel and P.W. Fischer; U.S. Patent 3,075,913; January 29, 1963; assigned to Union Oil Company of California have found that, in a process where clay-containing bituminous solids are reduced to a fluid pulp by agitation in the presence of aqueous sodium silicate at slightly elevated temperatures and the resulting pulp is mechanically treated, for example, by gravity settling or by centrifuging, to separate a solids fraction, a hydrocarbon oil fraction, and an aqueous sodium silicate fraction which is recycled to the pulping step, the buildup of clay in the sodium silicate recycle stream can be substantially prevented by introducing into the process system a small amount of an ionizable calcium compound. The calcium ions produced by the latter in aqueous solution serve to flocculate the clay which otherwise remains colloidally suspended in the circulating stream of sodium silicate, and allows such clay to be separated along with the nonclay solids in the separation step.

A number of other modifications of the above recovery process, using alkali metal silicates in the pulping operation are described by *J.E. Sherborne; U.S. Patent 2,921,010; January 12, 1960; A.E. Kelley; U.S. Patent 2,980,600; April 18, 1961; R.P. Vaell and P.W. Fischer; U.S. Patent 2,924,566; February 9, 1960; all assigned to Union Oil Company of California.*

Aqueous Extraction Using Seawater

J.H. Barkman, Jr., R.M. Jorda and J.B. Vogel; U.S. Patent 3,547,803; December 15, 1970; assigned to Shell Oil Company describe a process for recovering oil from a tar sand where pulverized tar sand is mixed with a liquid hydrocarbon diluent. The diluted tar sand is then mixed with a first volume of hot brine at a controlled pH. The materials which float in the first volume of hot brine are isolated and mixed with a second volume of hot brine containing a surfactant. The second volume of hot brine is also at a controlled pH. Finally, any hydrocarbon materials that are substantially free of solids and aqueous liquids and float in the second volume of brine are isolated. The process provides a method of separating or breaking oil from bituminous sands by aqueous extraction using neutral or acidic brine or seawater. Referring to Figure 3.11, tar sand ore is pulverized as by crushing or grinding in any conventional manner at a pulverizing station **10**.

FIGURE 3.11: AQUEOUS EXTRACTION OF TAR SAND USING SEAWATER

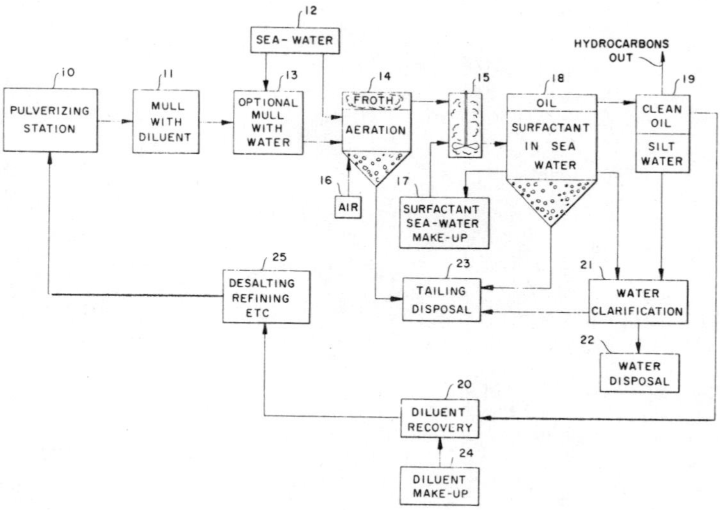

Source: J.H. Barkman, Jr., R.M. Jorda and J.V. Vogel; U.S. Patent 3,547,803; December 15, 1970

The pulverized ore is next mulled at mulling station **11** with a suitable hydrocarbon diluent in a conventional manner for a period long enough to allow the tar and diluent to mutually dissolve. The tar sand is substantially uniformly diluted to an oil phase exhibiting a viscosity of less than 100 cp at a temperature of less than 200°F.

At this point, if desired, the ore-diluent mixture may be mulled at station **13** with hot liquid brine or seawater, received from a seawater source **12**, having a pH of not more than 7 and a volume sufficient to extend between materials which float and sink in the brine or seawater so as to change the slurry characteristics of the ore. The resulting ore-diluent mixture (or slurry) at station **13** is introduced into a flotation stage or process comprising one or more flotation cells (such as skimming stage **14**, countercurrent washer **15**, and aeration cell **16**) which may employ in cell **16** aeration from any convenient air source or other conventional techniques as indicated to create a froth or emulsion upon the introduction of brine or seawater into the cell.

The froth or emulsion is then treated by countercurrent washing at washer **15** with a preferably hot liquid brine or seawater solution of a detergent or surfactant received from a surfactant-seawater makeup station **17**. The surfactant used is one that exhibits a significant amount of solubility in both brine and oil, the second volume of brine also having a pH of not more than 7, and a volume sufficient to extend between materials which float and sink in the brine or seawater.

The washed froth-surfactant slurry from washer **15** is separated, by conventional settling, into a solid and aqueous-free oil, surfactant in seawater and tailings mixture at station **18**. The oil is removed and, at station **19**, any hydrocarbons that are as substantially as volatile as the hydrocarbon employed as the diluent are evaporated and the remaining oil is recovered by passing it to a diluent recovery station **20**. The silt-water, separated from the solid-free oil at station **19**, is clarified with water at station **21** and the water and tailings are disposed of by conventional disposal means.

Thus, the water is passed to a water disposal station **22** and the tailings to a tailings disposal station **23**. The seawater removed from the oil-surfactant-seawater slurry at station **18** is also clarified with water at station **21** while the tailings are disposed of to station **23**, indicated in the drawing. The surfactant in the seawater is passed from station **18** to surfactant makeup station **17**. Finally, at the diluent recovery station **21**, any necessary diluent is added from a diluent makeup station **24** while any surplus diluent is passed to a processing station such as a desalting and refining station **25** for further processing by conventional means. The necessary diluent for mulling station **11** is supplied from the diluent recovery station **20**. Tailings from skimming stage **14** are also passed to tailings disposal station **23**.

Thus, as can be seen from the above, a clean hydrocarbon product with good recovery yields is obtained at the diluent recovery station **20**. The diluent from station **20** contacts the tar sand at mulling station **11** before any water is added. This is an important feature of the process since, if the diluent is added after water is introduced, recovery may be reduced. Once the diluent has been added, and well mixed, water may be added from station **12**, as desired, to improve the slurry characteristics. It is also important that the surfactant used to upgrade the froth or emulsion produced in the flotation stages is not added at an earlier stage in the process, such as during the mulling process at station **11** and **13** or the skimming stage **14**. In these cases, such an addition would be ineffective for improving the froth quality and may actually interfere with the flotation process by causing an increase in the amount of solids being floated and in the volume of froth due to inclusion of excess air. Suitable surfactants for use in the froth upgrading of the process are described in U.S. Patent 3,553,100.

Such surfactants may comprise an aqueous solution of an anionic surface-active material which is a free acid or salt of a complex organic phosphate ester (as for example, Gafac RE-610 and described in U.S. Patents 3,331,896 and 3,168,478). Further, any suitable closely related compound, as for example, any surfactant material in which ethylene oxide groups, acyl radicals and from about 10 to 22 carbon atoms are contained in each of the molecules of a surfactant that exhibits a significant amount of solubility in both oil and brine in addition to having a pH of not more than 7 and a volume sufficient to extend between materials which float and sink in the brine may be used.

The addition of water or brine from water source **12** to station **13** is optional since certain tar sand ores, such as those from the Edna, California region, as well as other regions, commonly contain from 3 to 10% water by weight in their natural state. The addition of more water, especially before contact with the hydrocarbon diluent, either reduces the recovery of tar from the sand or, if it is a very small amount, has no effect. It is believed that, in the former case, it hinders adequate solution of the tar and diluent by filling and blocking the pore volume. The diluent is generally required to effect complete separation of tar, although, with some tar sands, it may be possible to obtain separation without diluent by careful operation of the flotation process.

In related work *R.M. Jorda, J.H. Barkman, Jr. and H.F. Young; U.S. Patent 3,553,100; January 5, 1971; assigned to Shell Oil Company* describe a process for breaking or separating an emulsified mixture of liquid hydrocarbon, insoluble fluid and finely divided solid material. The emulsified mixture is mixed with a volume of hot brine having controlled pH and containing a surfactant material in which ethylene oxide groups, acyl radicals, and from 10 to 22 carbon atoms are contained in each of the molecules of a surfactant that exhibits a significant amount of solubility in both oil and brine. Hydrocarbon materials that are substantially free of solids and aqueous liquids and float in the brine are isolated and the hydrocarbon components of the isolated materials are recovered.

Utilization of Methane as Flotation Gas

R.M. Butler, J.L. Tiedje and J.A. Bichard; U.S. Patent 3,203,888; August 31, 1965; assigned to Esso Research and Engineering Company describe a method of securing the separation of bitumen particles from sand particles which, in essence, comprises the utilization of a hydrocarbon gas as, for example, methane which will function to change the apparent density of the bitumen particles to less than the density of water. The process is particularly adapted for use in conjunction with a water technique conducted at ambient temperatures where 40 to 400% by weight of water is added to the tar sands. It has been found that when tar sand is intermittently mixed with sufficient cold water, such as 200 weight percent of water at 70°F, a mixture containing small particles of bitumen and sand is obtained.

Under these conditions, the sand particles in such a dilute suspension are essentially bitumen-free. This is the case even when, under certain mixing operations, a relatively small amount of the water as, for example, from 30 to 50 weight percent is added and mixed under a shearing action and then the remaining quantity of water added to the slurry in order to dilute the slurry. This latter technique helps to break down any large particles in the sand.

However, in such a process as described, the dilute cold suspension in general tends to produce a settled layer containing both the particles of bitumen as well as the particles of sand. The reason there is no sharp separation of sand from the bitumen is that the bitumen has a density slightly higher than that of water and, thus, does not float. For example, specific gravities of Athabasca bitumen vary from 1.007 to 1.022 at 77°F. Even higher apparent specific gravities may be obtained if some fine clay is contained within the bitumen particle.

By this process, the apparent densities of the bitumen particles are lowered to less than that of water, thereby producing a very effective separation between the sand particles and the bitumen particles. Figure 3.12 specifies that natural tar sands are introduced into zone **3** and mixed with water which is introduced by means of line **2**. In zone **3**, the mixture is subjected to a mixing-shearing action. The slurry is then passed by means of pump **4** through line **5** into a gas absorption zone **14**. In this zone, the slurry is thoroughly mixed with the gas as, for example, methane, or ethane, which is introduced by means of line **6**. A preferred adaption is to maintain the zone **14** at a pressure in the range from 5 to 50 lb/in^2 gauge. Thorough mixing is secured in zone **14** by means of stirrer or equivalent means **15**.

FIGURE 3.12: UTILIZATION OF METHANE GAS TO CHANGE THE APPARENT DENSITY OF BITUMEN PARTICLES

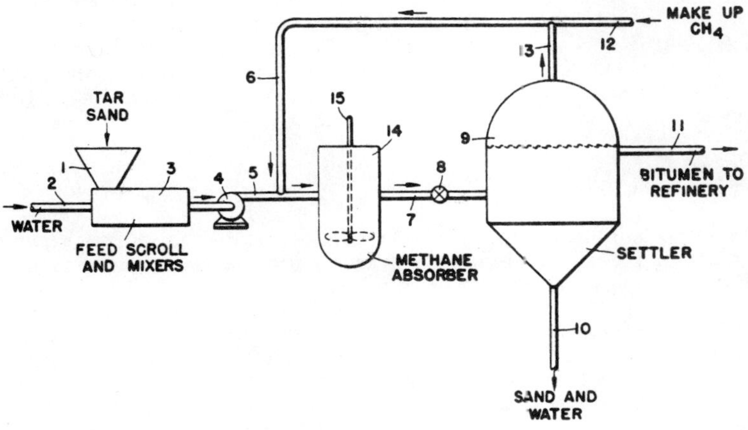

Source: R.M. Butler, J.L. Tiedje and J.A. Bichard; U.S. Patent 3,203,888; August 31, 1965

By contacting the tar sands with a normally gaseous hydrocarbon under moderate elevated pressure, the tar sand tends to dissolve the hydrocarbon gas, thereby changing its apparent specific gravity. The mixture is then passed by means of line **7** through valve **8** into a settler **9** which is maintained at a pressure below that of the pressure in zone **14** as, for example, from 5 to 20 lb/in^2 less. In zone **9** the hydrocarbon gas as, for example, methane comes out of solution and forms gas bubbles attached to the bitumen particles. This decreases the average specific gravity of the bitumen particles and causes the same to float.

Thus, the bitumen particles may be removed as a bitumen phase by means of line **11** while the sand and water phase is removed as a bottoms stream by means of line **10**. Gas particles are removed overhead by means of line **13** and repressured as a recycle stream. Makeup gas may be added by means of line **12**. A further advantage of the process is that, even if clay is present in the bitumen particles, these clay-containing particles can also be made buoyant by the employment of adequate hydrocarbon pressure. In general, it is preferred that the pressure in zone **14** be in the range from 20 to 30 lb/in^2 gauge, preferably 25 lb/in^2 gauge. It is preferred that the pressure in zone **9** be 10 to 15 lb/in^2 gauge below the pressure existing in zone **14**.

In a process described by *M.C. Lowman, Jr. and E.J. Fisch; U.S. Patent 2,871,180; January 27, 1959; assigned to Shell Development Company* an aqueous pulp of the bituminous sand is supplied to a vertical extraction zone and a liquid low molecular weight paraffinic solvent flows in an upward direction of the vertical extraction zone in countercurrent direction to the downwardly descending sand. During the countercurrent movement of the aqueous pulp and solvent, there is formed a deasphalted oil and solvent phase, an asphaltene phase diluted with a lesser portion of the solvent, a water phase and a substantially oil-free sand. The several phases and sand are then separated into distinct layers with the asphaltene-enriched layer being disposed below the point of introduction of the hydrocarbon solvent and the water layer immediately below the asphaltene layer. The sand settles to the bottom of the water layer. A deasphalted-oil enriched solvent layer forms above the point of introduction of the aqueous pulp.

Addition of Granulated Carriers During Mixing

J.A. Bichard; U.S. Patent 3,153,625; October 20, 1964; assigned to Esso Research and Engineering Company describes an integrated process where in an initial operation, the concentration of the bitumen is increased on a certain quantity of sand and then this bitumen-rich sand is handled in a coking operation such as a fluid type coking operation. The carrier particles or granulated carriers may be nonporous as for example stones, but are preferably a hydrocarbon adsorbent as for example coke particles such as are secured in a fluid coking operation. In essence, the operation comprises mixing the carrier which preferably comprises coke particles with the tar sands which mixture is then treated with an optimum amount of water and mixed so as to cause separation of oil-free sands and the enrichment of the remaining sands.

Vacuum Separation Technique

R.A. Hemstock; U.S. Patent 3,139,397; June 30, 1964; assigned to Esso Research and Engineering Company describes a recovery technique which employs a vacuum separation step. The following examples illustrate the process.

Example 1: A 100 grams of a sample of tar sands (dry basis) containing 7 weight percent connate water and 13 weight percent oil was mixed with 30 weight percent water for half a minute to effect disintegration of the tar sands into sand and very small oil particles distributed in water. This was carried out at 45°F. The mixture was transferred to a flask containing excess water and the pressure was reduced. When the pressure reached 200 mm of mercury, oil floated to the surface. At lower pressure, increasing amounts of oil floated to the surface. Complete separation was obtained, with clean sand remaining at the bottom of the vessel, in less than half a minute from starting the evacuation. Very small oil particles that floated to the water surface immediately coalesced, forming large agglomerates, thus reducing the water content of the oil mass. These agglomerates were skimmed off the water.

Example 2: The sand reduction in this operation can be compared with that from mixing with water and mechanical separation. This is shown in the following table. From the table it is apparent that a better oil recovery is secured in a shorter time period when utilizing a vacuum separation technique. Good sand reduction is accomplished at lower temperatures and with lower amounts of water in premixing and minimum premixing time. The tar sands used contained 7 weight percent water and 13 weight percent oil.

	This Process	Sand Reduction [3]
Temperature, °F	45	70
Premixing Time, Minutes	0.5	5
Premixing Water, Wt. percent [1]	30	75
Sand Removed, Wt. percent [1,2]	66	80
Oil Recovery, Wt. percent	90	88

[1] On tar sands (dry basis).
[2] Also contains oil not recovered in oil phase.
[3] Use of mechanical separation at atmospheric pressure.

Use of Shearing-Mixing Zone to Obtain Bitumen-Rich Sand

J.A. Bichard and J.W. Wunder; U.S. Patent 3,159,562; December 1, 1964; assigned to Esso Research and Engineering Company describe an integrated process for the recovery of oil from tar-sands where in one operation or an initial phase, at relatively low temperatures, the process comprises removing substantially oil-free sand from a quantity of tar sands as mined, thereby increasing the concentration of the tar or bitumen on the remaining sands. The bitumen-rich sand is then further handled in a secondary phase utilizing steam and relatively high temperatures in order to produce an oil phase containing a relatively small amount of sand.

Referring to Figure 3.13, tar sands as mined are introduced into a shearing-mixing stage or zone **3** of the sand separation phase by means of line **1**. Water is introduced by means of line **2** and mixed with the sands. This water may be introduced directly into the shearing-mixing zone **3**. As a specific example, the composition of the sand introduced into zone **3** comprises 6% by weight connate water, 15% by weight of bitumen, 80% by weight of sand and 5% clay. The clay content varies from 0 to 30% in these sands but, in the specific example given, the clay comprises 5% by weight of the total.

FIGURE 3.13: TWO PHASE SEPARATION PROCESS

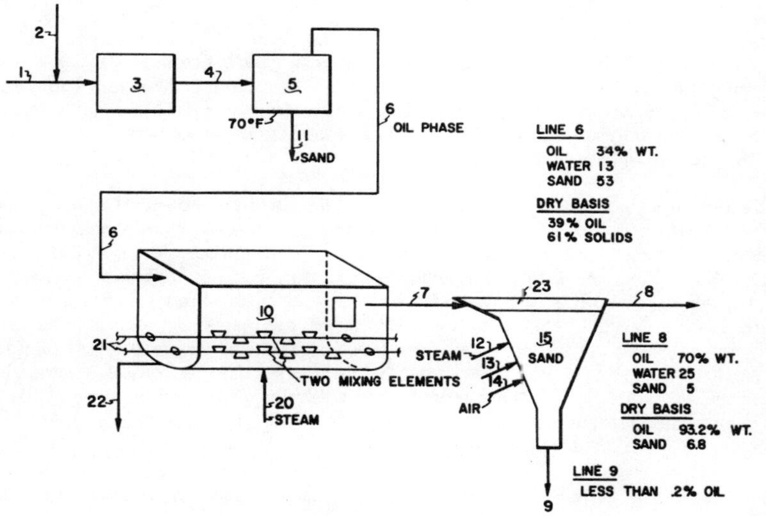

Source: J.A. Bichard and J.W. Wunder; U.S. Patent 3,159,562; December 1, 1964

The mixing operation in zone **3** is carried out under conditions where a shearing thrust is imparted to the sands being mixed with the added water. Under these conditions, substantially oil-free sands separate and the entire mixture is then passed into the second stage or zone of the sand separation phase which preferably comprises a mechanical separation zone **5**. The sands are introduced into zone **5** by means of line **4**.

This mechanical separation zone may comprise any suitable mechanical means for separating the substantially oil-free sands from the bitumen-rich sands. For example, the mixture introduced by means of line **4** may be sieved in order to separate the oil-free sand which oil-free sand is removed by means of line **11**. The bitumen enriched sands in a plastic physical state are removed by means of line **6** and passed into the second phase of the integrated process which comprises a relatively high temperature operation. Stages **3** and **5** are preferably conducted at 32° to 110°F, preferably 50° to 90°F. It is also preferred that the amount of water utilized in stages **3** and **5** be 40 to 400% of water by weight, preferably **40** to 150% of water by weight based upon the tar sands being processed.

Under the conditions of the process, the oil phase removed from stage **5** by means of line **6** contains from 30 to 50% of oil. This phase is introduced into separation zone **10** which preferably comprises a pug mill where steam is injected by means of line **20**. Zone **10** may be any desirable type of pug mill as, for example, one which contains mixing elements **21** at the bottom with paddles in order to remove the sand countercurrent to the introduction of the feed stream. The sand stream is removed by means of line **22**.

The temperature in zone **10** is maintained in the range from 140° to 210°F, preferably at a temperature in the range from 185° to 195°F. The amount of steam introduced into zone **10** is adjusted so that there exists in this zone 20 to 160% of water, preferably 30 to 100% of water by weight based upon the oil present. The mixed oil phase is withdrawn from pug mill **10** by means of line **7** and introduced into a separation zone **15**.

In this zone an oil phase **23** formed on the top which is removed by means of line **8**. Sand collects in the lower area of zone or cell **15** and is removed by means of line **9**. In accordance with this example, steam and/or water is introduced into cell **15** by means of lines **12** and **13** and air by means of line **14** in order to bring the sand in a fluidized condition. A plurality of addition points may be used.

Example: In one operation, a sand containing 6% by weight of connate water, 15% by weight of bitumen, 80% by weight of sand and 5% clay was processed in zones **3** and **5** at a temperature of 70°F in a manner as described. The oil phase removed from zone **5** comprised 34% by weight of oil, 13% by weight of water, and 53% by weight of sand. This feed was introduced into zone **10** at 190°F under conditions where approximately 40 to 100% by weight of water is present.

The streams were then introduced into zone **15** maintained at 190°F and an oil phase removed by means of line **8** and a sand phase by means of line **9**. Under these conditions of operation, the oil phase removed by means of line **8** comprised 70% by weight of oil, 25% by weight of water, and 5% by weight of sand on a wet basis. The oil content was 93.2% by weight, while the sand was 6.8% by weight on a dry basis. In the same operation, the sand removed by means of line **9** contained less than 0.2% by weight of oil. Thus, overall, the process is concerned with a two-phase operation where in the first phase tar sands are subjected to shearing-mixing agglomeration action at relatively low temperatures, and in a second phase the oil phase is treated at a relatively high temperature utilizing injected steam and air. Under these conditions of operation, the oil phase is relatively pure and substantially no oil is lost in the sand phase.

In related work *J.A. Bichard, R.M. Butler, J.R. McEachern and J.W. Wunder; U.S. Patent 3,152,979; October 13, 1964; assigned to Esso Research and Engineering Co.* describe an integrated process for the recovery of oil from tar sands, which utilizes a shearing-mixing stage and then employs a particular technique in a secondary separation stage. In the latter stage, air is utilized for the more efficient separation of the oil phase.

Hydrogenation of Emulsion

E.W. White; U.S. Patent 3,291,717; December 13, 1966; assigned to Cities Service Athabasca, Inc., Imperial Oil Limited, Atlantic Richfield Company and Royalite Oil Company Limited, Canada has found that water can be effectively separated from bituminous emulsions by a relatively mild hydrogenation of the emulsion. Further, such a mild hydrogenation serves as a convenient point for further hydrogenating and separating the bitumen to more useful products such as those having an API gravity above 15 or above 20. Hydrogenation of the bituminous emulsion of froth destroys many of the surfactants and reduces the specific gravity, viscosity and surface tension of the bitumen. Also it is believed that hydrogen sulfide released from hydrogenated sulfur compounds in the emulsion decreases the pH of the mixture which also tends to favor separation.

The process comprises feeding the products of a heated bituminous emulsion to a reactor maintained under suitable hydrogen pressure and reaction conditions for reacting hydrogen with such feed and particularly the bitumen when it is desired to effect hydrocracking in addition to merely breaking the emulsion. The major portion of the water vaporizes in the reactor whereas the bitumen remains in the liquid phase. During hydrogenation much of the surfactant content of the emulsion is destroyed. The oil is then separated from water by various techniques such as cooling the effluent from the reactor to a temperature below 400°F and flashing to reduced pressure.

The hydrogen which is supplied to the reaction system can be substantially pure hydrogen or of a gaseous stream containing other constituents such as carbon oxides, nitrogen, methane, ethane, and steam. The hydrogen-containing gas introduced to the reaction zone is adjusted to provide sufficient hydrogen for reaction of about 600 scf per barrel of bitumen feed to 1,500 scf per barrel of bitumen feed. Preferably, the hydrogen consumption for hydrocracking of the bitumen is from 800 to 1,000 scf per barrel of bitumen feed. When it is principally desired to merely make a separation of the emulsion into the oil and water phase with minimum cracking of the feed, the hydrogen consumption of the feed can vary from 600 to 800 scf per barrel of bitumen in the feed.

The hydrogen partial pressure in the reactor can vary over a wide range such as 800 psig to 2,000 psig and preferably 1,000 psig to 1,500 psig. The total pressure in the reactor will be substantially greater than that of hydrogen, particularly due to vaporization of water in the reactor although the total pressure preferably varies from 1,000 to 3,500 psig. The temperature in the reactor can vary from 700° to 900°F and preferably from 775° to 880°F. Since water is the more easily vaporized major constituent of the feed and for the most part vaporizes in the reactor, the lower and intermediate portions of the reactor are enriched with oil.

Hydrogenation of the bitumen is in the liquid phase, preferably without the addition of a catalyst. However, conventional hydrogenation catalysts can be employed. After hydrogenation the water is separated from the oil by cooling and flashing the effluent. Water can then be recovered as overhead vapor or condensed as a bottom liquid layer, depending on temperature and pressure of the effluent.

Hydraulic Cyclone

M.R. Tek and S.J. Marwil; U.S. Patent 2,910,424; October 27, 1959; assigned to Phillips Petroleum Company describe the use of a hydraulic cyclone in the separation and recovery of oil from oil sands. The process comprises introducing a suspension or slurry of the oil sand into a hydraulic cyclone, subjecting the slurry to the forces developed in the cyclone, withdrawing a stream consisting essentially of sand from one end of the cyclone, and withdrawing a stream of oil and water from the other end or base portion of the cyclone. The stream of oil and water is then processed for the recovery of the oil.

Example: 120 lb/min of Athabaska oil sand having an oil content of about 14% by weight is crushed to the extent that the particle size is less than 60 mesh. The crushed sand is introduced into a mixing tank agitated with a motor driven stirrer and is there mixed with 60 gal/min of water introduced at 170°F to produce a slurry or suspension containing about 19.3% by weight of solids. 80 cc/min of Aerosol MA are introduced into the mixing tank to aid in the formation of the suspension or slurry. The residence time in the tank is about 21 minutes.

The suspension or slurry is pumped at a rate of 66 gal/min and a pressure of 60 lb/in^2 gauge through a ¾ inch i.d. inlet orifice and tangentially into the base portion of a 4 inch diameter cyclone with a 10 inch cone section which gradually and uniformly tapers to a ⅝ inch underflow or apex nozzle. The cyclone has a ¾ inch i.d. overflow orifice axially positioned in the base portion or feed section. A stream containing about 3% by weight oil and about 97% by weight water together with a small amount of finely divided sand is removed through the overflow orifice. The remainder of the slurry introduced into the cyclone is removed through the underflow or apex orifice.

The overflow stream of oil and water is filtered to remove the sand and then passed to a phase separation vessel where a surface active agent of the emulsion breaker type is introduced at a rate of about 30 cc/min. A stream of oil in an amount of about 15 lb/min or 900 lb/hr is recovered from the phase separator.

Sound Wave Induced Separation

A.G. Bodine; U.S. Patent 3,123,546; March 3, 1964 describes a process and apparatus for separating the petroleum constituent from tar sand mixtures by use of high intensity sound waves. One typical treatment apparatus constitutes a batch container, with the walls functioning as a sound wave generation and/or transmission system. A liquid may simply be added to the tar sand in the container. The added water, or other liquid, is apparently helpful in that it provides a fluid medium of intermediate density and impedance. It is additionally very helpful and important in that it very greatly increases the degree of acoustic coupling between the sound wave source, e.g., the wall of the container, and the mixed sand and petroleum material. In the operation, assuming the tar-sand material to be placed in a container, and intense sound waves passed through, the oil quickly migrates to the surface of the sand.

With liquid added to the batch, the material disintegrates as the sound wave action drives the liquid into the material, and the oil leaves the sand more completely, undoubtedly as a result of the sound waves being transmitted more effectively to and through the mass. The oil rises to the surface of the water (if water is used as the added liquid), and the sand collects in the bottom of the container. The sand settlement occurs no matter what the nature of the added liquid may be. Complete details of the apparatus are provided.

Mechanical Separation Technique

R.T. Eyre; U.S. Patent 2,790,750; April 30, 1957 describes mulling the bituminous sand with a minor amount of water until the pulp has reached a certain state. The pulp is then vigorously mixed with a large volume of water and the mixture is fed into an inclined conduit which functions as the separating zone, the conduit being inclined at an angle of 28° to 35°. The dispersion of mulled tar sands in water is fed at a low temperature, below 50°F, into the conduit which is maintained filled with liquid, and the conduit is subjected to vibrations to aid the flow of material down the inclined bottom of the conduit. This causes the mixture to separate effectively into a lower layer of sand particles, an intermediate layer of oil and an upper water layer containing the clay or silt which was present in the tar sands charged. The lower layer of sand is trapped and drawn off, while the oil and water layers are also separately withdrawn from the separation zone.

OTHER PROCESSES

Extraction Apparatus with Endless Perforate Conveyor

A process described by *T.A. Pittman and J.L. Woods; U.S. Patent 3,856,474; December 24, 1974* is directed to the recovery of bitumen or oil from tar sands or oil sands. An endless perforated conveyor belt is used and a spray technique employed such that a maximum of recovery can be had with the minimum use of solvent. Spray pools are recovered below the endless conveyor belt and are cascaded such that the most oil-rich pool occurs most proximate the forward end of the conveyor and nearest to the collector in which the solution of the pool is pumped.

Variable weir features are utilized to effect the cascading necessary. The system is so arranged that clean solvent need only be used to spray the last sector of a run. Other spray compartment areas simply have recovered solutions recirculated through the spray mechanism. A portion of the recovered petroleum product may be used as an energy source for supplying heat, refrigeration and other power where needed.

In the accompanying diagrams, Figure 3.13a is a side elevation, partially cut away, of the apparatus of the process. Figure 3.13b is a fragmentary perspective of a portion of the conveyor belt utilized in Figure 3.13a. Figure 3.13c is similar to Figure 3.13b but illustrates an optional or alternate belt that can be used. Figure 3.13d is a transverse section taken along the line **4–4** of Figure 3.13a. Figure 3.13e is a fragmentary perspective illustrating the weir structure. Figure 3.13f is a schematic illustrating the manner in which power is supplied to various motors and pumps in the system.

In Figure 3.13a apparatus **10** includes an enclosure housing **11**. The same provides a complete enclosure for the structure provided endless conveyor belt **14**, but also includes a slot **D** for permitting passage therethrough of spent ore to endless conveyor **83** driven by motor **M1**. Such spent ore of the process about to be described may be used for further processing or simply for disposal. Disposed within housing **11** is a frame structure **12** which includes upstanding end roller yoke supports **13** at either end of the structure. Two such support yokes respectively journal discharge end roller **15** and feed end roller **16**.

FIGURE 3.13: TAR SANDS OIL EXTRACTION PROCESS

a.

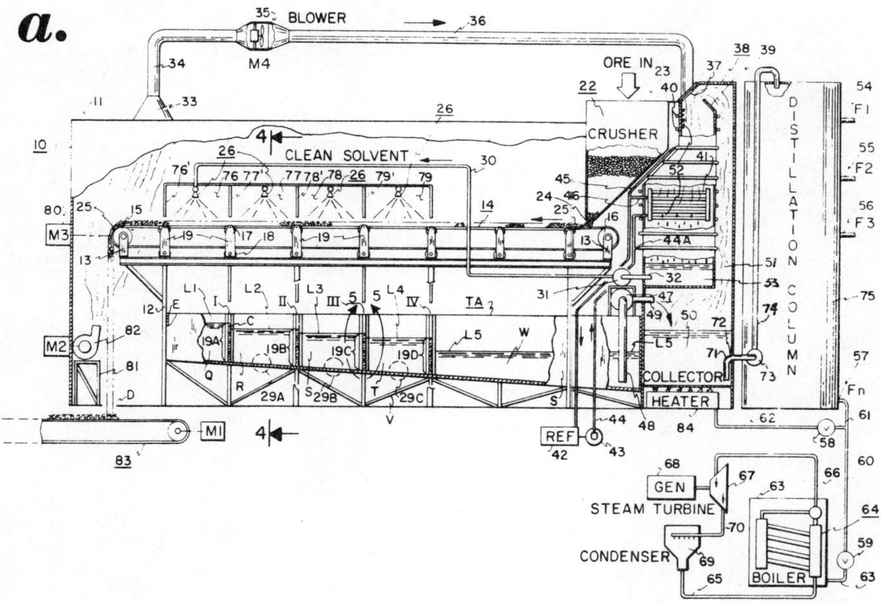

b.

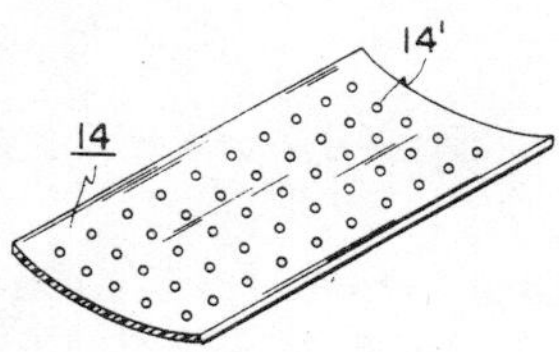

c.

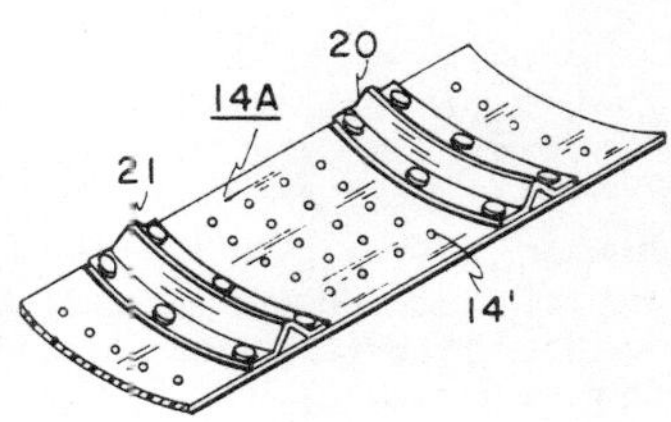

(continued)

FIGURE 3.13: (continued)

d.

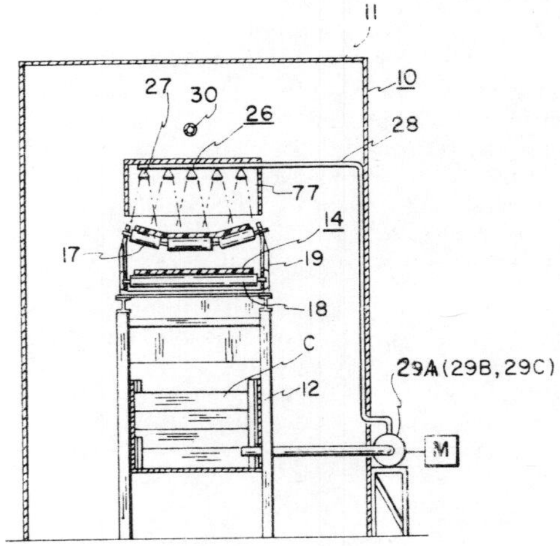

e.

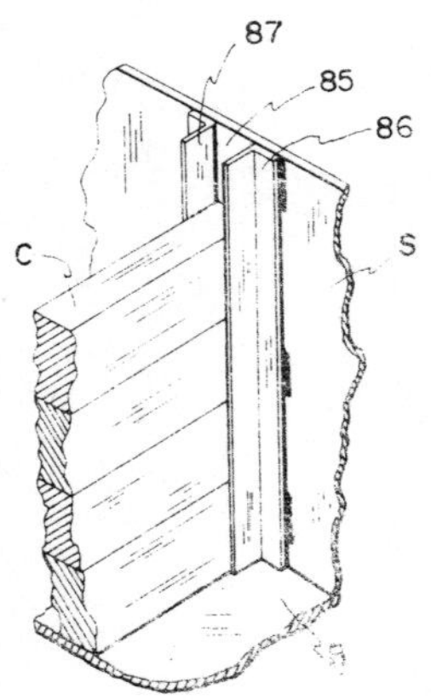

f.

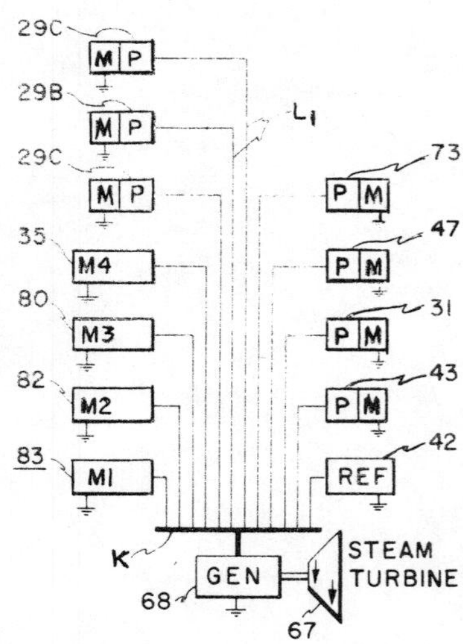

Source: T.A. Pittman and J.L. Woods; U.S. Patent 3,856,474; December 24, 1974

Also included with respect to frame **12** is a series of upstanding yokes or U-shaped members **19**, the same journalling idler rollers **17** and **18** at each respective location. Accordingly, the endless conveyor belt travels about the idler rollers and also the end rollers **15** and **16** in the manner shown to convey crushed ore **25** from right to left.

The conveyor belt **14** is perforate and may be of any suitable metalic, web, or elastomeric material. In any event, the belt will have the series of perforations **14'**. As an alternate to the belt **14** of Figures 3.13a and 3.13b, there may be provided a substitute belt **14** employing spaced cleats **20** and **21**, where this is deemed desirable for ore conveyance. A solution-collection tank **T** is provided and is preferably supported by the frame **12** having support bracing **V**. The solution collection tank includes closed opposite ends **E**, closed opposite sides **S**, and bottom **B** all being integrally formed. Figure 3.13e illustrates that welded to the sides **S** will be a series of angles **86** and **87** to provide for a slot **85**. Opposed slots are provided to receive one or more weir members **C**. These may be unitary or may be stacked vertically as shown in Figure 3.13e. Multiple element weirs as shown in Figure 3.13e are preferred in some applications since the fluid level may be incrementally adjusted as appropriate for each collection section.

At this point, it is appropriately mentioned that bitumen-solvent solution pools will establish fluid levels **L1–L5** within respective solution compartments or volumes **Q**, **R**, **S** and **T**. The final compartment or volume at **W** directly supplies the suction conduit **48** of pump **47**. In returning to the input end of the endless conveyor at roller **16**, it is seen that the apparatus is supplied a fixedly supported crusher **22** having inlet opening **23** and outlet opening **24**. Crushed ore proceeds, hence, from the interior of the hopper to appear as a layer **25** on the upper side of the enless conveyor **14**. As the layer of ore, preferably not more than 1½ inches thick, and preferably crushed to pass a minus twenty screen mesh, proceeds from right to left on the conveyor belt, the same will pass underneath sprays as are given by spraying means **26** within the several compartments **76** through **79**. This spraying means is shown in detail in Figure 3.13d, where a respective spraying means is supplied at **29** to which conduit **28** is connected.

A series of spray nozzles **27**, of standard form, are secured to conduit **28** of the means **26** such that these sprays spray downwardly upon the crushed ore of conveyor belt **14** in Figure 3.13d. It will be observed that each of the compartments formed by the intercompartment partitions, i.e., weirs **19A** through **19D** and compartment walls **76′** through **79′**, will have an independent spraying means **26**. Thus, each of the compartments includes its own respective pump **29A** through **29C** and respective conduit (**28**) and spray means. Accordingly, each compartment has its own spray circuit.

Collector unit or chamber **38** includes a vaporization baffle **39**; disposed within the interior of unit **38**, preferably, is a condenser structure **41**. With respect to collector unit **38**, the same receives pressure conduit **49** of pump **47**, thereby receiving bitumen or oil dissolved within the solvent within the collector chamber as indicated at **50**.

A coolant conduit **40** may be disposed about the condenser chamber of structure **38**. Thus, the cooling means associated with the structure proximate **53** may be performed either exterior to the structure by virtue of coolant conduit **40**, or may be performed interior of such structure by a condenser **41**. Such cooling effect may be performed by circulating tap or stream water, or there may be provided a refrigerant and compressor **42** which, with refrigerant pump **43**, completes the circuit as by supply conduit **44**, supply conduit stub **44**, return conduit **45**, and return conduit stub **46**. These are all connected together in a conventional manner as shown in Figure 3.13a.

The purpose for the baffle **39** is to direct ascending vapors **51** from the pool **50** upwardly and to the right so that the same falls down as condensation or collects upon condenser **41**. A visual indication of such condensed vapors is given at **52**, which condensation settles at **53** for subsequent reuse. Conduit **32** serves as an outlet for the condensed solvent at **53**, and pump **31** serves to receive such solvent and supply the same under pressure via conduit **30** to the several spray means **26**.

The outlet of the collector unit **38** in Figure 3.13a receives fluid at **50** and routes the same, via pump **73**, aperture **72** and conduit **74** to the top of the conventional refinery, fractionation, or distillation column **75**. Such structure **75** includes outlets **54** through **56** for recovering desired fractions **F1** through **F3** in the usual manner and, in the usual manner, and, in addition, supplies a fraction **Fn** at conduit outlet **57**. This fraction will be in the heavier region, such as conventional refining bottoms or heavy fuel oil, which may be conducted by connection **57** and conduit **60** to and through manual control valves **58** and **59**. Conduit **62** interconnects valve **58** with heater **84**, and valve **59** is connected to the boiler as indicated by conduit **63'**. Conduit **65** interconnects the boiler **63** with condenser **69**. Conduit **66** interconnects steam turbine **67** with the boiler **63**. Generator **68** is mechanically coupled to steam turbine **67** in the manner indicated, and conduit **70** serves a spent steam return to condenser **69**, such condensation to be fed by pump or other means to the boiler in a usual manner.

With respect to the tank **TA**, it is noted that the same is divided into compartments by the weir structures **19A** through **19D**, these being made up of one or more weir members **C**, see again Figure 3.13e. The weirs are constructed such that each successive weir, going from left to right, is at a progressively lower height. Thus, there is a series of cascading pools having liquid levers **L1** through **L5** as indicated. The purpose for this is to make the pools successively richer as the collection chamber is approached.

It is noted that crushed ore therefore is received by the crusher at **25** and is conveyed through the system from right to left. The clean solvent which is conducted by conduit **30** is sprayed onto the crushed ore at the respective compartments, i.e., at **76** through **79**, and such solvent carries with it the oil or bitumen of the crushed ore. The solvent and entrained oil passes through the apertures or holes of the conveyor, the same falls into the pools of compartments **Q** through **W**. The fact of independent circulatory systems for each of the compartments is to be noted.

As to nomenclature with reference to Figure 3.13a, **M** and **P** within blocks, see Figure 3.13f, indicate motors, pumps and motors driving otherwise identified pumps. It is seen in Figure 3.13a that the clean solvent is routed through conduit **30** solely to the foremost compartment at **76**, whereas the solution below each of the prior stages is recirculated in such stages for further recovery of bitumen from the crushed ore as the same passes through the route shown. By such method the quantity of clean solvent needed is greatly reduced, and simply performs the final step of cleansing any remaining ore or bitumen from the processed crushed shale at its last stage. The solution pools at **S** and **T**, for example, both containing oil, may be recirculated through new ore passing into these chambers so as to enhance recovery.

The purpose for the cascading pools by virtue of the weir construction is simply to ensure that the progressively dirtier solvent, that is, the solvent which includes the most oil or bitumen, runs progressively rearwardly to pool **W** forming fluid level at **L5** and leading to the recovery chamber. Thus, the pool having level **L1** will have the least oil entrained whereas the pool at **L5** will have the most oil entrained. Accordingly, only a small amount of solvent is needed for spray appreciation, i.e., at the last stage of the recovery process.

Accordingly, the oil and solvent mixture is taken at conduit **48** and is exhausted into the collector by virtue of pump **47**. In the collector at structure **38**, the solvent will gradually vaporize under ambient temperature conditions, say from 60° to 100°F. This is particularly true where one of the three following solvents is used: methylchloroform, trichloroethylene, and perchloroethylene. These solvents are highly preferred since they have low heats of vaporization, low specific heats, are not flammable.

Where it is desired to speed the vaporization process of the solvent within the solution at **50**, then heater **84** may be supplied with an appropriate fuel fraction from column **75**, this is indicated in Figure 3.13a. Cooling of the collector structure at **38** is accomplished by either the interior condenser or by exterior coolant pipes, the same activated by refrigerant **42** of any conventional form in combination with refrigeration pump **43** in the usual manner.

In addition to supplying heat and coolant, the fraction taken at **Fn**, **57** in Figure 3.13a, may also be used to actuate a steam turbine and electrical generator, see **68** in Figure 3.13a, which will supply power to the various motors and pumps used in the system, see Figure 3.13f. These elements will be supplied power by leads L_1 connecting to the bus bar **K** of the electrical circuit of Figure 3.13f. Where it is desired, a motor-driven blower at **82** be provided so as to blow any vaporized solvent or fumes up the stack including section **33** and conduit **34** and **36**. This is performed by blower **35** which uges such fumes back into the collector structure at **38**. In being positioned above the slot **P**, the blower **35** not only dries the final run of spent ore but also eliminates essential loss of the solvent either in vapor or liquid form.

Use of Soluble Oils

A process described by *F.H. Poettmann and J.T. Kelly; U.S. Patent 3,392,105; July 9, 1968; assigned to Marathon Oil Company* comprises contacting the oil sands with soluble oils, then diluting the tar sand-soluble oil mixture with a solvent-diluent, and separating the sand from the resulting relatively low viscosity solution. The term soluble oil refers to solutions of surfactants in nonpopular solvents with the ability to emulsify water. The process involves a combination of separation and dilution steps which are to be carried out aboveground. Comminuted oil sands are contacted with a soluble oil to form a slurry in a mixer unit; the slurry is then mixed with a solvent-diluent and separated. This process is illustrated with reference to Figure 3.14.

FIGURE 3.14: EXTRACTION PROCESS USING SOLUBLE OILS

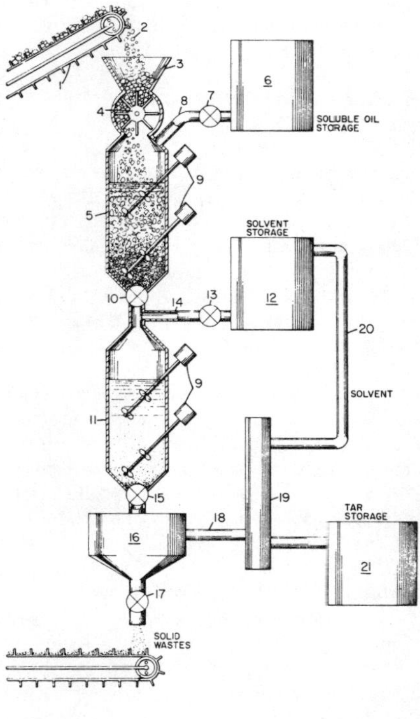

Source: F.H. Poettmann and J.T. Kelly; U.S. Patent 3,392,105; July 9, 1968

In the figure illustrated above, conveyer **1** passes comminuted oil sands **2** into the hopper **3**. Comminuted oil sands **2** pass through the star feeder **4**, or other suitable valving mechanism, into slurry preparation tank **5**. Soluble oil from the storage tank **6** passes through valve **7** in line **8** to slurry preparation tank **5** where it is rapidly mixed with oil sands **2** by stirring devices **9**. Normally, from at least about 0.1 to 10, preferably at least from 0.5 to 15, volumes of soluble oil per volume of oil sand produce a desired slurry.

A mixture of four parts tar sand and one part of a soluble oil comprising straight run gasoline, isopropanol (about 4%), and water (about 25%) creates a loose slurry on mixing. After a slurry of suitable viscosity, preferably from about 500 to 50,000 cp, is prepared, it passes through valve **10** into dilution tank **11** where it is mixed with solvent. Solvent, from solvent storage **12** passes through valve **13** in line **14** into dilution tank **11** where it serves to further reduce the viscosity of the slurry. A mixture of four parts of the above loose slurry and one part straight run gasoline gives a solution which appears to have a viscosity approximating that of kerosene.

Normally, from at least 0.5 to 15 volumes of solvent are added per volume of slurry. The amount of solvent added depends upon the viscosity of the solvent, the temperature of the solvent and the slurry, etc. Solvents such as straight run gasoline, liquefied petroleum gases, xylene, isopropanol, dimethylformamide, furfural, phenol, etc., can be used.

When diluted to a desired degree, for example, from 30 to 1,000 cp, the diluted mixtures pass through valve **15** into separator **16**. Separator **16** is depicted as a cyclone separator but can be any desired liquid-solid separator. On separation, the sand and other solid wastes pass through valve **17** to a waste disposal means. The liquids pass from separator **16** through line **18** into distillation tower **19** where the light, low-boiling solvent is distilled from the mixture. The solvent is then recycled through line **20** to solvent storage **12**. The bituminous bottoms are then removed to storage tank **1**. From storage, the bituminous materials are transported to market, etc.

Addition of Hydrocarbon Diluent

W.H. Price; U.S. Patent 2,965,557; December 20, 1960; assigned to Cities Service Research and Development Company describes a process by which crude oil and silt may be separated from bituminous sands with essentially no loss of crude oil with waste sand. Accordingly, crude oil and silt may be separated from bituminous sands by mixing the bituminous sand with a small amount of hydrocarbon diluent and introducing the mixture into a separation zone in which a lower fluidized layer of sand and an upper liquid layer of oil and silt are formed. Additional hydrocarbon diluent is introduced into the lower layer of sand to aid in the separation of the original mixture into these upper and lower layers and each of the layers is separately removed from the separation zone.

In a continuation *W.H. Price; U.S. Patent 3,070,541; December 25, 1962; assigned to Cities Service Research and Development Company* has found that it is desirable to remove some or all of the silt prior to passing the tar sand mixture to the processing unit. Thus, for instance, if the mixture is to be passed to a coking unit, removal of a majority of the silt before passing the mixture to the coker will substantially lessen the load on the coking unit and will also make a larger proportion of the coke produced available for supplying heat or for other purposes.

In the process, a mixture containing crude oil and silt separated from bituminous sand in a sand separation zone and preferably containing liquid hydrocarbon diluent is passed to a first silt settling zone wherein it is allowed to settle to form a lower layer of silt, crude oil and hydrocarbon diluent and an upper layer of hydrocarbon diluent and crude oil. The upper layer is withdrawn from the first silt settling zone and preferably passed to a fluid coking unit for conversion of crude oil into liquid hydrocarbon products and coke while the lower layer containing silt is withdrawn from the first silt settling zone and passed to a second silt settling zone.

Additional hydrocarbon diluent is introduced into the second silt settling zone, preferably together with the silt, and the resulting mixture is allowed to settle to form a lower layer of silt substantially free of crude oil and an upper layer of hydrocarbon diluent substantially free of silt. Hydrocarbon diluent is preferably withdrawn from this upper layer and recycled as the original hydrocarbon diluent contained in the mixture of silt and crude oil passed to the first silt settling zone. Silt is withdrawn from the second silt settling zone and passed to a silt washing zone to remove the remainder of the hydrocarbon diluent therefrom.

In the silt washing zone the silt is contacted with water to form a lower layer containing silt and substantially free of hydrocarbon diluent, an intermediate layer of water and an upper layer of hydrocarbon diluent. Silt is withdrawn from the lower layer of the silt washing zone and discarded while water is withdrawn from the intermediate layer and hydrocarbon diluent is withdrawn from the upper layer of the silt washing zone and recycled with the hydrocarbon diluent from the second silt settling zone. A portion of the liquid hydrocarbon product from the fluid coking zone is passed to the second silt settling zone as fresh hydrocarbon diluent, while the remainder of the liquid product from the fluid coking zone may be used in any suitable manner. Likewise the coke formed in the coking zone may be disposed of in any suitable manner.

J.D. Frame, J.D. Haney and E.W. White; U.S. Patent 3,041,267; June 26, 1962; assigned to Cities Service Research and Development Company describe a process which mixes the tar sand with a hydrocarbon diluent in sufficient amount to form a pumpable slurry. This is accomplished by utilizing the diluent in a volume ratio of tar (bitumen) to diluent of from 1:2 to 1:5. The tar sand slurry after gentle mixing to insure substantially complete solution of the tar or bitumen in the diluent and recycled oil, is delivered to the first stage in a two step separating and disengaging treatment.

In the first stage where separation and disengagement of tar and diluent from sand is accomplished, the sand diluent mixture is eased gently into a water wash tank that includes at the inlet end an inclined surface. The sand, in the presence of water moves at a slow rate, less than the rate of gravitational free flow down the inclined or sloping surface during which movement a substantial portion of the diluent tar mixture is disengaged and separated from the sand particles. As the sand moves down the sloped surface it assumes a natural angle of repose at which time, and while in movement, the sand is gently fluffed or fluidized by water. The fluidization of the sand is conducted in such a manner as to continuously move the sand away from the area of the wash tank to a sand outlet. In this fluidized zone the second stage of the disengagement and separation of hydrocarbons from sand is accomplished.

The gentle fluffing of the sand in this stage effects a substantially complete disengagement of hydrocarbons (diluent tar mix) from the sand particles. After disengagement a hydrocarbon water mixture is withdrawn from the water wash tank and delivered to conventional separatory means with separated water being recycled to the wash tank. The recovered oil product after separation is processed according to conventional hydrocarbon processes such as hydrocracking, visbreaking or similar processing techniques.

Use of Propane

R.C. West; U.S. Patent 3,131,141; April 28, 1964; assigned to Jersey Production Research Company describes a process for the recovery of oil from tar sands utilizing a process which eliminates the use of water, thereby eliminating the problems of sludge separation, oil-water emulsions. An integrated operation is utilized employing a hydrocarbon fraction in the gas oil boiling range and a light hydrocarbon such as propane and butane, where by a particular technique the bitumen is recovered and the gas oil and light hydrocarbon efficiently recovered.

Referring to Figure 3.15, bituminous sands at atmospheric temperatures are introduced into heating zone **1** by means of feed line **2**. In zone **1** the sands are heated to 100° to 140°F. These sands are then introduced into the upper area of a first countercurrent treating zone **4** by means of line **3**. In essence, countercurrent treating zone **4** comprises a bituminous sand solvent contactor maintained at a pressure sufficient to keep hydrocarbons in the liquid phase. A relatively heavy solvent which comprises a conventional light gas oil is introduced into the lower end of zone **4** by means of line **5**. Any suitable contacting or distributing means may be employed in zone **4**.

FIGURE 3.15: NONAQUEOUS PROCESS EMPLOYING PROPANE

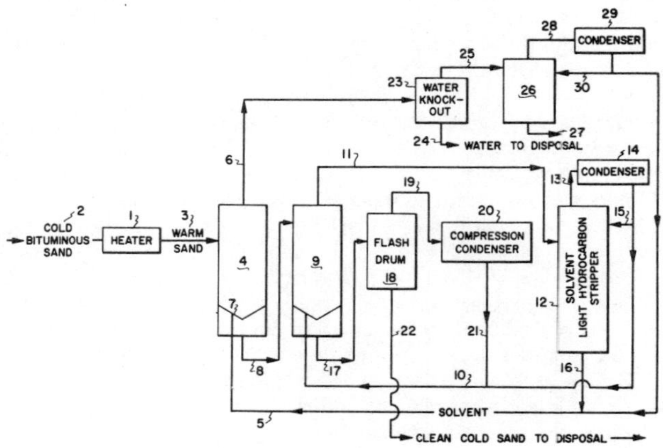

Source: R.C. West; U.S. Patent 3,131,141; April 28, 1964

The solvent is preferably a gas oil boiling at 400° to 700°F, preferably 425° to 600°F. The API gravity is 28° to 35°, preferably 32°API, while the Saybolt viscosity at 100°F is 60 to 100 seconds, preferably about 80 seconds. A bitumen gas oil solution is removed from the top of zone **4** by means of line **6** and handled as hereinafter described. A sand-gas oil mix falls through a screen or separating element **7** and is withdrawn from the bottom of zone **4** by means of line **8**. In the operation, the crushed sand is moved forward by a rock pump, screw conveyor, or any suitable means.

This sand-gas oil mix is introduced into the top area of a second countercurrent treating zone **9** by means of line **8**. In this zone, the sand is contacted with a light hydrocarbon solvent so as to produce clean sand. A light hydrocarbon solvent such as a liquefied normally gaseous hydrocarbon as, for example, liquid propane or butane is introduced into the bottom of zone **9** by means of line **10**. Zone **9** is maintained at a sufficient pressure to keep the liquefied normally gaseous hydrocarbons in the liquid phase. The light gas oil and the light hydrocarbon solvent is removed overhead from zone **9** by means of line **11** and introduced into a light hydrocarbon stripping zone **12**. In this zone, the light hydrocarbon, as for example propane, is removed overhead by means of line **13**, passed through condenser **14** and recycled to zone **9** by means of line **10**. The light gas oil free of propane is removed from the bottom of zone **12** by means of line **16** and recycled to zone **4** by means of line **5**.

The sand and light hydrocarbon solvent is removed from the bottom of zone **9** by means of line **17** and passed at a reduced pressure into flash drum **18**. Under these conditions, the light hydrocarbon, as for example propane, flashes from the sand and is removed overhead from flash drum **18** by means of line **19**. This stream is passed into compression condenser **20** where the propane is liquefied and recycled to the system by means of line **21**. Clean sand is withdrawn from flash drum **18** by means of line **22** and withdrawn from the system and subsequently handled as desired.

The bitumen and solvent removed by means of line **6** overhead from initial zone **4** is passed to a water knockout drum **23** where connate water separates and is removed from the system by means of line **24**. The solvent plus the bitumen is removed from knockout drum

23 by means of line **25** and passed to a solvent or light gas oil stripper **26**. In this zone, the bitumen is removed as a bottom stream by means of line **27** and further handled as desired, as for example by cracking, coking and the like. The light gas oil solvent is removed overhead by means of line **28**, passing through condensing zone **29** and recycled to the system by means of line **5**. A portion of this stream may be recycled to zone **26** by means of line **30**.

Solvent Extraction

A process described by *C.M. Gable, E.A. Duncan, Jr. and E.R. Freitas; U.S. Patent 3,475,318; October 28, 1969; assigned to Shell Oil Company* involves extracting tar low in asphaltenes from a tar sand that contains asphaltenes by treating tar sand with a saturated hydrocarbon solvent having 5 to 9 carbon atoms per molecule and steam stripping the solvent-containing sand after the solvent-tar liquid has been drained off. Steam stripping preferably is accomplished in two stages, the first stage effected by passing steam into a solvent-containing sand bed until just prior to steam breakthrough and the second stage effected by countercurrent contact between a moving bed of sand and a flowing stream of steam.

The process has the obvious advantages of providing deasphalting and extraction in one step which eliminates or greatly simplifies a subsequent flash distilling step in which undesired asphaltenes are removed. This process also has the advantages of not forming water-tar emulsions as found in the recovery of tar by water addition. Large quantities of solvent are not required as in the slurrying operations and only a single solvent is used for product extraction, thereby essentially eliminating the problems arising with complex solvent systems. For example, solvent recovery by steaming is a relatively simple operation and provides a means for substantially complete solvent removal.

Example 1: A tar sand containing 6.2% by weight tar having an asphaltene content of about 31% by weight basis tar was extracted with n-heptane, a nonaromatic solvent. The tar sand was placed on a filter, thereby forming a bed initially 9.5 inches deep. A pressure drop of 7.5 psi was maintained across the bed by means of applied vacuum. When extracted in two stages with n-heptane at a solvent to original tar ratio of 4.2 per stage, a tar recovery of 65% was obtained. A similar two-state extraction with benzene gave a 97% tar recovery. The difference in recovery indicates that 33% of the tar remained on the sand after the n-heptane extraction, in good agreement with 31% asphaltene content of the tar.

Example 2: In separate experiments with extracted tar, insolubles were determined at several solvent to tar ratios in several heptane-toluene solvent mixtures.

Solvent, percent weight			
Toluene	n-Heptane	Solvent/Tar	Insolubles
10	90	4/1	25.5
19.5	80.5	4/1	19.8
36	64	4/1	9.8
49	51	4/1	2.0
10	90	20/1	29
36	64	20/1	17

From these data it follows that by suitable selection of solvent composition, any desired rejection of asphaltenes may be obtained.

W.E. Savage and H.A. Cheney; U.S. Patent 3,553,099; January 5, 1971; assigned to Shell Oil Company describe a process for recovering tar from tar sand by solvent extraction. The process is initiated by subjecting tar sand to solvent extraction in an expanded, or mobile, bed within an extraction zone. Preferably, the tar sand is ground and premixed with solvent before being introduced into the extraction zone; and in the extraction zone, it is maintained as an expanded or mobile bed by an upwardly moving stream of solvent. The term mobile bed is intended to mean a bed of solid particles in an upwardly moving liquid stream wherein the combined effect of moving liquid and buoyancy cause the particles to be supported primarily by the liquid instead of being supported by resting on each other. A liquid hydrocarbon phase consisting of tar dissolved in solvent is removed from the mobile bed through a passageway that starts within the bed, passes laterally from it, and terminates above the upper level of the bed. It has been found that a mobile particle bed acts as a self-cleaning filter which removes substantially all of the fines from a liquid stream.

The liquid stream removed from the mobile bed in the extraction zone may be subjected to a final particle separation before it is fractionated to remove solvent from the tar product, but the removal of the small amount of entrained fines from the product stream is a feasible process step in the process because the small amount of entrained fines does not require treating large amounts of solids. For example, the fines may be removed readily by subjecting the liquid stream to centrifugal separation in apparatus such as a hydroclone, or even filtration may be employed because the small solids content of the large volume liquid stream does not result in rapidly clogging filters.

As a specific example, a cement mixer is employed as a low energy first-stage mixing device. Ground tar sand and toluene containing some dissolved tar are charged to the mixer. Enough liquid is charged to the low energy mixer to produce a flowable slurry of tar sand in toluene and the slurry is passed into an expanded bed of tar sand in an upwardly flowing solvent stream. Fresh toluene is added at a rate of 0.25 gallon per pound of tar and the upwardly flowing liquid stream is divided between the portion that recirculates to the low energy mixer and the product portion that is withdrawn first laterally and then vertically so that it is removed from the mobile bed laterally and withdrawn from above the upper level of the bed.

The product stream consists of a tar-solvent liquid containing about 0.1 pound of fines per gallon of liquid. This stream is subjected to final fines removal by centrifugal separation in a series of hydroclones, it is washed with water and the resultant hydrocarbon stream is fractionated conventionally into a tar fraction and a solvent fraction, the latter being returned to the extraction zone as fresh solvent and the former being recovered as the product from the process.

From the lower portion of the mobile bed in the extraction zone a mixture of sand and hydrocarbon phase is withdrawn and introduced into the center portion of an elutriation zone wherein a mobile bed of solid particles is maintained in an upwardly flowing liquid stream. The elutriation zone is charged at a lower portion with 0.1 gallon of water per pound of sand, the water containing enough sodium hydroxide to produce a 0.3 molar solution.

A mixed phase system is withdrawn from the upper portion of the elutriation zone and passed to a phase separation zone wherein hydrocarbon phase is withdrawn from the lower portion. Fines are entrained in both the hydrocarbon phase and in the water phase, but the distribution of fines is much heavier in the water phase due to preferential wetting. The hydrocarbon phase is returned to the extraction zone at an intermediate point and the water phase is returned to the lower portion of the elutriation zone. A mixed water and sand phase is withdrawn from the bottom of the elutriation zone and separated into water and sand phases, the former being recirculated to the elutriation zone and the latter being discarded.

The underflow from the hydroclone centrifugal separators consists of a small liquid stream densely loaded with fines particles and it, too, is returned to the extraction zone. Examination of the sand indicates that 99.7% of the total solids charged are recovered in the sand phase and it contains 0.6% total hydrocarbon and only 0.1% solvent. Examination of the hydrocarbon phase indicates that 95% of the tar in the tar sand is recovered as product.

Sonic Energy

In a process described by *T.J. Bulat, J.R. Logan and P.F. Kusy; U.S. Patent 3,017,342; January 16, 1962; assigned to The Bendix Corporation* oil bearing media is immersed in a body of solvent for the oil. The solvent is sonically activated, advantageously to cavitate the liquid solvent, for rapid separation of the oil from the media. The media is then removed to a second body of liquid where any oil and solvent remaining with the media is substantially separated therefrom. The second liquid is advantageously a nonsolvent for the oil and solvent and has a specific gravity different from the oil and solvent and they will be separated by gravity.

Residing in the same container, one of these liquids, either the solvent or the second liquid, will float on the other. Advantageously, the solvent has the lower specific gravity whereby it will float on the second body of liquid and the media may be moved, as by being dropped, through both bodies of liquid.

The combination of a solvent floating on a nonsolvent, the whole being sonically activated, has several advantageous results. The crude oil contains a wide range of fractions united in solution and is soluble in light petroleum fractions such for example as kerosene, benzene, fuel oils and others, and both the oil and these solvents are insoluble in and lighter in weight than certain other liquids such, for example, as water. The oil bearing media is moved first through the solvent and then the nonsolvent; the latter being heavier provides the immediate advantage that gravity may be utilized to move the oil bearing media through the liquids.

The media may simply be dropped into the solvent and allowed to free fall through the solvent and nonsolvent. The oil will be dissolved in the solvent so that as the particles of media move into the nonsolvent layer they carry with them not oil but only a coating of solvent. The solvent layer will contain some dissolved oil but is substantially a single substance which can be removed from the media particles by a sonic action, control of the acidity of the nonsolvent liquid and other simple expedients. Once removed from the media particles, the solvent floats upward through the nonsolvent to rejoin the body of solvent floating above.

In related work *J.R. Logan; U.S. Patent 2,973,312; February 28, 1961; assigned to The Bendix Corporation* also describes the use of sonic energy to activate the solvent and sand solution.

Solvent Addition with High Energy Shear Treatment

A process described by *E.L. Claridge, G. Edwards and J.T. Smith; U.S. Patent 3,553,098; January 5, 1971; assigned to Shell Oil Company* is based on preparing tar sand to have certain critical physical characteristics and then subjecting each particle of the tar sand to a high energy shearing treatment in an aqueous phase where a critical amount of kinetic energy is expended on each particle of tar sand. When these critical relationships are employed, there is substantially complete recovery of tar and solvent as a gravitationally separated hydrocarbon phase; and substantially all of the liquid losses in the discarded sand fraction involve the inexpensive and readily replaced aqueous phase.

The process is initiated by preparing tar sand to be in a specific predetermined condition by mixing it with a solvent and, if necessary, adjusting the temperature of the mixture. The objective of this first step in the process is to produce a hydrocarbon phase, which includes the tar and the solvent, having a maximum viscosity of two poises and preferably a viscosity of less than one poise. Such a hydrocarbon phase can be prepared by employing a light hydrocarbon solvent such as gasoline, which will produce a hydrocarbon phase when mixed with the tar that has the required viscosity at ambient temperature.

When heat energy is available, the hydrocarbon phase at proper viscosity may be produced with a heavier solvent such as gas oil if the mixture is heated to at least 50°C. The preferred mode of operation will depend upon the conditions that prevail. If heat energy is available, reducing viscosity through temperature elevation may be preferred because a less volatile solvent can be employed; but if processing is at a mine site where heat is not available inexpensively, a lighter hydrocarbon, which reduces the viscosity of the resultant hydrocarbon phase without heat, is preferred. Mixing of tar sand and hydrocarbon solvent is accomplished in a low energy mixer such as a muller or rotary drum. Mixing of hydrocarbon and tar sand may be in the presence of water or other material to aid in handling. For example, added water may be included in the hydrocarbon-sand mixture to produce a slurry for transporting the mixture to the next step in the process.

The mixture of hydrocarbon solvent and sand, with or without added water, is subjected to treatment that produces a predetermined amount of shearing on each particle of sand. This treatment involves introducing each particle of sand into a flowing aqueous stream in such manner that there is sufficient difference in velocity between the sand and the aqueous stream to produce a Reynolds number in the system including a sand particle and the surrounding water of at least 150. This portion of the process should not be confused with stirring or mixing because the criteria are quite different. This portion of the process requires that the sand and water have greatly different relative velocities so that a rapid acceleration of each sand particle is effected to produce strong shear forces at the surface of the sand particles.

Flow may be quite smooth, for example, when the sand-solvent mixture is introduced into the throat of a venturi in which there is rapid flow of water. Stirring in its usual sense is not indicated because in ordinary stirring both the solid phase and the liquid phase might be at high velocity, and acceleration of the solid particles by the liquid phase might be quite small.

In the sense of accelerating a sand particle in an aqueous phase, the Reynolds number found to be critical is a special case of the Reynolds number limited to the system including the sand particle and the surrounding aqueous phase without regard to the conduits in which the system is contained. The dimensionless group known as the Reynolds number usually relates the diameter of the conduit, the density of the liquid medium, the viscosity of the liquid medium, and the linear flow rate of the liquid medium in feet, pounds, and seconds as dimensions to determine the degree of turbulence in the system.

In the special system under consideration, the diameter involved is the diameter of the sand particle, and the linear flow rate involved is the difference between the velocity of the sand particle and the velocity of the surrounding aqueous phase. In such a limited system, the change from laminar to turbulent flow occurs at particle Reynolds numbers between 2 and 5 rather than at the usual Reynolds numbers between 2,000 and 4,000 when measuring turbulence in fluid flowing through a cylindrical pipe.

The tar separated as set forth above remains tacky and is capable of recontaminating the sand, particularly any sand grains which were originally tar-wetted on any part of their surface. To avoid recontamination, it has been found that an adequate volume of aqueous phase must be present. Recontamination is substantially avoided if the volume ratio of water to the initial sand charge, exclusive of solvent, is maintained at at least four and preferably ten. In other words, a minimum of four volumes of water must be employed for each volume of sand measured before solvent is added to it; and when such volumetric proportions are observed, it has been found that tar recovery is excellent, usually involving more than 90% and frequently more than 99% of the tar in the original sand. A concomitant effect is that solvent loss to the tailings fraction is negligible.

The aqueous phase may be ordinary water, seawater, effluent from water treating plants, or water from other sources; but it is preferably fresh water having the pH adjusted to 8 or higher. Although water with a pH in excess of 8 is not essential for effecting the separation of tar from sand, water with high pH is beneficial in reducing interfacial tension between the liquid phase and the solid phase so that tar-wetted fine sand grains separate more readily from the hydrocarbon phase and are more easily recovered with the tailings fraction. Since it is more difficult to achieve high-shear conditions with small sand grains (fines) than with large ones, and the net density of grains with adherent tar and hydrocarbon solvent relative to water tends to be lower for fine grains than for coarse grains, separation of tar from large grains is usually readily achieved even at pH values below 8. For fine grains, the removal of tar, while not adding greatly to total tar recovery, may add substantially to the recovery of fine sand in the tailings. This of course also greatly reduces the contamination of recovered tar by fine sand. Therefore, the high pH water is not required so much to promote the recovery of tar as to promote the inclusion in the tailings of the fine sand fraction as clean sand. High pH water thus reduces the load on subsequent separation equipment which is required to remove suspended fines from the hydrocarbon phase and the aqueous phase.

Screen Separation

In a process described by *E.W. White; U.S. Patent 3,068,167; December 11, 1962; assigned to Cities Service Research and Development Company* hydrocarbon oil is recovered from bituminous sand by the process which includes the steps of mixing the sand with liquid hydrocarbon diluent to form a slurry, passing the slurry through a first separation screen to thereby separate a first portion of sand particles too large to pass through the openings of the screen and containing entrained hydrocarbon liquid; passing the remainder of the slurry through a second separating screen having smaller openings than the first screen to separate from the slurry a second portion of sand particles too large to pass through the openings of the second screen containing entrained hydrocarbon liquid, separately mixing each of the thus separated portions of sand with water to form slurries of water, hydrocarbon liquid and sand, and treating the last mentioned slurries for removal of hydrocarbon liquid.

Freezing Technique Employing Propylene

M.A. Bergougnou and T. Kalina; U.S. Patent 3,114,694; December 17, 1963; assigned to Esso Research and Engineering Company describe a technique for reducing the tar sands to a relatively low temperature in a manner to freeze the bitumen and water in the sands. The mixture is then readily fractured by grinding and the smaller and lighter bitumen particles are separated. An efficient operation is utilized for cooling the tar sands to the required low temperature, -50° to -60°F. The process involves cooling the sands by direct contact with boiling propylene. One source of make-up propylene is available from the off-gas from a fluid coker associated with the recovery of oil from tar sands.

Centrifugal Separation

J.L. Tiedje, J.A. Bichard and R.M. Butler; U.S. Patent 3,161,581; December 15, 1964; assigned to Esso Research and Engineering Company describe an anhydrous separation technique, using a continuous solid bowl centrifuge. In essence, the operation comprises mixing the tar sands with a quantity of a suitable solvent and then introducing the mixture at an intermediate point in the centrifuge. Gravitational forces remove the hydrocarbons outwardly toward the wide end of the bowl and the sand is moved by a screw conveyor up the incline toward the narrow end. Solvent introduced at the narrow end gives the sand a countercurrent wash and an efficient recovery of the oil from the sand is secured.

In Situ Process Employing Gamma Radiation

S.L. Ruskin; U.S. Patent 2,906,680; September 29, 1959, assigned to Union Carbide Corporation has found that electronic and ionic influences exerted by gamma radiation profoundly affected petroleum and reservoir gases leading to the formation of reaction products of petroleum and gaseous hydrocarbons. The carrying out of these reactions in the oil-bearing natural compositions in the earth by gamma radiation in the range of 100 million roentgens to 400 million roentgens not only promotes the formation of these valuable hydrocarbon products but markedly increases the flow of petroleum from the rock compositions. Thus a sample of steel gray oil-bearing rock, having the appearance of dry concrete and extremely hard in composition, from which earth formation no petroleum could be made to flow, after irradiation with 100 million roentgens changed in color to a brownish shade and readily crumbled in the hand under moderate pressure, releasing the trapped oil which could be taken up with absorbent paper. A source of gamma radiation is encased in a specially constructed canister so that the radiation source is completely shielded. The dimensions of the canister are such as not to exceed the dimensions of the average oil well pipe. The radiation source is permanently fixed in the canister bottom while the shielding cylinder around the canister is readily separable from the canister Thus the radiation source may be lowered into the well for the desired period and then re-shielded before removal. The source and canister are housed within a lead cylinder so that the movements of the canister and source are at no time exposed to possible contact. Referring to Figure 3.16, the part intended to be lowered into the well includes a canister or cylindrical casing **2** in each end of which is secured a block **4** of lead or other shielding material. Between the lead blocks is a source **6** of gamma radiation, such as cobalt 60. In the upper lead block is secured a staple **8** by which a chain or cable can be secured to the canister.

FIGURE 3.16: IN SITU IRRADIATION PROCESS

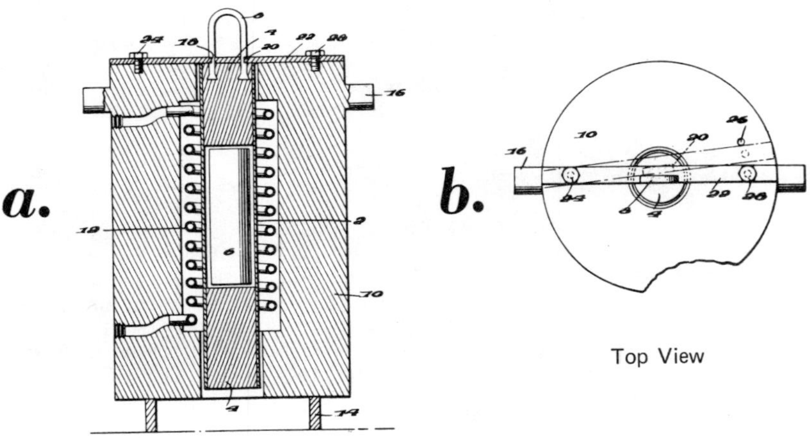

Top View

Cross-Section of Shielded Radiation Source

Source: S.L. Ruskin; U.S. Patent 2,906,680; September 29, 1959

The canister occupies the center of a lead cylinder **10**. With this cylinder may be a coil **12** for passage of a regulating fluid. This block can rest on a well pipe **14** so as to allow the canister to be lowered into and pulled out of the well without exposing the operator to the radioactive material. The cylinder **10** can be handled by projection **16**. Staple **8** has notches **18** in its legs, which can engage parts **20** of a lever **22** pivoted at **24** on the top of the cylinder **10**, and swingable between the solid and broken line positions shown in Figure 3.16b, being limited in the latter position by stop **26**. In this latter position, canister **2** is free to move downward, but cannot move upward because lever **22** still overlies it. In the solid line position, in which lever **22** can be releasably held by removable bolt **28**, parts **20** engaged in notches **18** support the canister within the shielding cylinder **10**.

To further promote the effects of the irradiation of the oil-bearing strata, one may cause an intermediate chemical reaction to occur as a result of the irradiation with 100 million roentgens whereby the products of the chemical reaction affect the flow of petroleum and/or the composition of the oil-bearing rock, shale or sands. By irradiation with 100 million roentgens to 400 million roentgens in the presence of glycerol, the glycerol is converted to acrolein, glyceraldehyde, glyceric acid and further breakdown products. This leads to a softening of the silicate structure of the oil-bearing rock, shale or sand and an aggregation of the petroleum molecules facilitating the flow of petroleum in strata having nonproducing oil wells. Thus when a limestone core from a nonproducing oil well is immersed in glycerol and irradiated with 100 million roentgens to 400 million roentgens or more, the white or gray color of the limestone darkens perceptibly. The limestone tends to disintegrate and soften and is more easily fractured. When Athabasca sand is so treated, the petroleum separates from the sand and discolors the conversion products of the glycerol. Agitation of the sands promotes the separation of the petroleum.

Example 1: A light colored 8 inch core from an oil well that has ceased producing oil was immersed in glycerol and irradiated with 200 million roentgens. The core darkened in color and developed a strong odor of petroleum. The compression test gave the following results:

Cores	No. 1 light, control	No. 2 dark, 200 million R.
Capping material	Sulphur	Sulphur
Height after capping	2.437	2.437
Diameter (in.)	2.437	2.437
H/O	1	1
Correction factor	.85	.85
Area (in.)	4.66	4.66
Total load	7,260	5,820
P.s.i.	1,558	1,250
Corrected p.s.i.	1,325	1,064

The glycerol became syrupy and the presence of acrolein and glyceraldehyde could be readily detected. The glycerol also appeared to have polymerized.

Example 2: A three inch limestone core was irradiated with 100 million roentgens while immersed in water. The core darkened in color and became granular with an increased odor of petroleum.

Example 3: A three inch limestone core was irradiated with 100 million roentgens in the presence of an atmosphere of acetylene. A marked increase in darkening and odor of petroleum developed. The core could be readily crushed by the hand at the edges.

Substituted Nitrobenzenes in Water Flooding

D.G. Feuerbacher and T.E. Sample, Jr.; U.S. Patent 3,554,286; January 12, 1971; assigned to Texaco Inc. describe a process for the recovery of hydrocarbons from subterranean hydrocarbon-bearing formations by treatment with an aqueous medium containing a base and a substituted nitrobenzene. It has been found that when an aqueous medium containing a base, e.g., an alkali metal hydroxide, and a substituted nitrobenzene, such as o-nitroanisole, is contacted with a hydrocarbon-bearing formation, mobility of the hydrocarbons from the matrix occurs together with the formation of an oil-in-water emulsion, thereby enhancing the recovery of the hydrocarbons. It has also been found that a nonionic emulsifying agent may be incorporated in the aqueous medium so as to facilitate the formation of an oil-in-water emulsion. Furthermore, the use of a solubilizing agent in aqueous medium may in addition be preferred to improve the solubility of the substituted nitrobenzene.

The method can be utilized as an improved water flood where the aqueous medium is employed as the flooding agent. Alternately, the method may be utilized by the injection into a subterranean hydrocarbon-bearing formation via an injection well of a slug of the aqueous medium containing a base and a substituted nitrobenzene, and then injecting a drive agent, e.g., H_2O as water or steam, so as to drive the slug toward a producing well, from which the hydrocarbon is produced.

As an example of the process, test runs were made in which an aqueous medium composed of a 0.1 N aqueous solution of sodium hydroxide, and containing 500 ppm of the o-nitroanisole (ONA) was passed through a pack of 50 to 60 mesh glass beads containing a heavy crude (11°API) at an oil saturation of approximately 32%. Initially, three pore volumes of the displacing medium were passed through the pack during which time the displaced oil and effluent displacing medium were collected and the amount of displaced or extracted oil determined. The flood was then continued until either no more oil was displaced or until a total of twenty pore volumes of the displacing medium (including the first three) had been passed through the pack. The following table shows that improved oil recovery was obtained and that the presence of the substituted nitrobenzene, o-nitroanisole (ONA), resulted in about fifteen times more recovery than that obtained using the sodium hydroxide solution in the absence of a substituted nitrobenzene. Furthermore, the use of a sodium hydroxide aqueous solution containing either nitrobenzene or anisole did not show improved oil recovery.

In addition, other substituted nitrobenzenes that demonstrated improved oil recovery over that obtained using a sodium hydroxide aqueous solution included o-nitrophenol and nitrated mixed (m-,p-) cresols. The latter agent is a mixture of mono-, di-, and trinitrated meta- and para-cresols. In runs which were performed where a nonionic surfactant was incorporated into the solution containing sodium hydroxide and o-nitroanisole, an additional improvement in recovery was noted.

Additive	Substituted nitrobenzene conc. (p.p.m.)	Oil recovered (grams)[1]	Total oil recovered (grams)	Total, percent recovery
(1) Dist. water		None	None	0
(2) 0.1 N aqueous NaOH		0.19	0.24	3.1
(3) Nitrobenzene	500	0.08	0.08	0.9
(4) Anisole	2,000	0.11	0.11	1.6
(5) ONA	500	0.48	3.04	42.6
(6) N-95 [2]	5,000	None	None	0
(7) ONA [3] plus N-95	([4])	1.33	4.14	44.0
(8) ONA [4] plus N-95	([4])	1.18	3.74	49.6
(9) ONT plus 5% isopropanol	2,000	0.25	3.22	41.6
(10) o-Nitrophenol	2,000	0.46	0.56	6.7
(11) Nitrated mixed (m,p) cresols	2,000	0.54	0.70	8.0

[1] After 3 pore volumes through pack.
[2] Jefferson Chemical Co. Surfonic N-95.
[3] 5,000 p.p.m. o-nitroanisole (ONA) plus 1,000 p.p.m. N-95 as an emulsion.
[4] Saturated.
[5] Repeat of above run.

NOTE.—In Runs 3 through 11, the additives were contained in 0.1 N aqueous NaOH.

The results also demonstrate that a solubilizing agent such as isopropyl alcohol, can be used in combination with the substituted nitrobenzene such as o-nitrotoluene (ONT) to increase the solubility of the substituted nitrobenzene and thereby increase oil recovery. The tests demonstrated that the preferred location of the substitute relative to the nitro group is in the ortho position.

An injection well was drilled into a subterranean hydrocarbon-bearing formation through which a slug of the aqueous medium containing the base and the substituted nitrobenzene as described above was injected into the formation. The slug is then followed by a subsequently injected aqueous drive agent, e.g., water. The size of the slug injected may vary within relatively wide limits, and will depend on a number of conditions, including the thickness of the formation, its characteristics, and the conditions for the subsequent injection of the aqueous drive medium. The aqueous drive agent may be H_2O in the form of either water or steam, the temperature of which may range up to 500°F.

In the passage of the aqueous medium through the subterranean hydrocarbon-bearing formation, hydrocarbons are desorbed from the formation, forming an oil-in-water emulsion which is then produced at production wells. Separation of the hydrocarbons from the emulsion is accomplished by one of several known emulsion breaking techniques.

Alkali Metal Hydroxides and Amine for Water Flooding

D.G. Feuerbacher; U.S. Patent 3,581,823; June 1, 1971; assigned to Texaco Inc. describes a process for the recovery of hydrocarbons from subterranean hydrocarbon-bearing formations by the treatment with an aqueous medium containing an alkali metal hydroxide and a low molecular weight amine. Thus, it has been found that when an aqueous medium containing an alkali metal hydroxide, and a low molecular weight amine, such as n-butylamine, is contacted with a hydrocarbon-bearing formation improved recovery or extraction of low-gravity hydrocarbons from sands is obtained. Illustrative of the process, test runs were made in which an aqueous medium composed of a 0.1 N aqueous solution of sodium hydroxide, and containing 0.5% by volume of n-butylamine was passed through a pack of 50 to 60 mesh glass beads containing a heavy crude (11°API) at an oil saturation of approximately 30%. Initially, three pore volumes of the displacing medium were passed through the pack during which time the displaced oil and effluent displacing medium were collected and the amount of displaced or extracted oil determined. The flood was then continued until either no more oil was displaced or until a total of 20 pore volumes of the displacing medium (including the first three) had been passed through the pack. The following table shows that improved oil recovery was obtained. The presence of n-butylamine resulted in about six times more recovery than that obtained using the sodium hydroxide solution in the absence of a low molecular weight amine. In a second test, the presence of trimethylamine resulted in about eight times more recovery than that obtained using the sodium hydroxide solution alone.

Additive	Substituted nitrobenzene conc. percent by vol.	Oil recovered (grams), after 3 pore volumes thru pack	Total oil recovered (grams)	Total percent recovery
1. Distilled water		None	None	0
2. 0.1 N aqueous NaOH		0.19	0.24	3.1
3. n-Butylamine*	0.5	0.50	1.78	20.3
4. Trimethylamine*	1.0	1.40	1.48	25.3

*In Runs 3 and 4, the amines were contained in 0.1 N aqueous NaOH.

In a preferred case, an injection well is drilled into a subterranean hydrocarbon-bearing formation through which a slug of the aqueous medium containing the alkali metal hydroxide and the low molecular weight amine as described above is injected into the formation. The slug is then followed by a subsequently injected aqueous drive agent, e.g., water. The size of the slug injected may vary within relatively wide limits, and will depend on a number of conditions, including the thickness of the formation, its characteristics, and the conditions for the subsequent injection of the aqueous drive medium. The aqueous drive agent may be H_2O in the form of either water or steam, the temperature of which may range up to 500°F.

In the passage of the aqueous medium through the subterranean hydrocarbon-bearing formation, hydrocarbons are desorbed from the formation, forming an oil-in-water emulsion which is then produced at production wells. Separation of the hydrocarbons from the emulsion is accomplished by one of several known emulsion breaking techniques. Another example of the process is illustrated by its application to the recovery of hydrocarbon materials from tar sands utilizing the aqueous medium in an extractive technique in combination with techniques for the mining of the tar sand, and techniques for the recovery of the oil from the tar sand following its extraction.

pH-Sensitive Aqueous Surfactant Solutions

A process described by *H.L. Greenwald; U.S. Patent 3,108,059; October 22, 1963; assigned to Rohm & Haas Company* involves recovery of oil by the use of an aqueous surfactant solution which causes an oil-water emulsion to be produced. The emulsion is readily broken so that the surfactant can be efficiently recovered in a single phase and reused and the oils economically obtained.

The process makes possible high recovery of all components by controllably changing the hydrophile-lipophile balance of a surfactant during the process. The controlled change of this balance is brought about by the use of pH-sensitive emulsifiers. The oil is either emulsified under alkaline conditions and the emulsion later broken by adding acid to make the solution acidic, or emulsified under acid conditions and broken by adding alkali to make the solution alkaline. In actual practice, it may be desirable in some cases to effect a closer control of the acidity or alkalinity of the treated solution in order to maximize the recovery of surfactant. This variation of pH effectively controls emulsion stability and the concentration of the surface-active agent in each phase. In the oil recovery process the mined oil-bearing formations are treated in a reservoir such as an open pit. The steps comprise agitating portions of the mined formations with an alkaline aqueous solution of a pH-sensitive surfactant to emulsify the oil in the mixed formations, the surfactant being selected from the class consisting of:

(a) the ethylene oxide adducts of a range of amines represented by the formula $C_{11\text{-}24}H_{23\text{-}49}NH_2$ where the number of ethylene oxide units is from 5 to 50 per amine;

(b) $C_{11\text{-}24}H_{23\text{-}49}NH(C_2H_4O)_nSO_4Na$, where $n = 3$ to 50; and

(c)

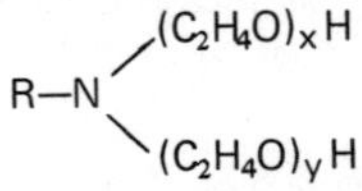

in which R is a dehydroabietyl group and $x + y = 15$ to 50.

The solids are allowed to settle and the emulsion is subsequently broken by addition of an acidic substance so as to make the oil-water system acidic and thereby concentrate the surfactant in a bulk phase where it is available for further use, while concentrating the oil in another phase.

TAR SANDS RETORTING AND REFINING PROCESSES

RETORTING AND COKING

Simultaneous Separation and Cracking

V. Haensel; U.S. Patent 2,733,193; January 31, 1956; assigned to Universal Oil Products Company describes a continuous fluidized method for carrying out simultaneously the separation and cracking of oil from oil-bearing sands. The process is accomplished by contacting the oil sands material with heated reactivated catalyst particles and recirculated heated oil sand particles, which in turn are separated and returned in a controlled operation.

The method for effecting the separation and catalytic cracking of oil, from oil containing sands, comprises, introducing the oil sands in particulated form to a distillation and cracking zone together with a heated stream of particles consisting primarily of subdivided catalytic material, effecting a fluidized contacting between the oil sands and the heated catalyst-containing particle stream within this cracking zone, discharging resulting hydrocarbon vapors from the upper portion of the distillation and cracking zone and withdrawing contacting solid material from the lower end.

The process continues by passing the withdrawn and contacted particles, with a coke deposition thereon, to a burning and classification zone, introducing an oxygen-containing stream into the latter zone and effecting a burning of at least a major portion of the carbonaceous deposit from both the catalyst and sand particles while fluidizing and passing the oxygen-containing stream through the particles in a manner effecting a classification thereof into an upper fluidized bed of primarily catalyst particles and a lower fluidized bed of primarily sand particles.

The resulting combustion gases are removed from the upper portion of the burning and classification zone while discharging a continuous stream of sand particles into the lower portion of the zone, and continuously withdrawing hot catalyst particles from the upper fluidized bed in the burning and classification zone and returning them to the distillation and cracking zone, together with fresh oil sand particles.

By effecting the simultaneous processing steps within both the cracking zone and the burning and classification zone, a relatively compact unit is provided and a minimum of transfer lines is necessitated to carry out the continuous fluidized operation. In other words, only two major contacting zones are utilized, with the separation and cracking of oil vapors being effected in a single confined conversion zone, and reactivation and heating together with the elutriation and classification of particles being effected within a single confined regenerating and classification. The latter zone being operated to effect substantially distinct fluidized beds of catalyst and sand particles and is equipped with separate draw off means, so that selected quantities of catalyst or sand may be discharged or withdrawn from the chamber and selectively returned to the conversion zone.

Preferably a catalyst having a density and particle size substantially different from that of the sand particles in the oil and tar sands is utilized, such that the classification of particles within the reactivation and separating zone effects the maintenance of two distinct layers or beds within the lower portion of that zone. The cracking catalyst may be any of the known types of catalyst such as silica-alumina, silica-magnesia, silica-zirconia, or a natural catalyst such as Superfiltrol, with the catalyst particles being maintained in a finely divided state so that they readily fluidize and maintain a light fluidizable characteristic.

Internal Combustion Retorting

F.J. Sanders; U.S. Patent 3,130,132; April 21, 1964; assigned to The Standard Oil Co. describes a process which involves a retorting method for recovering oil from oil-bearing minerals so that the oil may be effectively removed from the solids and, furthermore, to avoid subjecting the product oil vapors to severe refluxing conditions with consequent over-heating and loss in yield.

The process also provides a retorting apparatus of compact and simple design adaptable to the continuous passage of oil-bearing solids downward through the vessel, thereby attaining a high throughput unit capacity. The process for recovering oil from oil-bearing minerals is made self-supporting in heat requirements for retorting by utilizing direct internal combustion of part of the hydrocarbons present on such solids without appreciably affecting the potential yield of valuable liquid and vapor oil products.

Referring to Fgiure 4.1, the reference numeral **1** refers generally to an elongated upright retorting vessel comprising a metal shell **2** suitably insulated with a refractory lining **3**. A charge hopper **4** of any suitable construction is disposed at the top of the retort and is so adapted as to maintain a continuous feed of solid material into the top of the retorting vessel **1**.

The upper portion **5** of the retorting vessel **1** is substantially cylindrical in shape with the wall diverging outwardly in a downward direction to a maximum diameter for the vessel somewhat below the combustion zone of the unit so as to diminish the tendency of bridging or agglomeration of the particles within the combustion zone. The lower portion **6** of the retorting vessel forms a frustum of a cone and converges into a central opening **7** in which is disposed a suitable valve means **8**. The rate of flow of solids through the retort will, of course, be controlled by regulating the discharge of spent solids from the valve.

A series of conduits enter the retorting vessel at various levels as shown, extending horizontally across the unit. These conduits may be supported in any suitable manner within the retort, such as being directly supported from the walls of the retort. At the uppermost level is the oxygen-containing gas conduit **9**.

In order to achieve a uniform distribution of gas within the retort, conduit **9** is adapted with communicating circular tubes **10** which are arranged in a common horizontal plane with conduit **9** and are provided with openings **11** to supply gas in a downward direction. At a level lower in the retort is the product outlet conduit **12** adapted with communicating circular tubes **13**. The topside of tubes **13** have slotted openings **14** to effect the removal of combustion gases and products from the retort.

Depending upon the size of the retorting vessel, conduits **9** and **12** each may be adapted with more or less than the two communicating circular tubes shown. It is only important that the structural arrangement associated with conduits **9** and **12** will provide for a uniform distribution of oxygen-containing gas from conduit **9** to the combustion zone and the even removal of products from the retort to the product outlet conduit **12**.

FIGURE 4.1: RETORT PROCESS

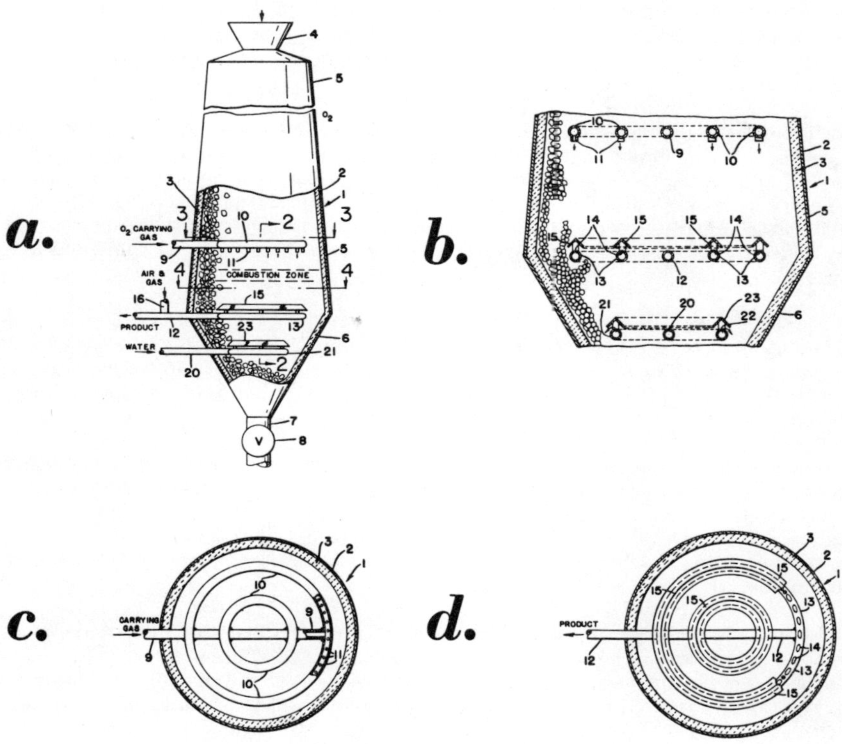

(a) Overview of Apparatus
(b) View Aong Line **2–2** of Figure 4.1a
(c) Section Along Line **3–3** of Figure 4.1a
(d) Section Along Line **4–4** of Figure 4.1a

Source: F.J. Sanders; U.S. Patent 3,130,132; April 21, 1964

Deflectors **15**, which may conveniently be angle bars positioned as inverted Vs, are mounted above each of the tubes **13** and overlying the tubes sufficiently to provide an angle of repose for the solids immediately above the tubes to prevent the entrainment of granular solid particles in the gases removed from the un t. Deflectors **15** may be mounted on the tubes in any suitable manner, such as by welding directly to the tubes, or if desired, the deflectors may be supported directly from the walls of the retort.

The level of conduit **9** in the retort is not critical but its position will be determined by the relationship that must be observed between the height of the retort bed over the conduit to the distance in the retort between conduits **9** and **12**. For a fixed distance between the levels of conduits the pressure drop from conduit **9** to conduit **12** will be inversely proportional to approximately the square of the distance from conduit **9** to the top of the retort.

It therefore becomes obvious that in order to have the gases which are supplied through conduit **9** flowing substantially downward in the retort toward conduit **12**, it is necessary to maintain the height of the retort above conduit **9** several times the distance between the levels of conduits **9** and **12**. Consequently, in order to supply sufficient air in a downward direction within the retort for the effective removal of products by conduit **12** without the need for excessive pressures on the gas in conduit **9**, it is desired to have the distance between conduits **9** and **12** to the height of the retort above conduit **9** maintained in a ratio of at least 1:4 and preferably 1:5 to 1:6.

For example, if the retort height is to be 30 ft, conduit **9** would be located at a level 22 ft from the top of the retor and conduit **12** would be located 26 ft from the top of the retort so that the height above conduit **9** to the distance between conduit **9** and conduit **12** would be in the ratio of 5.5:1. Of course, a ratio of these distances less than 1:4 can be used if it is desired to have a retort of relatively short over-all length; then a corresponding adjustment of pressure on the oxygen-carrying gas supplied through conduit **9** is made.

Conduit **20** is positioned in the lower portion **6** of the retorting vessel **1**. Conduit **20** supplies a cooling medium to the retort and is adapted with communicating circular tube **21** having openings **22** so as to introduce the cooling medium uniformly to this level in the retort. Deflector **23** is provided in the manner shown to protect the openings **22** from being clogged by solids.

The operation of the retort is described in particular reference to the processing of tar sands. At the start of the operation, let it be assumed that the retorting vessel **1** has been filled from hopper **4** by gravity flow with tar sands crushed to a suitable particle size and that star valve **8** is closed. Hopper **4** contains an additional supply of crushed tar sands particles so that once the valve is opened the particles will pass continuously downward in an uninterrupted column.

The size of the tar sands particles is not critical, but it is preferred that the particles will pass a 2" mesh sieve while being substantially free of silt and fines which might be entrained in the product gases and vapors leaving the retort. It may be desirable in initially starting the operation of the unit to fill the lower portion of retorting vessel **1** with sand such as flint shot Ottawa ranging from 20 to 50 mesh in particle size, and filling the balance of the unit with the tar sands particles.

The retorting operation is started by igniting the hydrocarbons present on a horizontal strata of tar sands particles extending across the unit immediately above the level of conduit **9**. To accomplish this, a mixture of a combustible gas and air may be temporarily injected through conduit **12** from conduit **16** and burned within the retort to heat the hydrocarbons to a temperature above their ignition temperature, which is approximately 500° to 600°F. When the combustion process is under way, the supply of gas is stopped while the supply of air is continued through conduit **12** to support combustion.

As an alternate method of initiating the combustion process, an air gas burner may be temporarily placed within the retort just above conduit **12** to again heat the hydrocarbons across this level of the retort to a temperature above their ignition temperature. When the combustion process is under way, the air gas burner may then be removed from the side of the retort so as not to later interfere with the sustained operation of the process and air is temporarily supplied through conduit **12** to support combustion.

The combustion zone thus created is permitted to move upward a short distance from the original ignition level. It is most desirable that the combustion zone attains uniform depth across the bed as well as uniform temperatures. Thermal indicators well-known to the art may be positioned within the retorting vessel to observe the conditions of the combustion zone so it will be possible to adjust air flow rates as required to reach desired conditions within the unit.

When the combustion zone reaches a level at least one-half the distance between the level of conduit **12** and the level of conduit **9** and has built up a uniform temperature across the bed in the range of 650° to 1100°F, the temporary supply of air through conduit **12** is stopped and a supply of air is introduced through conduit **9**. This supply of air may be preheated if desired.

During the initial ignition period, the products of combustion move upward in the retort and are cooled by the unheated particles above the combustion zone. In this comparatively brief period of time, a small band of oil contained on the particles above is displaced by such gas flow; and when the reverse flow of air from conduit **9** reaches this area, these liquid products, together with products of combustion, flow downward in the retort toward product outlet conduit **12**, and the combustion process continues with relatively even temperatures throughout the combustion zone. After the flow of air has been reversed, the temperature in the combustion zone is continued in the range of 650° to 1100°F for sustained operation.

At temperatures below 650°F the viscous oil present on the tar sands is not effectively removed from the mineral particles. At temperatures substantially greater than 1100°F the combustion reaction progresses too vigorously and the oil burns in place, giving off substantially all combustion gases with very little yields of valuable oil products.

Valve **8** is now opened and the rate of discharge of residual solids from the retort is regulated so as to maintain the level of the combustion zone constant for the continued operation of the unit, and the rate of feed of fresh particles to retorting vessel **1** from hopper **4** is equal to the rate of discharge.

It is desirable that not more oxygen be introduced by way of conduit **9** than will be consumed in the reactions at the combustion zone. In this manner, no further combustion of the oil products can occur while these products pass downward in the retort toward conduit **12**. Therefore, it will be obvious that when using air as the oxygen-containing gas that if the rate of air supplied through conduit **9** to attain the desired temperature level in the combustion zone is inadequate to effectively force the oil products and the combustion products to flow downwardly in the retort, then some extra noncombustible gas such as nitrogen must be added with the air to cause such a result.

The heat from the combustion zone is sufficient to educt the hydrocarbons from the oil-bearing solids immediately above the combustion zone as vapors and liquid. The actual distribution of product between the vapor and liquid states will depend in great part on the temperature level maintained in the combustion zone. The products are forced across the combustion zone and pass downwardly in the retort through the residual solids to the open areas found around tubes **13** due to deflectors **15**.

The liquid and vapor oil products entering these open areas are readily collected in tubes **13** by means of slots **14**, together with the gaseous combustion products and steam which is rising from the lower portion of the retort. The products so entering the tubes are withdrawn from the retorting vessel by means of conduit **12** into a recovery system forming no part of this process where they are to be cooled, condensed, and put in storage. The products, while passing through the residual solids below the combustion zone, do not cool or reunite to the solids in any appreciable degree since these solids are still hot from leaving the combustion zone. A slight vacuum of approximately 100 to 300 mm may be placed on conduit **12** if desired to facilitate the removal of the products and combustion gases therethrough to the recovery system.

The residual solids flowing below conduit **12** are cooled by the introduction of water through conduit **20** and tube **22**. The water is converted to steam on the hot solids which forms a seal against vapor leakage into outlet pipe **7**. Some of the steam produced rises upwardly in the retort to be collected in tubes **13** and is quite helpful in stripping the descending solids of residual oil products and forcing these residual products upwardly in the retort to where they are also collected by tubes **13** and by conduit **12**.

H.F. Tse; U.S. Patent 3,518,181; June 30, 1970; assigned to Sun Oil Company describes an improvement in pyrolytic methods of treating bituminous tar sands. Specifically, the process comprises preheating the tar sands to between 200° and 380°F and feeding the preheated sands, while within this temperature range, into the pyrolytic treating zone. It has been found that tar sands, when heated above 200°F, flow readily and are easily handled and transported but that when heated above 380°F the sands become viscous again and tend to reagglomerate. The process can be applied to thermal cracking, catalytic cracking and coking among others.

Recycle of Hot Decoked Feed Solids

C.M. Fitch; U.S. Patent 3,267,019; August 16, 1966; assigned to Signal Oil and Gas Co. describes a process which includes the recycling of hot decoked feed solids to a point, prior to the still or retorting zone, where the recycle solids contact and are mixed with fresh feed sand or shale, initiate cracking and materially assist in the particle size reduction. In such initial mixing zone, the feed solids can also be ground, pulverized, etc., to free-flowing particles, preferably of substantailly the same particle size as the recycle solids.

The recycle solids prevent balling up or clogging of pulverizing equipment in the mixing zone due to plasticized feed. Such recycling of the decoked solids to an initial mixing (and, preferably, particle size reducing) zone provides improved heat transfer efficiency and results in improved recovery of oil in the subsequent still, holding or retorting zone, and at a relatively low temperature. Large quantities of fluidizing gases are not needed, so that recovery of the oil is simplified.

As one particular example of the process, tar sand in up to steam shovel sized chunks is passed through a pulverizer in which it is intimately mixed with hot decoked sand having a temperature of 1200°F. The weight ratio of the recycled sand to the fresh feed sand is 5.7:1. The hot recycle sand heats the feed sand to 950°F in the pulverizer, aids in the breakdown of large lumps of the feed in the pulverizer and also initiates cracking of the oil in the feed.

The pulverized particulate hot mixture is then dropped directly from the pulverizer into a rotary retort and is passed through for a 7 minute retorting time at 950°F, the hot mixture being tumbled or cascaded approximately 8.75 times per minute. Hydrocarbons vaporize and are stripped from the mixture by means of countercurrently flowing purge gas, preferably superheated steam, although recycled retort gas or any hydrocarbon gas can be used. The purge gas-vaporized hydrocarbon effluent mixture passes to a hydrocarbon recovery unit, the purge gas, when a hydrocarbon purge gas is used, being recycled back to the retort for repassage through it countercurrent to the solids flow, and the hydrocarbons, including oil and gas, being recovered and passed to storage, to a product pipeline, etc.

The stripped solids mixture passes from the retort to a rotary coke burner in which the residual hydrocarbons are burned, heating the solids sufficiently (to 1200°F) so that they can be utilized as hot recycle solids, that is, so they can be passed back to the pulverizer unit. Some of the hot decoked solids are discarded, after they pass through an air preheater. The air heated thereby is used in the coke burner. In discarding the hot solids, such solids go into a unit where they are directly contacted with water to generate steam. The steam is used to provide pneumatic conveying of the excess solids out of the system and into a waste area.

Compacted Tar Sands

P.E. Chaney, R.W. Ince and C.M. Mason; U.S. Patent 3,487,002; December 30, 1969; assigned to Sun Oil Company describe a method of recovering oil from mined tar sands which involves forming compacted tar sands pieces by special conditioning treatment that provides low internal permeability and then retorting in fixed bed form. The conditioning treatment can involve rolling of preformed pellets, compaction in a mold or pressure extrusion. Substantial collapsing of the bed during retorting is avoided.

The improvement provided by the process in the retorting of bituminous sands is achieved by virtue of the manner by which the bituminous sands are prepared for retorting, and the manner of handling the material during the retorting operation. The conditioning treatment can be carried out in several ways, but in all cases the tar sands are compacted in a manner such that air or gas is squeezed out of the pores and the internal permeability of the tar sands material is reduced to within the range of 0 to 2,000 millidarcies. The tar sands pieces so compacted can then be retorted in fixed bed form, and the pieces will substantially retain their shapes during retorting so that collapsing of the bed is avoided.

The process involves first forming the bituminous sands into pellets or chunks having sizes which are more or less uniform in all dimensions and which are mainly in the range of ¼ to 6", more preferably 1 to 4". The pellets then undergo treatment in a conditioning step that is essential for imparting to the pellets weight-supporting characteristics necessary for effecting the retorting step successfully. This conditioning step involves subjecting the pellets to a rolling action that effects a continuous gentle deformation, whereby the sand grains within each pellet become more densely packed and the pellets assume a smoother and glossier appearance than they had before rolling.

This rolling action causes a marked decrease in the permeability of the individual pellets. After the pellets have been rolled sufficiently, they are then capable of forming a bed which can be subjected to retorting conditions without exhibiting any substantial tendency to collapse. During the retorting the pellets retain their shapes and do not tend to stick together appreciably. By passing sufficient gas heated to above 700°F, preferably to within the range of 900° to 1500°F, through the bed, typically 80% by weight of the bituminous material can be driven out and recovered as oil. The rest of the bitumen or heavy oil is converted to normally gaseous hydrocarbons and carbonaceous material which remains as a residue in the pellets.

Compaction and Retorting

J.D. Bennett; U.S. Patent 3,623,972; November 30, 1971; assigned to Sun Oil Company describes a retort system for retorting bituminous materials which includes a horizontally moving bed of such materials which are compacted and then perforated prior to entry into the retort. Such perforating provides vertical flow channels through the compacted bed of material to permit a controlled distribution of heat and burning within the retorting apparatus.

The retorting apparatus is comprised of two zones, the first being a retorting zone for educting volatile hydrocarbon materials from the horizontal bed. The second zone, into which the bed passes after reduction of the hydrocarbons, comprises a burning zone which burns the remaining coke within the materials. Heat from such burning operation is transferred to the retort zone for providing retort heat. A heat exchanger also utilizes heat from the educted hydrocarbons for heating the burning zone.

Referring to Figure 4.2, an apparatus is shown for forming vertical flow paths through a bed of bituminous sand material to be retorted. Bituminous materials which are mined from the earth are placed on top of the conveyor **12**. The conveyor has side members **14** which form a trough for holding the materials in place on the conveyor. The conveyor is constructed in hinged sections to permit movement of the bottom and side members over a pulley.

The materials are moved on the conveyor until they are aligned with a compacting and perforating device **16** which is positioned above the conveyor. The compacting and perforating device **16** is comprised of a body platen **20** having downwardly extending perforating members **18** which are formed in the shape of a truncated cone. Upon lowering the compacting and perforating device **16** onto the conveyor, the lower face of the platen **20** engages the materials on the conveyor and compacts these materials into the conveyor trough formed by the conveyor **12** and sidewalls **14**.

The platen is sized to be received on the conveyor between the sidewalls so that the top of the compacted material is below the top of the sidewall members **14** in the conveyor. The perforators **18** are arranged to be aligned with openings **22** in the conveyor. During the perforating operation the platen is lowered onto the material being carried by the conveyor. The material is thus compacted between the platen and the conveyor trough.

The perforators form holes or conically shaped openings through the compacted material to provide communication between the top and bottom sides of the compacted bed of materials on the conveyor. The shape of the perforating members causes the tar sands to be laterally displaced and compacted upon entry of the conically shaped members into the material.

Compacting of the material about the opening provides a strong structure which will retain its shape and thereby maintain an opening while the material is being moved through the retorting process. Figure 4.2b shows a cutaway view of a portion of such material as it appears after the compacting and perforating operation.

FIGURE 4.2: HORIZONTAL BED PROCESS FOR TAR SAND RETORTING

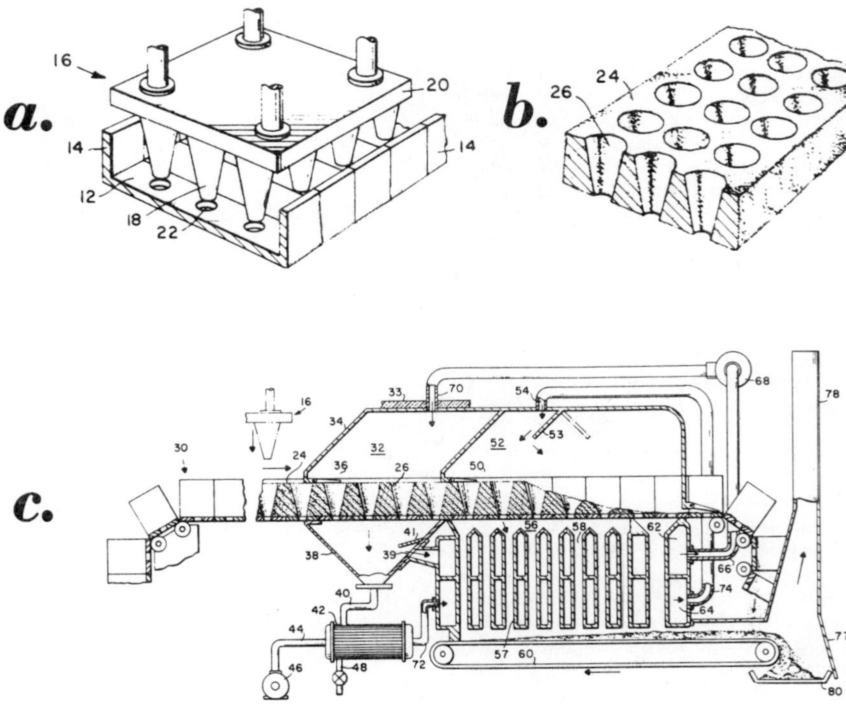

(a) Compacting Apparatus
(b) Horizontal Bed of Compacted Material
(c) Retorting System

Source: J.D. Bennett; U.S. Patent 3,623,972; November 30, 1971

The bed of materials **24** has conically shaped openings **26** formed therein with the large portion of the cone at the top of the bed and the small portion of the truncated cone at the bottom of the bed. The openings are spaced in the bed to provide an approximately equal thickness of material in all directions for purposes to be hereinafter described. The thickness of the bed, and size and spacing of the openings will be determined by the rate of movement of the bituminous materials through the retorting system.

The compacted material forming the bed together with the conical perforations create a bed of material which is characterized by a substantially greater strength than the unconsolidated materials which are originally placed on the conveyor belt. The conical openings provide a strong unsupported sidewall portion which is designed to maintain its form throughout the retorting operation, while being subjected to movement of the conveyor and movement of gaseous fluids therethrough during the retorting and burning operations.

Such openings provide a means of passing gaseous fluids through the beds and at the same time expose a maximum amount of cross-sectional area of the bed to such gases, leaving an equal volume of such materials unexposed between the opening to effect a uniform exposure of such materials to the atmosphere within the retorting zones. By maintaining a proper feed of materials through the retorting system, such materials are uniformly treated and therefore a maximum amount of hydrocarbons may be educted together with a maximum amount of thermal energy which is derived from the materials during passage through the burning zone.

Referring to Figure 4.2c, a conveyor **30** is shown for moving a continuous bed or briquet **24** of bituminous materials into a retorting apparatus. While the conveyor is shown passing over pulleys at the ends of the retorting mechanism, additional pulleys are provided for diverting the conveyor outwardly around the retorting mechanism to produce a continuous conveying system. The apparatus includes a first retorting zone **32**, provided with a housing **34** enclosing the zone. Insulating materials such as at **33** are used throughout the construction of the retort system to maximize the utilization of heat energy generated in the system.

A swinging door seal **36** rides lightly along the top surface of the bed of material to provide a sealed enclosure to the interior of the retort zone. The door seal is sized to swing downwardly between the sidewall members **14** on the conveyor. A pipe **70** communicates with the top of the housing **34**. Hot gases are forced through pipe **70** into the retort zone to bring the zone **32** to an approximate temperature range of from 1000° to 1300°F. Such heated gases move about and through the bed of materials to raise the temperature of such materials to a retorting temperature of approximately 750° to 900°F.

Such heated gases pass through the openings **26** in the material and openings **22** in the conveyor into a funnel shaped enclosure **38** positioned below the conveyor beneath the retort zone. Enclosure **38** has an opening **39** in its sidewall to permit passage of such hot gases from the enclosure. The enclosure also provides a means for catching educted hydrocarbons which are driven from the bed of material during heating in the retort zone.

A cover **41** prevents educted hydrocarbons from entering the opening. The funnel shaped housing terminates in an outlet conduit **40** at the lower end thereof. The conduit **40** provides means for moving the educted hydrocarbon fluids into a heat exchanger **42**. The heat exchanger may be of any known type where heat is conducted through tubular members from one medium to another. In this case the heat exchanger provides a means of transferring heat energy in the educted hydrocarbons to a gas, being blown through conduit **44** by means of blower **46**. The hydrocarbon materials are thus cooled and leave the heat exchanger through conduit **48**. The educted hydrocarbons enter the exchanger from conduit **40** at 200° to 600°F. Air entering the heat exchanger through conduit **44** will absorb some of this heat energy, before exiting the heat exchanger through conduit **72**.

The moving bed of bituminous materials leaves the retort zone **32** and moves through a swinging door seal **50** into a second, burning zone **52**. Heated air is supplied to the burning zone **52** by means of a pipe **54**, which communicates with the top of the burning zone. A defuser **53** is positioned below the opening of the pipe into the burning zone. The defuser provides a means for regulating the distribution of hot gases within the burning zone. The heated air entering the burning zone causes the fixed carbon or coke within the sand to be burned.

The coke serves as a bonding material to hold the sand in its compacted and perforated shape. As the coke burns, the face of the exposed material becomes unconsolidated to the extent that the burned face sluffs off and drops by gravity through the openings in the conveyor. The openings in the material are thereby enlarged as the burning takes place until the material has substantially completely burned. The velocity and turbulence of hot gas moving through the openings **26** in the material will carry the material out the openings **22** in the conveyor. Means may be provided at the end of the conveyor for sweeping any remaining sand from the conveyor before the conveyor exits the retort.

A heat exchange chamber **56** is formed below the conveyor belt, beneath the burning zone. The heat exchange chamber is provided with baffles **57** which form vertical channels **58** interconnecting the upper and lower portions of the chamber. The channels provide a means for the unconsolidated sand to fall to the bottom of the chamber, where they collect on a conveyor belt **60** for removing spent sand to the exterior of the retort system. Upper and lower transverse flow ducts **62, 64** respectively, are formed within the baffles **57** positioned within the heat exchange chamber.

The flow ducts establish a means for transferring heat energy from the burned sand to gases passing within the ducts **62** and **64.** Additional baffles or surface area may be provided in the chamber to facilitate such heat exchange. The upper duct **62** is connected at its left end, as viewed in Figure 4.2c, with the opening **39** in the funnel shaped enclosure **38** beneath the retort zone. The opposite end of the duct **62** is connected by means of a conduit **66** with a blower **68** for moving gaseous fluids between the duct **62** and the pipe **70** communicating with the top of the retort zone **32.**

The flow path just described creates a continuous circulation of gases within the retort zone. The movement of such gases through duct **62** adds heat energy to the gases, and thereby maintains the temperature within the retort zone at a level which will insure retorting of the volatile constituents from the bituminous material.

Such temperature may, for example, be in the range of 1000° to 1300°F. In the event that fogging occurs in the enclosure **38**, a defogging device may be placed in the entry duct **62** to prevent the fog from getting into the heat exchanger and forming coke deposits.

The lower transverse duct **64** within the heat exchange chamber communicates at its left end with a conduit **72** leading from the heat exchange device **42**. The lower duct provides a means for passing gaseous fluids which emit from the heat exchanger **42** through the heat exchange chamber **56** for supplying additional heat energy to such gaseous fluids. The lower duct **64** connects at its other end with conduit **74**, which in turn communicates with conduit **54** and the upper end of the burning zone within the retort system.

Gases entering the burning zone through conduit **54** may have an approximatel temperature range of from 500° to 1000°F. Such heated gas flowing into the burning zone maintains the burning zone at a temperature which will insure burning of the coke or fixed carbon residue within the bituminous materials. The burning of such coke or fixed carbon materials provides additional heat to gaseous fluid flow through the heat exchange chamber to maintain temperatures within the retorting system. The burning coke may raise the temperature in the burning zone to an approximate range of from 1300° to 1700°F.

The hot gases which are moved into the upper end of the burning zone by means of the blower **46,** pass downwardly through the burning sands, the conveyor, and outwardly through the vertical flow paths **58** within the chamber. Such gases then continue along the path taken by the spent material conveyor belt into a plenum **77** at the end of such conveyor, where they are permitted to escape through a flue gas chimney **78.**

Additional blowers other than those shown may be provided throughout the system to provide for necessary heat transfer. It is anticipated that a substantial amount of the heat energy will have been removed from such escaping gases so that the temperature of the gas escaping through chimney **78** will be very low and contain relatively few unburned hydrocarbon elements. A transversely moving conveyor **80** at the bottom of the plenum **77** receives sand from conveyor **60** for removal to a disposal site.

Summarizing the operation just described, mined bituminous tar sand materials are loaded on a conveyor **30** where they are compacted and perforated by means of an apparatus as shown in Figure 4.2a for performing this operation. Such perforated and compacted material is then moved into a first sealed retort zone **32** where heated gases are passed downwardly through openings **26** within the material to heat the material and permit the flow of volatile hydrocarbon fluid into the funnel shaped enclosure **38** below the retort zone.

Next, the retorted bituminous materials are passed into a burning zone **52** where the remaining fixed carbon or coke is burned. While such coke is burning, a gaseous fluid is passed through the burning zone to move the heat energy evolving therefrom into a heat exchange chamber **56** below the conveyor **30**.

A pair of transverse heat exchange ducts **62, 64** within the chamber **56** move gaseous fluids through the chamber to remove the heat energy from the burning zone gas for: (1) heating the circulating fluid through the retort zone; and (2) supplying additional heat to the burning zone.

As the coke is burned from the bituminous material in zone **52**, the sands containing such coke become unconsolidated and are permitted to drop through the openings **22** within the conveyor belt and vertical channels **58** within the chamber **56** onto a spent material conveyor belt **60**, which removes such unconsolidated sands from the retort zone to the disposal conveyor **80**.

Slurry Process

H.V. Rees; U.S. Patent 2,793,104; May 21, 1957 and U.S. Patent 2,885,275; May 5, 1959; both assigned to Texaco Development Corporation describes a process for the generation of a mixture of carbon monoxide and hydrogen suitable as a source of gaseous fuel or as a source of feed gas for the synthesis of hydrocarbons from oil-bearing minerals. In the process, the oil-bearing mineral in particle form is mixed with a liquid which may be converted to vapor form on heating. The suspension is passed under conditions of turbulent flow into a tubular heating zone where it is heated to a temperature at least sufficient to convert all of the suspending liquid to vapor.

Part of all of the resulting residual solid may be separated from the resulting vapors. Liquid condensed from the vapors may be used in the preparation of the slurry. The residual solid may be separated and discarded, or where combustible matter is still associated with the solid, used for fuel or for the generation of carbon oxides and hydrogen. If desired, the dispersion of residual solid may be passed directly to a combustion zone or gas generator without separation. Residual unvaporized oil or carbonaceous solid associated with the mineral residue is useful for the production of a mixture of carbon monoxide and hydrogen which may be used as fuel gas or as a source of feed gas for the synthesis of hydrocarbons.

Important advantages result from the method of recovering hydrocarbon oil from minerals in accordance with the process. As a dispersion or slurry, the oil-bearing mineral is readily transported and subjected to elevated pressures. Since the slurry may be handled in a manner analogous to handling of a liquid, troublesome lock hoppers and similar devices which are normally required for handling solid materials are eliminated and replaced simply by the slurry preparation equipment and pumps. The slurry may be made up some distance from the processing site and transported thereto by a pipeline. The quantity of material fed to the process may be accurately metered as a slurry of relatively constant controlled composition.

Retorting and Coking Process with Hot Sand Recycle

C.H.O. Berg; U.S. Patent 2,905,595; September 22, 1959; assigned to Union Oil Company of California describes a tar sand treating process in which a recycled stream of hot sand is utilized to heat the heavy hydrocarbon materials in the bituminous sand to produce a coked sand and to produce a coker gasoline and gas oil which are directly useable as feed stocks to existing refinery processes.

Referring to Figure 4.3 the three essential elements of the apparatus of this process include coking kiln **10**, coked sand furnace **12**, and coker distillate fractionating column **14**. Sisquoc tar sand is introduced at a rate of 10,000 tons per day by means of conveyor **16** into the upper or sand inlet end of rotary coking kiln **10**. Simultaneously recycle sand removed from the bottom of coked sand furnace **12** through line **18** is combined and introduced at a rate of 20,000 tons per day and at a temperature of 1445°F. The coking kiln **10** is approximately 16 feet in diameter, 90 feet long, is rotated at a rate of about 3 rpm, and has a total slope of 3.1 feet.

FIGURE 4.3: TAR SAND DISTILLATION PROCESS

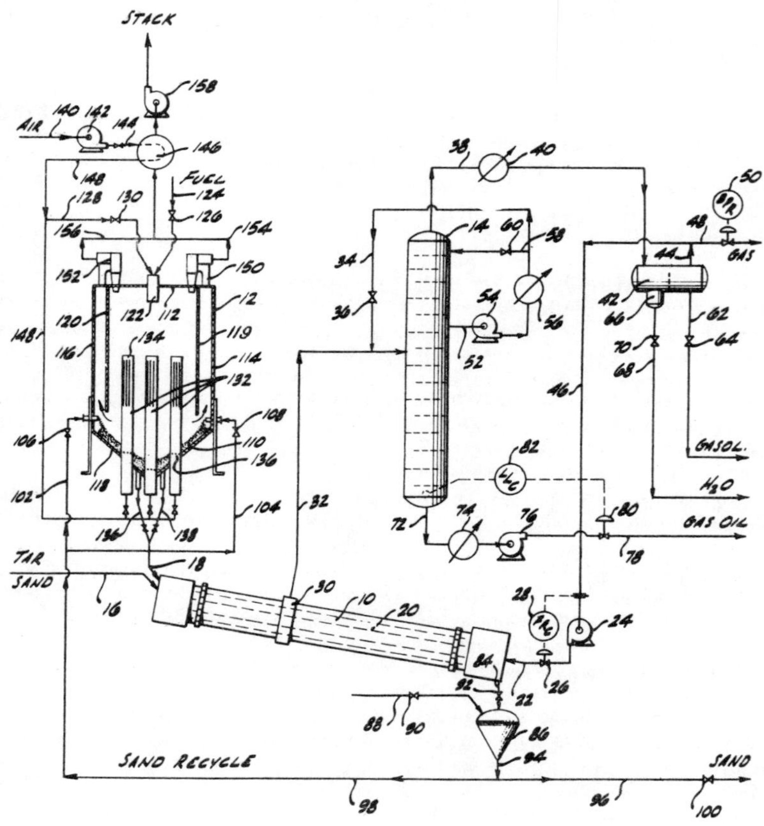

Source: C.H.O. Berg; U.S. Patent 2,905,595; September 22, 1959

It is provided with a plurality of longitudinal baffles of flights **20** by means of which the sand is mixed forming a uniform mixture at 900°F. The sand mixture progresses downwardly through coking kiln **10** which provides a residence time of approximately 1 hour. A stripping gas comprising the coked noncondensable gas product is introduced into the bottom of coking kiln **10** by means of line **22** under the influence of blower **24** at a rate of 2,490 Mscf/day (1 Mscf = 1,000 standard cubic feet) controlled by valve **26** under the influence of flow recorder controller **28**. The stripping gas sweeps the coker distillate upwardly through the kiln and it is removed from disengager **30** by means of line **32** at a temperature of 900°F.

This total coker distillate is combined with 9,180 barrels per day of 200°F quench oil flowing through line **34** controlled by valve **36** and the mixture is introduced at 500°F into coker distillate fractionation column **14**.

The overhead vapor is removed from column **14** through line **38** and is cooled and partly condensed at 100°F in cooler **40**. The mixed overhead phases are introduced into vapor liquid separator **42** wherein the noncondensable gases which are in vapor form at 100°F and approximately atmospheric pressure are removed at a rate of 5,980 Mscf/day per line **44**. Of this produced coker gas approximately half is passed through line **46** by means of blower **24** previously described and is introduced as a stripping gas into the bottom of coking kiln **10**. The remainder is removed through line **48** at a rate controlled by back pressure regulator **50**. This gas may be used for fuel as subsequently described in the coked sand furnace **12**.

A side cut is removed from the column above the coker distillate inlet through line **52** at a temperature of 400°F. This material is pumped by means of pump **54** at a rate of 43,880 barrels per day through side cut cooler **56** where it is cooled to 200°F. This cooled stream includes the 9,180 barrels of quench oil referred to previously, and the remainder is introduced through line **58** at a rate of 34,700 barrels per day controlled by valve **60** as a reflux into the top of column **14**.

The coker gasoline is removed from separator **42** through line **62** at a rate of 853 barrels per day controlled by valve **64** actuated by means of a conventional liquid level controller. Water is removed from trap **66** through line **68** at a rate of 570 barrels per day as controlled by valve **70** actuated by a conventional differential liquid level controller.

From the bottom of distillation column **14** is removed the heavy or gas oil fraction of the total coker product, which boils above 400°F. This material flows through line **72** through bottoms cooler **74** at a rate of 5,360 barrels per day. It is pumped by pump **76** through line **78** at the rate given controlled by valve **80** which in turn is actuated by liquid level controller **82**. The coker gas oil thus produced constitutes a feed stock for a catalytic desulfurization unit to produce a low sulfur cracking stock.

The coked sand product is removed from the bottom of coking kiln **10** through line **84** at the rate of 28,806 tons per day. This material is either intermittently or continously introduced into coked sand induction zone **86**. A conveyance gas under pressure is introduced thereinto through line **88** at a rate controlled by valve **90**. With the solids inlet sealed by element **92**, which may constitute either a solids feeder or a valve adapted to seal line **84** against a counterflow of gas, the conveyance gas depresses through lines **94**, **96** and **98** concurrently with an elongated gas permeable moving mass of coked sand. The gas flow is controlled at a rate sufficient to overcome opposing forces of gravity and friction acting on the sand so as to permit solids movement.

Through line **96** controlled by valve **100** a flow of coked sand is produced from the process at a rate of 8,756 tons per day. Through line **98**, which later divides into a plurality of parallel lines **102** and **104**, each in turn controlled respectively by valves **106** and **108** which serve to restrict the discharge of solids from the outlet of lines **102** and **104**, a sand recycle flowing at a rate of 20,150 tons per day is introduced into a plurality of points along each side near the bottom of coked sand furnace **12** forming a moving bed **110** there.

By restricting the discharge of solids from the outlet of lines **96, 102** and **104** without causing any substantial restriction on the discharge of gas therefrom, the sand is caused to move through the lines in the form of a nonfluidized dense or compact continuous moving mass of solids. The quantity of gas required to convey the solids is roughly less than 10% of that required to convey the same solids at the same rate as a gas lift suspension in a conventional pneumatic conveyor, and the conveyor conduit is approximately 80% smaller in diameter.

Coked sand furnace **12** is shown in Figure **4.3** as it appears in vertical cross-section. It is actually an elongated furnace having a flat or curving top **112**, flat parallel sides **114** and **116** and a trough shaped bottom **118**. The slope of the trough is 45° or any slope sufficient to cause downward flow of the solids there. Extending longitudinally along each side of the furnace and spaced apart are bridge walls **119** and **120** extending downwardly from roof **112** to a point just above the sand inlets and forming flue gas removal flues.

The coked sand furnace is 20 feet wide, 36 feet high, 100 feet long, and is provided with a series of ten surface combustion gas burners **122** disposed in a row along the top. Fuel gas is introduced through line **124** at a rate of 1,300 Mscf/day controlled by valve **126** together with air flowing through line **128** at a rate of 37,700 Mscf/day controlled by valve **130**. The combustion of this fuel supplies approximately 35% of the heat required in the process, the remaining 65% being generated within furnace **12** through combustion of the coke deposited on the coked sand. The fuel and air rates are controlled to produce coke free sand at the desired rate and temperature.

Disposed in three parallel rows along the length of furnace **12** and extending upwardly through trough shaped bottom **118** are gas lift tubes or zones **132**. These tubes are 10 inch nominal iron pipe size conduits approximately 35 feet high disposed vertically in three rows each containing 65 tubes, being spaced apart 1.5 feet. The upper part of each of these lines is provided around its periphery with six slots **134** each 10 feet long and 2 inches wide. Provided around the periphery of each of these tubes at a point just above the trough shaped bottom of the kiln are apertures **136**.

Moving bed of coked sand **110** flows downwardly and enters these apertures; the sand is suspended in an upwardly flowing stream of preheated 1000°F combustion air to form a sand suspension. The lift gas velocity is 25 feet per second in the lift lines. The suspension velocity is decreased in the slotted portion of the combustion tubes and the gas and sand move radially outward through the slots and the sand falls by gravity back onto moving bed **110**. The recirculation rate of the sand through the combustion tubes **132** may be varied at a rate sufficient to adequately reduce the coke content of the sand.

In the process of this example this recycle rate was controlled at a value of 27 to 1, that is, with a coked sand feed rate of 20,150 tons per day, the internal sand recirculation rate was 540,000 tons per day. The hot spent sand is withdrawn from the bottom at points along the bottom of the furnace section through lines **136** and **138** and passed through line **18** into coking kiln **10**. Combustion air is introduced at 116,000 Mscf/day through line **140** via blower **142** and through line **144** to preheater **146**; then heated from 80° to 1000°F and passed through line **148** at 78,300 Mscf/day into the bottom of vertical sand combustion tubes **132**. The coke from the coked sand grains is suspended and burned. The flue gas produced by sand combustion and by fuel combustion in burners **122** passes downwardly therethrough under the lower edge of bridge walls **118** and **120** and then upwardly between these bridge walls and the outer walls of the furnace.

Here part of the suspended sand grains are separated and the gas passes through a plurality of dust collectors **150** and **152** disposed along the edge of the top of furnace **12**. Here the dust is separated from the flue gases and is returned to the furnace and the dust free flue gas passes through lines **154** and **156** through air preheater **146**. Herein the gases are cooled from 1445°F and discharged by means of blower **158** to the atmosphere at a temperature of 550°F.

The furnace is constructed of mild steel and has 6 inches of internal insulation. The gas lift tubes exposed to 1000°F gas and sand are fabricated of type 314 stainless steel containing 25% chromium, 20% nickel and 2.5% silicon. The raw tar sand used as a raw material in the above example analyzed 1% water, 12.1% hydrocarbon material and 86.9% silica sand by weight. The product obtained in the process includes a total condensable coker distillate amounting to 72.8% of the hydrocarbon contained in the raw tar sand feed and the average gravity thereof is 19.1° API if the coker gasoline and the coker gas oil are combined. The coke production is 18.3% of the hydrocarbon feed and approximately 70% of this is burned to supply heat to the process. The gas production is 8.9% by weight of the hydrocarbon content of the raw tar sand fed to the process.

The process as described above successfully treats hydrocarbon saturated sands to produce a coker distillate hydrocarbon without the use of aqueous phases or flotation reagents. It is characterized by producing directly a distillate product rather than a heavy residual oil which often is heavier than water and thus difficult to treat in aqueous systems. Frequently the raw tar sand contains a hydrocarbon phase containing substantial quantities of hydrocarbon derivatives of sulfur. In some cases the hydrocarbon phase will analyze between 4 and 6% by weight of sulfur. In the combined tar sand retorting and hydrocarbon coking process a substantial desulfurization is effected with the production of hydrogen sulfide in gas phase and a coker distillate containing less than 3% by weight of sulfur.

Coking a Mixture of Tar Sand and Froth Product

I. Steinmetz; U.S. Patent 3,466,240; September 9, 1969; assigned to Great Canadian Oil Sands Limited, Canada has found that bitumen extracted from tar sands by the hot water process can be mixed with fresh tar sands and used as a feed for direct coking. The proportion of added bitumen in the feed can be adjusted so that regardless of the proportion of bitumen in the fresh sands, the mixed feed will always contain the same amount of cokable material.

This process has several advantages over direct retorting alone. Firstly, a coker feed comprising a mixture of bitumen and sands has a lower mineral content than a feed consisting of only tar sands. This allows a more efficient utilization of heat since a great deal less siliceous and mineral material need be heated compared to that of a feed of only tar sands. Secondly the mixed feed gives improved combustion characteristics in a moving bed type of operation where it is essential that enough coke is generated to heat the sands so that they can be recycled hot back into the reaction zone. In short, by adjusting the proportions of bitumen and tar sands in the mixed feed of this process, the advantages of both an extraction process and a direct retorting process may be maximized and disadvantages minimized. The process is particularly advantageous in that it allows for a uniform coker feed regardless of the composition of the tar sands.

For example, with a high tar sand, the amount of bitumen in the feed mixtures could be reduced while with a low tar sand content, the bitumen content could be increased making possible a constant composition retort feed. The particular process used in the coking step is not critical.

Fluid Coking Processes

In a process described by *D.E. Blaser; U.S. Patent 3,261,775; July 19, 1966; assigned to Esso Research and Engineering Company* a feed extracted from tar sands and containing bitumen, solids and water is preheated and then is introduced into the top of a tower. The feed is then contacted countercurrently with flue gas and much of the water is stripped out of the feed. The resulting mixture is then passed to a conventional fluid coker which consists of a reactor and burner; within the burner may be two or three stages of cyclone separators.

The remaining water is vaporized out of the feed and much of the coke formed during the coking step is burned off the sand within the burner; the remaining coke serves to coat the solids. The diplegs of the cyclones within the burner empty into the burner itself or outside the burner. All or part of the fines may be rejected; rejected fines are disposed of by conventional means. The reactor cyclone diplegs discharge to the reactor stripper. In this manner most of the fine materials are carried to the burner.

Since fluid coking requires a fairly constant particle size, the efficiency of a fluid coker designed to process these bitumens would be greatly enhanced by providing these methods for separation of fine material. The solids and the remaining coke are transported to a second high temperature burner, where all the coke is burned and the sand is heated. The hot sand is now transferred to a steam-water drum. Contacting of the hot sand and water releases 15 to 45 psig of steam; this large amount of energy is recovered and may be used for any desired purpose. The sand is then removed in the form of a sand-water slurry.

The flue gas from the second, high temperature burner is passed to the first stage burner thereby combining the flue gases. The combined gases are then passed through one or more stages of cyclones where dust and fine coke is recovered. The flue gas may then be channeled through a waste gas turbine, driving an air blower which supplies air for both the first and second burner. From the turbine the hot flue gas passes to the lower portion of the tower where it serves to impart some heat to the feed and also strip out most of the water. Water is injected into the second stage burner to absorb the heat of combustion of the coke by forming steam and also provide additional heat carrier in the flue gas for stripping. Water may also be added to the first stage burner for the same reason.

In related work *J.W. Brown; U.S. Patent 3,278,412; October 11, 1966; assigned to Esso Research and Engineering Company* describes a modified system of fluid coking in which viscous crude petroleum oil is separated and cracked from a mixture with sand, clay and water.

According to the process a tar sand which may be an Athabasca tar sand is subjected to a froth flotation. That is to say, hot water is mixed with the tar sand; several phases are formed including an aqueous phase, a sand phase, and a froth. The froth forms because gas is liberated from the bitumen, the gas being predominantly ethane, methane and trapped air. The forth may contain up to 40 volume percent of gas. The water phase and the sand phase are withdrawn and discarded.

This froth is then subjected to a fluid coking process to remove oil and water and to form coke and lower boiling hydrocarbons from the oil within the froth. Extraneous hot gas or fuel may be added to the burner of the coking unit and this combined with the low coking temperature utilized prevents the coke from being entirely burned to provide heat for the coking process.

The lighter overhead hydrocarbon products are separated from the heavier products and removed or may further be put to use. The heavier, 900+°F fraction, is subjected to conventional process such as fluid coking in which gas and liquid product are produced along with a high value coke. However, the heavier fraction may be subjected to other processes such as visbreaking, for example.

The solids, including sand and other mineral matter such as clay, are coated with coke in the first stage, low severity or low temperature coking reactor during the coking step, and the coated solids are then trapped within the coking reactor. By trapped it is meant that the coke coating serves as a glue. Therefore, the smaller solids become stuck to larger particles and this serves to keep the smaller solids within the reactor rather than allowing them to escape through the stack. This is particularly true of very fine solid material within the 1 to 5 micron range; the cohesive force of the coke tends to keep them stuck to larger particles and, therefore, they tend to remain within the coking unit. As much as 50% of the coke formed during coking is maintained on the solids.

Fines from the cyclones and some of the coarser solids of the low temperature burner are regenerated in a fluid bed and used as cracking catalyst in the reactor. The clay in these particles becomes activated and acts as a cracking catalyst. Some of the stack solids from the high temperature burner may also be used. The net carbon or coke coated solids are removed from the burner of the fluid coker and are introduced into a high temperature burner. Because of the high temperature in this burner, and the fact that all the coke is burned, the coke coating is burned to CO_2 thereby producing a large amount of heat. This heat in the form of hot gas is passed into a boiler where it may be utilized for the production of steam.

Alternately, part of this hot gas may be passed through the coker burner to supply heat. The extreme heat causes some of the sand and clay to become sintered, that is to say agglomerated in large masses which are, therefore, easier to remove than smaller pieces. Alternately, the solids may be removed as slag where ash melting point is low. Solid particles are removed from the high temperature burner and introduced into a cooler. The heat liberated during this cooling process is captured in a boiler and may be used for production of steam.

In the case of a froth from an Athabasca sand, the protective coke covering on the solid particles eliminates problems of erosion of apparatus and entrainment of fine solid particles, which would otherwise be encountered if all the carbon were burned off the solids in a conventional fluid coker burner. The principle of operation of the burner is that at least 5%, and preferably 50% of coke is maintained on the solids to produce a protective covering during burning in the burner. If necessary, in cases where the heat load is unusually high in the reactor, supplementary fuel such as torch oil is burned in the burner in preference to the coke covering to maintain this protective coating.

In a process described by *M.F. Nathan, G.T. Skaperdas and G.C. Grubb; U.S. Patent 3,320,152; May 16, 1967; assigned to Pullman Incorporated* tar sands agglomerates are introduced into a feed preparation zone and mixed with relatively hot particulate material obtained from a reaction zone. The feed preparation zone is maintained under temperature conditions to convert the agglomerates and particulate material into a fluidizable mixture suitable for introduction as feed material into a fluid coking zone and at a pressure suitable for introducing agglomerates into the feed preparation zone.

Mixing in the feed preparation zone is preferably carried out in a fixed bed agitated by suitable method and means to insure adequate intermixing of the agglomerates and hot particulate material. The relatively hot particulate material can be obtained at a suitable elevated temperature, which is preferably between 400° and 1400°F, from a hydrocarbon conversion zone, e.g., the coking zone, or from a heat generation zone employed in the process.

The ratio of relatively hot particulate material to agglomerates feed is preferably maintained between 0.5:1 and 2.5:1 in order to regulate the temperature of the bed in the feed preparation zone at the desired level. Hot particulate material, such as sand, is a preferred hot contact material, since it serves as a fluidization promoting diluent for the sand-oil mixtures contained in the agglomerates feed as well as a heating medium to reduce the viscosity of the heavy hydrocarbons in the feed.

The temperature in the feed preparation zone is generally maintained at a level which is sufficiently high to cause reduction in the viscosity and expel water from the agglomerates feed such that the agglomerates are triturated, and yet, preferably, below the level at which any substantial percentage of the entering hydrocarbon material is converted to gaseous material which is not condensable under atmospheric pressure conditions, e.g., dry gas.

It is important to minimize dry gas production at this point in order to minimize the cost of recovering hydrocarbon materials leaving the feeder as vapor. For these purposes, the preferred temperature range in the feed preparation zone is maintained between 250° and 600°F, and most preferably, between 275° and 450°F.

The feed preparation zone is preferably maintained under essentially atmospheric pressure and most preferably under a subatmospheric pressure to prevent loss of gaseous material at the location where agglomerates are introduced. It has been found that agglomerates feed can be converted into a fluidizable mixture employing relatively hot particulate material as the contact agent employing an agglomerates residence time in the feed preparation zone preferably between 1 and 10 minutes.

The fluidizable mixture produced in the first zone by the action of the relatively hot contact material mixed with the agglomerates feed is withdrawn from the feed preparation zone into a pressure developing zone for passage into a hydrocarbon conversion zone. The conversion zone preferably constitutes a coking zone containing a dense bed of fluidized solids material operating at an elevated temperature and pressure substantially above the conditions in the feed preparation zone.

Preferably, the feed preparation zone is situated at an elevation above that of the thermal coking zone such that the fluidizable mixture can be passed to the upper portion of a standpipe communicating between such zones for passage by gravity into the fluidized bed of the thermal coking zone. Suitable agitation means is provided preferably within the standpipe in order to prevent bridging of the standpipe.

The hydrocarbon conversion zone employed to convert the hydrocarbon material contained in the fluidizable mixture of tar sands and particulate materail produced in the feed preparation zone comprises a fluidized bed of particulate material maintained under thermal coking conditions. In the coking zone, heavy oil adhering to the solid particles undergoes pyrolysis, evolving lighter hydrocarbon vapors and leaving carbonaceous residue denoted as coke on the solids.

Temperatures between 850° and 1100°F and pressures between atmospheric and 40 psig are preferable. Particulate material bearing coke at the surface is passed from the coking zone into a heat generation zone in which coke is burned thereby raising the temperature of the particulate material to an elevated level. Heated particulate material at a temperature preferably 200° to 300°F above the temperature in the conversion zone and in the range from 1050° and 1400°F is passed from the heat generation zone into the coking zone to supply heat required for the coking operation.

Removal of Suspended Material by Filtration

In a process decribed by *D.S. Pasternack; U.S. Patent 3,365,384; January 23, 1968; assigned to Canadian Patents and Development Limited, Canada* the high viscosity residue of water washed dehydrated tar sands is subjected to controlled thermal-cracking at atmospheric pressure or under vacuum. This is followed by filtration of the hot residual oil through a filter at a suitable temperature to prevent congealing of the residual oil and permit rapid and complete filtration. Desirably the filter temperature is low enough so that additional thermal-cracking does not take place in the pores of the filter, e.g., less than 650°F. The filtration may be carried out either under vacuum or under pressure. Preferably the feed material is subjected to controlled thermal-cracking at atmospheric pressure (or under a slight vacuum provided that excessive frothing does not occur), with simultaneous removal of the hydrocarbon vapors and gases which are produced, until an adequate amount of thermal-cracking has occurred.

An adequate amount of thermal-cracking has occurred when the hot residual oil will filter rapidly and completely by conventional means through a hot filtering medium and still provide a low ash filtrate. Generally some advantage in separating the solids is obtained when distillate formation is within the range 20 to 60% by weight of the pure oil in the feed. This corresponds to free carbon formation of from 0.2 to 3.0 weight percent based on pure oil in the feed. The following examples illustrate the process.

Example 1: Dehydrated Athabasca oil sands oil (obtained by a warm water washing method) was heated slowly at atmospheric pressure from 650° to 775°F in a period of 3.5 hrs until about 40 weight percent of distillate was obtained. The hot residual oil was then filtered rapidly and completely by passing under a vacuum of 27 inches, through a hot filtering medium composed of a No. 1 Whatman filter paper on a Buechner funnel suitably preheated and kept hot during the filtration.

The temperature of the filter medium was about 600°F. The solids in the filter cake consisted of 78% clay and 22% free carbon, and were oil wet. A low ash filtrate, with an ash content of 0.11 weight percent was obtained. (Coke produced from this filtrate by heating in a Ramsbottom furnace at 1020°F for 20 minutes had an ash content of 0.28%.)

Example 2: The same dehydrated Athabasca oil sands oil was heated comparatively rapidly at atmospheric pressure from 650° to 820°F in a period of 40 minutes until 40 weight percent of distillate was obtained. The hot residual oil was then filtered rapidly and completely similarly to Example 1. The solids in the filter cake consisted of 85% clay and 15% free carbon, and were oil wet. A low ash filtrate, with an ash content of 0.12 weight percent was obtained. (Coke produced from this filtrate by heating in a Ramsbottom furnace at 1020°F for 20 minutes had an ash content of 0.45%.) Examples of preferred variants of the process (using a diluent) are as follows.

Example 3: The warm unfiltered residual oil from Example 1 was diluted with two to three volumes of warm diluent (the diluent was the portion of the distillate from the thermal-cracking with a boiling range from 325° to 525°F.) The diluted residual was filtered rapidly and completely while under a vacuum of 27 inches, through a filtering medium composed of a No. 1 Whatman filter paper on a Buechner funnel, giving retention of all suspended solids (including the precipitated material).

The filter temperature was 250°F. The diluent-free filter cake consisted of 15% clay, 4% free carbon, and 81% diluent insoluble material. Upon removal of the diluent by distillation, the extract showed an ash content of 0.011 weight percent. (Coke produced from this extract by heating in a Ramsbottom furnace at 1020°F for 20 minutes had an ash content of 0.07%.

Example 4: The warm unfiltered residual oil from Example 2 was diluted with two to three volumes of warm diluent. The diluent was the portion of the distillate from the thermal-cracking with a boiling range from 325° to 525°F. Rapid and complete filtration was obtained using a vacuum of 27 inches, through a filtering medium composed of a No. 1 Whatman filter paper on a Buechner funnel, giving retention of all suspended solids (including precipitated material).

The filter temperature was 250°F. The diluent-free filter cake consisted of 20% clay, 4% free carbon, and 76% diluent insoluble material. Upon removal of the diluent by distillation, the extract showed an ash content of about 0.016 weight percent. Coke produced from this extract by heating in a Ramsbottom furnace at 1020°F for 20 minutes had an ash content of 0.11%.

HYDROGENATION

Upflow Liquid Phase Hydrogenation System

According to a process described by *R.H. Wolk and M.C. Chervenak; U.S. Patent 3,842,122; October 15, 1974; assigned to Hydrocarbon Research, Inc.* the recovery of valuable liquid hydrocarbons from tar sands bitumen by hydroconversion in an upflow reactor is improved when downstream equipment, including liquid-gas separators, is operated under selected temperature and liquid velocity conditions to prevent coke formation and precipitation.

Thus, it has been determined that there is a critical combination of the temperature of the liquid phase in the reactor effluent vapor-liquid separator and the downflow liquid velocity in that separator that must be followed if coke formation in that separator is to be avoided. This correlation is based on a plot of the logarithm of the linear velocity against a linear temperature range which indicates that coke-free and coke-forming combination conditions exist. Some variation in the location of the line separating the coke-free and coke-forming zones will exist with bitumens of different characteristics as well as variations in the hydrogenation reactor conditions.

FIGURE 4.4: LIQUID PHASE HYDROGENATION SYSTEM

a.

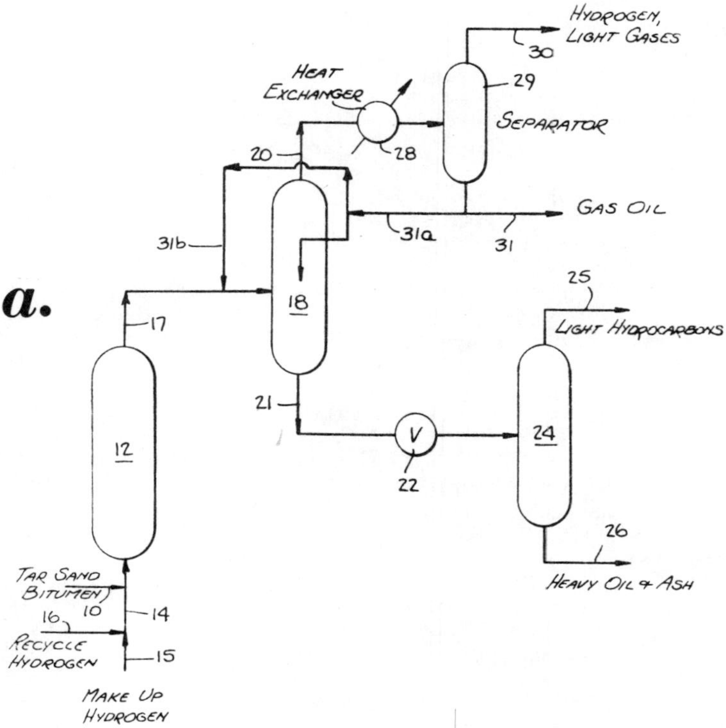

b.

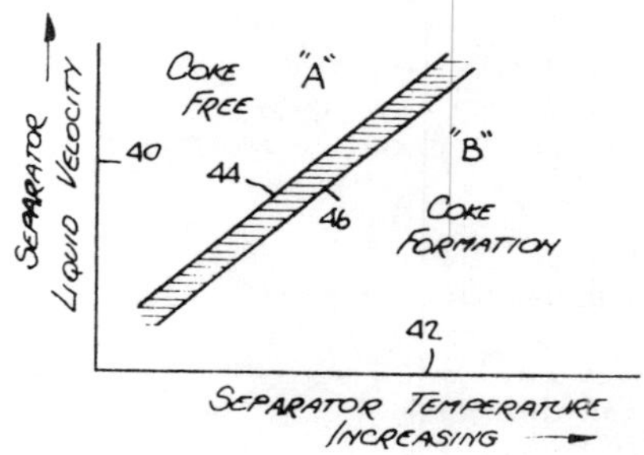

(a) Flow diagram for treating certain tar sands
(b) Graph of the coke-free versus coke-forming combination of conditions

Source: R.H. Wolk and M.C. Chervenak; U.S. Patent 3,842,122; October 15, 1974

As shown in Figure 4.4a, a preheated natural tar bitumen feed including from 0.5 to 4% of silt, inorganic sand or the like, the particles of which are generally less than 325 mesh in size, is introduced through line **10** to a reactor **12** through line **14**. A hydrogen rich gas (more than 60% hydrogen) is also introduced to the feed through line **14**. This will include, as needed, a fresh make-up hydrogen in line **15** and a recycle hydrogen in line **16**.

Conveniently, all of the liquid with the ash and the gas are taken overhead in line **17** and passed to the separator **18**, which is at substantially reactor pressure. There will be a phase separation of gas and vapors which are removed overhead in line **20** and the liquid and ash which are removed as bottoms through line **21**. The liquid is then passed through pressure reducing valve **22** to low pressure separator **24**.

The separator chamber **24** is preferably operated at essentially atmospheric pressure to permit further removal of hydrogen and light hydrocarbons at **25**, with the heavy oil containing the ash being removed at **26**. The vapors and gases that pass overhead at **20** from the high pressure separator **18** may be cooled at **28** and passed to separator **29**, from which the gases such as hydrogen, methane, naphtha, etc. may be removed through overhead line **30**. The hydrogen may be recovered by conventional processing and reintroduced in line **16**.

The liquid fraction removed at **31** as bottoms from separator **29** is a gas oil boiling range material. This gas oil fraction may be injected into the liquid phase in separator **18** through line **31a** to cause rapid cooling of the liquid phase. Alternately, it could be injected into line **17** through line **31b** to quench the entire reactor effluent to the desired temperature. The amount of gas oil used is a function of the temperature at which it is injected and the desired temperature for the operation of the separator.

Under normal conditions, it is found that the hydrogen partial pressure in reactor **12** should be in the range of about 200 to 1,500 psi, hydrogen partial pressure. Temperatures will range from at least 700°F but not to exceed about 950°F. Conversion of the 975+°F boiling range material to the material boiling below 975°F will be in excess of 50%.

A total of eight experiments were run in which data on coke laydown in separators was obtained on three different bitumens. Three different bitumen feeds, designated 1, 2, and 3, were obtained from two different sources designated A and B. Coking in the separation zone was not encountered with feed 3B which contained 3.4 weight percent ash in Experiment I (see table on following page).

Attempts to duplicate these experimental results in Experiment II were not successful with bitumen feed 1A which was obtained by a different extraction technique and contained 0.7 weight percent ash. The effect of the ash concentration on the coke laydown was evaluated in Experiment III.

Feed 2A, containing 3.1 weight percent ash, was run and coke was obtained in the separation zone. This result indicated that the difference in the ash left in the bitumen after its extraction from the tar sand was not by itself sufficient to prevent coke laydown. Experiment IV was then run with a hydrogen partial pressure in the reactor of 1,200 psig instead of the 750 psig used for the first three experiments.

All of these four experiments were at 68 to 73% conversion. The increase in pressure did not prevent coke laydown in the separation zone. Experiments V through VIII were made at various separator temperature and liquid velocity combinations in an attempt to define operable conditions.

A plot of all the temperature and liquid velocity data, presented at Figure 4.4b, shows that there are two distinct zones defining conditions for coke laydown and coke-free operation for the separator. The data suggest that high ash feeds may be less likely to lay down coke than low ash feeds.

However, in order to guarantee coke-free operation of the separator, it is necessary to stay above the minimum velocity and below the maximum temperature indicated by the lines in

Figure 4.4b. Certain bitumens recovered from Athabasca tar sand have essentially the following characteristics:

Bitumen number	1	2	3
Bitumen source	A	A	B
Gravity, ° API	9.1	5.6	5.5
Sulfur, w., percent	4.64	4.84	4.85
Ash, w., percent	0.79	3.09	3.38
Carbon, w., percent	83.35	81.8	80.8
Hydrogen, w., percent	10.53	10.39	10.29
Metals (V plus Ni), p.p.m.	210	198	253
Nitrogen, p.p.m.	3,900	3,700	4,100

Experiment	I	II	III	IV	V	VI	VII	VIII
Bitumen number and source	3B	1A	2A	1A	1A	1A	1A	1A
Reactor temperature, ° F	840	835	839	840	805	805	805	805
Hydrogen partial pressure in reactor, p.s.i.	750	750	760	1200	750	750	750	750
Conversion, volume percent (disappearance of 975° F., plus feed)	72	68	73	72	58	58	58	58
Separator temperature, ° F	700	675	800	800	550	650	700	800
Separation zone condition	Clean	Coked	Coked	Coked	Clean	Coked	Clean	Clean
Liquid phase velocity in separation zone, ft./sec.	0.002	0.002	0.004	0.004	0.002	0.001	0.50	0.80

The ordinate **40** is laid out on a log-log scale of superficial linear velocity of the downflowing liquid in separator **18** and the abscissa **42** is laid out as a linear temperature scale. The curves **44–46**, which are relatively close depending on the bitumen, formed a demarkation line between the area "**A**" which proved to be representative of coke-free conditions, and area "**B**", which generally showed objectionable coke formation.

It has been found essential to keep the effluent in the separator **18** down to a temperature between 500° and 800°F and to maintain the downflow linear liquid velocity between 0.006 and 1.0 foot per second. The lower temperature corresponds to the lower velocity as shown on the curves in Figure 4.4b. An approximate linear relation appears to exist between the lowest combination of conditions and the highest combination.

Hydrofining Process

J. Eng and W.R. Guggisberg; U.S. Patent 3,480,540; November 25, 1969; assigned to Esso Research and Engineering Company describe a hydrofining process which provides adequate desulfurization, denitrogenation and saturation of olefins for a blend of fractions derived from bitumen in a single unit employing the same conditions and catalyst for all. It has been found that when the bitumen fractions are blended in certain quantities and processed at certain hydrofining conditions, the fractions are each upgraded in the desired manner.

Generally, the process comprises separating the relatively solids-free bitumen into an overhead virgin fraction and a bottoms fraction, cracking the bottoms fraction, blending the virgin and cracked fractions, hydrofining the blend and recovering desired product fractions.

Referring to Figure 4.5, the bitumen derived from tar sands is fed by line **1** to distillation tower **2**. Suitable bitumens have distillation characteristics such that 50% of the material is distilled at a temperature in the range of 800° to 1100°F. The bitumens contain 1.5 to 6 weight percent sulfur, 0.1 to 1.0 weight percent nitrogen, 100 to 500 ppm metal, less than 5% inorganic solids and water and they have pour points in the range of 20° to 110°F. In tower **2** the bitumen is separated into a virgin overhead fraction having an end point of 500° to 1000°F, preferably below 700°F and a bottoms fraction boiling below 500° to 1000°F, preferably above 700°F.

The virgin fraction is passed by line **3** to hydrofining reactor **4** without any further treatment. The bottoms is passed by line **5** to a conversion unit, preferably a fluid coker **6** or a visbreaker. The conversion unit can be any suitable type of thermal-cracking or depolymerization reactor including a delayed coker, a fluid coker, a visbreaker or a hydrovisbreaker.

If desired, cracking processes involving inert solids, catalysts, hydrogen and steam can be used although conventional thermal-cracking processes are deemed more suitable for cracking the heavy bitumen bottoms.

In the case of fluid coking the coking zone is usually maintained at a temperature of 800° to 950°F. Typical fluid coking processes, thermal-cracking processes and visbreaking processes are described in *Hydrocarbon Processing*, September 1966, volume 45, No. 9, pp 191 to 194. Cracking severity can be controlled in accordance with the type of products desired. Coke is removed from the coker by line **7**. A gas and light ends fraction is removed from the coker by line **8**. A cracked fraction boiling in the range of C_5 to 1100°F, preferably 70° to 925°F, is passed by line **9** to line **3** where it is blended with the virgin fraction. The blend is hydrofined in reactor **4**. Any suitable hydrofining reactor can be used. Conventional hydrofining catalysts can be used.

The preferred catalyst is one containing 2 to 5 weight percent cobalt oxide, 10 to 25 weight percent molybdenum oxide and the balance silica stabilized alumina. Oxide catalysts are preferably sulfided in any conventional manner prior to use or in situ. The reaction conditions in reactor **4** depend in part on the characteristics of the components of the blended feed and on the type of products desired. In general the reaction temperature is 500° to 850°F, preferably 600° to 800°F; the reaction pressure is 400 to 2,000 psig, preferably 500 to 1,200 psig and the fresh feed rate is 0.2 to 3.0 v/v/hr. The hydrogen containing treat gas contains 60 to 100% hydrogen and it is supplied at the rate of 500 to 2,000 scf/bbl. Under these conditions hydrodesulfurization, hydrodenitrogenation and saturation of olefins and polyaromatics are the dominant reactions.

Reaction products are removed by line **10** and passed to separator **11**. A gas fraction containing hydrogen, H_2S and ammonia is passed overhead by line **12**. All or part of the gas is recycled to the reactor by lines **13** and **14**. Make up hydrogen is supplied by line **14**.

FIGURE 4.5: PROCESS FOR HYDROFINING BITUMEN

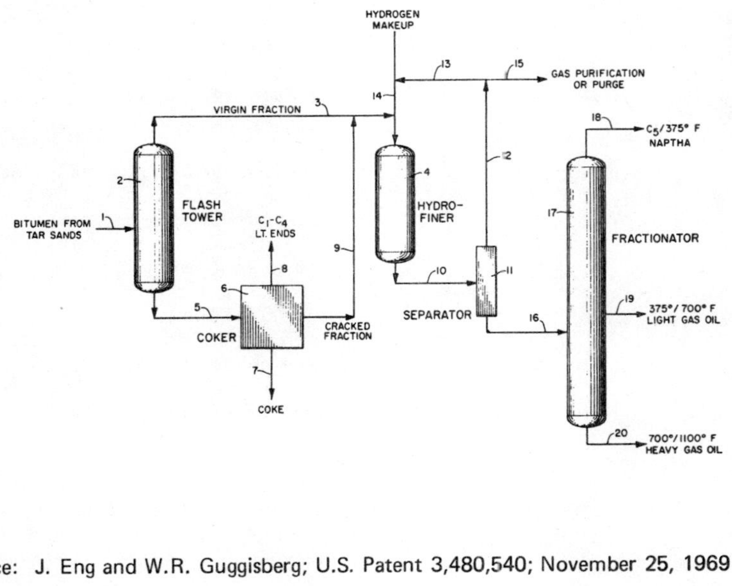

Source: J. Eng and W.R. Guggisberg; U.S. Patent 3,480,540; November 25, 1969

Part of the gas can be purged from the system or sent to purification by line **15**. Hydrofined oil is passed by line **16** to distillation tower **17**. In a preferred example, three fractions are recovered. A C_5 to 375°F gasoline fraction is recovered by line **18**. A light gas oil fraction boiling in the range of 375° to 700°F is recovered by line **19**. A heavy gas oil fraction boiling in the range of 700° to 1100°F is recovered by line **20**. The fractionation in zone **17** can be arranged in any suitable manner depending upon the number and type of products desired.

Example: Hydrofining of blended bitumen fractions is demonstrated by the following data. A blend of coker distillates and virgin distillates simulating a tar sand bitumen distillate blend was prepared. The blend contained 37 volume percent virgin gas oil from a tar sand bitumen and 63 volume percent of petroleum derived materials having characteristics similar to distillates derived from tar sands. The properties of the components of the blend are shown in Table 1 below. The simulated distillate blend and its fractions have the inspections shown in Table 2 below. 12,000 g of the total distillate blend was hydrofined in a reactor in the presence of a sulfided cobalt molybdate on silica stabilized alumina catalyst.

The catalyst contained 3.5 weight percent cobalt oxide, 12.1 weight percent molybdenum oxide, 2.0 weight percent silica and the balance was alumina. The pressure was 800 psig; the space velocity was 1.0 and the hydrogen rate was 2,000 scf/bbl. The data in Table 3 shows the results for various hydrofining temperatures. The data shows that the bromine number and the sulfur content of the total feed have been reduced to satisfactory values. Sulfur has been reduced from 2 to 0.3% or less depending on the reaction temperature.

The product naphtha fraction has a bromine number of below 1.0 which is desirable for gasoline components. The nitrogen and sulfur content of the naphtha is low enough for subsequent reforming and/or incorporation into mogas without sweetening. Light gas oils are often used as domestic heating oils and the light gas oil products are low enough in sulfur content for this use. The heavy gas oil product has a sulfur content of 0.5% or less making it suitable for use as industrial fuel oil in areas where sulfur contents must be very low.

TABLE 1: COMPONENTS IN SIMULATED ATHABASCA TAR SAND DISTILLATE BLEND[1]

	LV percent in Blend	Boiling Range, ° F.	Gravity API	Sulphur, Wt. percent	Nitrogen, Wt. percent	Bromine No.
Naphtha:						
Coker	14	C_5/350	62.9	0.36	0.006	73
Virgin	1	C_5/380	63.6	0.06	[2] 1	1
Light Gas Oil:						
Coker	12	355/680	24.0	1.43	0.046	32
Virgin	20	400/650	36.1	0.80	0.010	3
Heavy Gas Oil:						
Coker	16	650/1,000	23.3	2.50	0.520	
Virgin	37	650/1,000	15.4	3.00	0.118	8

[1] The heavy virgin gas oil was produced from Athabasca Tar Sand bitumen. Other components from topping and delayed coking unit when processing sour blend crude.
[2] Parts per million.

TABLE 2: INSPECTIONS ON THE SIMULATED TAR SAND DISTILLATE BLEND AND ITS FRACTIONS

		Fractions		
	Total Blend	Naphtha	Light Gas Oil	Heavy Gas Oil
LV, Percent on Blend	100	16.8	34.7	48.1
ASTM Dist. Range, ° F	C_5/1,000 Nominal	126/378	420/620	692/911 (95% pt)
API Gravity	28.3	58.5	32.5	16.5
Bromine Number	20.5	65.9	15.9	12.2
Sulphur, Wt. Percent	1.97	0.36	1.17	2.82
Total Nitrogen, p.p.m.	1,030	25	350	2,300
Basic Nitrogen, p.p.m.	291	25	139	533
Mercaptan Number	28.8	66.4	19.2	8.0

TABLE 3: HYDROFINING SIMULATED TAR SAND DISTILLATE BLEND

	Treat Temperature, ° F.								
	650			700			750		
Total Liquid Product:									
Bromine Number	2.1			2.1			2.5		
Sulfur, Wt. Percent	0.29			0.10			0.03		
	Fraction [1]								
	Naph.	Lt. G.O.	Hvy. G.O.	Naph.	Lt. G.O.	Hvy. G.O.	Naph.	Lt. G.O.	Hvy. G.O
LV Percent on Total Prod	17	39	44	18	39	43	21	43	36
API Gravity	58.9	33.9	22.2	59.4	34.1	22.9	58.5	33.4	23.8
Bromine Number	0.3	1.3	3.5	0.3	1.5	3.3	0.6	2.2	4.4
Sulfur, Weight Percent	[2] 16	0.10	0.55	[2] 17	0.05	0.19	[2] 16	0.01	0.04
Total N, p.p.m.	15	225	1,350	6	134	1,170	7	66	168
Basic N, p.p.m.	15	130	288	6	90	217	6	52	60
Mercaptan Number	0.2	0.2	0.6	0.1	0.[illegible]	0.2	0.2	0.1	0.2

[1] 15/5 distillation to nominal C_5/380, 380/650 and 650° F.+ vapor temperature fractions.
[2] Parts per million.

A process described by *E.T. Layng; U.S. Patent 3,151,054; September 29, 1964; assigned to Hydrocarbon Research, Inc.* is directed to the thermal hydrogenation of tar sands to convert the fines to a settled condition whereby a silt free product can be produced. The refined process includes the steps of passing the tar sand substantially in the liquid phase through a reaction zone in the presence of a hydrogen-containing gas under temperature in the range of 700° to 950°F and hydrogen partial pressure in the range of 200 to 1,000 pounds per square inch gauge and space velocity to provide a residence time in the order of ½ hour to 1½ hours whereby the silt becomes free settling and settling the effluent to remove a substantially silt free liquid and an ash containing residue.

Hydrogenated Thermal Tar as Hydrogen Donor-Diluent

A process described by *J. Stewart, S.C. Fulton and A.W. Langer, Jr.; U.S. Patent 2,772,209; November 27, 1956; assigned to Esso Research and Engineering Company* involves the use of an inexpensive hydrogen donor-diluent, such as a partially hydrogenated thermal tar obtained from catalytic cracking operations on hydrocarbon fractions such as petroleum gas oils in the recovery of oil from tar sands.

In the process it is preferred to use a donor-diluent having a boiling range within the limits roughly of 500° to 700°F. The lower boiling material makes it easier to strip the oil from the sand in later stages of the process. However, diluents boiling above 700° and up to 900°F may be used in many cases without difficulty. The diluent is preferably used in sufficient proportions to form a liquid or pumpable slurry with the sand. For this purpose, with high grade sands, the diluent should have a weight ratio of up to 5 to 1.

For sands of low bitumen or oil content, even larger proportions may be needed, up to 10 to 1 on the same basis. Using 5 to 10 parts of donor-diluent by weight for each part of tar or bitumen, the transferable hydrogen in the donor-diluent need not be very high. Hydrogenation to the extent of 200 to 500 scf of hydrogen per barrel of diluent is quite adequate.

In the process, sand is broken up and mixed with the donor-diluent in proportions suitable for forming a mobile slurry. The slurry is pumped through a fired coil and into a soaking and settling zone where the mixture is heated to a temperature within the range of 700° to 1000°F, preferably between 750° and 900°F. The material is held for a period sufficient to obtain good extraction, sufficient agitation being applied to accomplish this result. The usual residence time is from 10 minutes to 120 minutes. The process is preferably carried out under moderate pressure, e.g., up to 500 psig, a pressure of 200 psig being very satisfactory. However, the pressure need be no greater than is necessary to prevent loss of volatile ingredients.

The sand settles gradually to the bottom, as its oil content is extracted, and an oil phase forms at the top. The rate of sand settling and the completeness of oil extraction varies in part with the viscosity of the slurry and in part with the degree of agitation. The latter should be so controlled as to obtain maximum extraction of oil, consistent with good throughput. If the average residence time of the sand is within the limits described above and the sand slurry is free-flowing, extraction will be reasonably complete. With a residence time of 1 hour, for example, and a ratio of diluent to bitumen of 4:1, using a temperature of at least 750°F, and a diluent of boiling range about 500° to 700°F, 95% or more of the bitumen may be extracted.

In related work *J. Stewart, S.C. Fulton and A.W. Langer, Jr.; U.S. Patent 2,847,306; August 12, 1958; assigned to Esso Research and Engineering Company* describe the application of the above hydrogenated thermal tar donor-diluent process to the recovery of oil from shale.

Jet Fuels

A process described by *M.C. Kirk, Jr.; U.S. Patent 3,594,307; July 20, 1971; assigned to Sun Oil Company* involves producing a jet fuel having a luminometer number of at least 75. The process involves contacting a charge comprising a jet fuel range distillate from coked (thermally cracked) tar sands bitumen (such as that separated from Athabasca tar sands by the hot water process) or a straight run kerosene charge, or a catalytic gas oil charge, acyclic C_9 to C_{18} olefin charge (or a mixture of at least two such charges) with hydrogen in the presence of a hydrogenation catalyst formed from at least one member selected from the group consisting of nickel, cobalt, iron, molybdenum and tungsten and related oxides and sulfides on an inert porous carrier.

This is carried out at a temperature of 500° to 785°F, a pressure of 350 to 2,000 psig, a liquid hourly space velocity of 0.5 to 10.0 and a hydrogen circulation rate of 0 to 20,000 standard cubic feet per barrel of the charge. The resultant product is contacted with hydrogen in the presence of a catalyst which comprises a metal selected from the group consisting of nickel, cobalt, tungsten, ruthenium, rhodium, rhenium, osmium and the noble metal hydrogenation catalysts platinum and palladium, the catalyst being supported on a porous support (e.g., alumina, kieselguhr) at a temperature of 450° to 775°F, a pressure of 500 to 3,000 psig, a liquid hourly space velocity of 0.1 to 10.0 and a hydrogen circulation rate of 0 to 20,000 scf/bbl of the product of the first stage, the combination of conditions being selected to produce a jet fuel having an ASTM smoke point of at least 32 ± 3. The following example illustrates the process.

A straight run kerosene meeting the specifications for JP-5 and having the properties listed in the table below, under the heading charge, and containing 12.4% aromatics, was hydrodesulfurized in the presence of a sulfided catalyst comprising cobalt and molybdenum oxides on alumina (which catalyst was commercially available under the trade name Aero HDS-2). The hydrodesulfurization was conducted at 750 psig and 600°F at a liquid hourly space velocity of 2 and a hydrogen recycle of 5,000 scf/bbl of charge.

The hydrodesulfurized product was then charged to a second hydrogenation stage wherein the catalyst was Ni on kieselguhr. The second stage hydrogenation was conducted at 500°F and at 500 psig, at a liquid hourly space velocity of 0.75 with a hydrogen recycle of 10,000 standard cubic feet per barrel of feed. The product of the second hydrogenation stage contained only 0.05% by weight of aromatic hydrocarbons and had a smoke number of 35. Other properties of this two-stage product, are listed in the table under the heading JP-5A. From the California Research correlation, a smoke number of 35, for the second stage product, corresponds to a luminometer number of 82.

It can be seen from the charge properties listed in the table that this charge is a paraffinic kerosene. The table also lists, for purposes of comparison, runs made on the same straight run kerosene wherein only a single hydrogenation (or hydrodesulfurization) stage was used. Also shown, for comparison purposes, are similar runs made on propylene tetramer (which is a product obtained by the catalytic polymerization of propylene in the presence of a phosphoric acid on kieselguhr catalyst).

Preparation of Jet Fuels

	Charge stock							
	JP-5					Propylene tetramer		
	Operation							
	Deep hydrogenation of aromatics			Moderate hydrogenation of aromatics			Saturation of olefins	
	2 stages							
Catalyst type	CoMo[1]	Ni[2]	Ni-W	Ni-W		CoMo	CoMo	
Reactor conditions:								
Operating pressure, p.s.i.g	750	500	1,800		750	750	750	500
Gas recycle rate, s.c.f./bbl	5,000	10,000	5,000		0	0	0	3,000
Temperature, ° F	725	500	575		6[illegible]0	600	600	600
Liquid hourly space velocity	2	0.75	1		1.5	1	2	2
Inspection data	Charge							
Gravity, °API	43.9	45.3	44.8	44.4	44.1	52.1	54.2	54.1
Distillation (Engler) ° F.:								
10%	393	392	393	388	396	358	360	363
50%	419	418	418	419	418	365	370	371
90%	465	452	444	449	448	380	383	384
Aromatics, weight percent	12.4	0.05	4.0		9.8			
Olefins, weight percent						92.4	6.4	3.8
Freezing point, ° F		−54	−51	−58	−58		<−76	<−76
Flash point, (cc.), ° F	154	148			146	136	136	138
Estimated Luminometer No.[3]		82		69	62		95	94
Smoke point	24	35		30	27		40	39
Aniline point, ° F	149.5	164.6					175.0	176.4

[1] Desulfurization step.
[2] Deep hydrogenation step.
[3] Luminometer No. estimated from smoke point.

The hydrogenated propylene tetramer makes an excellent blending stock for incorporating with the two-stage hydrogenation products in order to make products having luminometer values above 85 (surprisingly, such hydrogenated acyclic olefins can be produced which have luminometer values of 100).

Two-Stage Deasphalting Process Using Controlled Solvent-Feed

D.L. Mitchell and J.G. Speight; U.S. Patent 3,779,902; December 18, 1973; assigned to Canada Cities Service, Ltd; Imperial Oil Limited; Atlantic Richfield Canada Limited; Gulf Oil Canada Limited, Canada describe a process for treating an asphaltene containing petroleum crude which may be contaminated by both metals and water. The petroleum crude may contain a considerable water concentration, being represented by a froth. The process comprises the introduction of a hydrocarbon solvent into the asphaltene containing petroleum crude so as to adjust the total asphaltene solvent power of the system. By asphaltenes is meant those hydrocarbons which are pentane insolubles. By solvent power of the system is meant the ability of the total hydrocarbon system, including solvent, to dissolve and contain the asphaltenes present.

In general, the process uses hydrocarbon solvents which may be selected from the group consisting of one or more paraffinic or isomeric hydrocarbons having from three to eight carbon atoms and saturated substituted cycloparaffins and saturated unsubstituted cycloparaffins having five or more carbon atoms. The solvent power of the system including solvent and feed is adjusted by altering the proportions of the solvent or the solvent type according to the required cotangent θ, as described hereafter, so as to yield a desired solvent power for the total system of solvent and petroleum crude. In general, in treating petroleum crude, for example an Athabasca bitumen, it is preferred that from 5.0 to 40.0 barrels of solvent are introduced in each step of the process per barrel of liquid to be treated.

In a two-step process, a solvent having a high solvent power is introduced to form two fractions. The first fraction preferably contains 10.0 to 25 weight percent of the asphaltenes and resins contained within the petroleum crude, at least 85.0 weight percent of the water and at least 98.0 weight percent of the mineral contaminants. A liquid fraction or heavy oil is formed which contains the remaining mineral contaminants and asphaltenes originally present in the petroleum crude.

Subsequently, the two fractions are separated and a second normally liquid or liquefied hydrocarbon solvent forming a low solvent power system is introduced into the liquid fraction, thereby forming two additional liquid fractions. A first additional liquid fraction is formed containing the remainder of the water, mineral contaminants and asphaltene content of the heavy oil and a second additional liquid fraction is formed consisting essentially of a mineral, water and asphaltene free heavy oil. These two additional liquid fractions are then separated to form the desired products of the process, that is, a heavy oil significantly free of water or mineral contaminants in a solvent phase and an asphaltene fraction of the bitumen substantially free of mineral and water contaminants.

Solvent power is generally measured as the solubility of an oil in a particular solvent system. As the asphaltenes are the least soluble components of the oil, it is necessary to define solvent power as the measurement of the critical solubility of the asphaltenes in given oil systems containing solvents so that deasphalting and demineralizing of asphaltene containing oils may be regulated. Conventional deasphalting processes for the deasphalting of heavy oils by pentane or other paraffinic solvents generally utilize the solvents singularly, wherein the asphaltenes are all concentrated in one deasphalting tower, but in most instances the asphaltenes recovered having a high mineral content.

It has been found that particularly useful solvent combinations may be formed by the mixing of one or more high power solvents with heavy oils in definite relationships, so that exact precipitation of asphaltenes in deasphalting tower occurs. Specifically, the yield of asphaltene precipitate obtained from the heavy oil is related to solubility of the normally precipitating asphaltenes in the solvent system.

The major influence on precipitation is governed by the solvent power of the single solvent or solvent blend, in conjunction with the heavy oil system, with the contribution of the solvent powers of the soluble oils and resins in the heavy oil being considered. Additional information concerning solubility characteristics may be found in *Regular Solution* by Hildebrand and Scott, Prentice Hall, 1963.

The relationship of asphaltene solvent power of the heavy oil-solvent precipitating medium may be expressed by the solubility parameter Cotangent θ (Cot θ). Cot θ is derived from experimental data in which the percent asphaltenes precipitated from a system for each solvent or solvent blend mixture with the heavy oil is plotted against the individual solvent concentrations or blend concentrations in the system, thereby deriving a definite linear relationship. Cot θ being the contangent of the angle θ formed between the X-axis or the axis containing the solvent concentration and the linear line formed by the experimental data. The Cot θ relationship and means for determining θ for deasphalting system is fully presented by J.A. Bichard, "Oil Solubility," *19th Canadian Chemical Engineering Conference and 3rd Symposium on Catalysis,* October 19-22, 1969, Edmonton, Alberta.

As originally conceived, the Cot θ relationship was intended to provide a means for insuring that undesirable precipitation would not occur. Using this concept, it was found that precipitation would be precluded by maintaining the Cot θ of the system at a positive value. It has been found that by use of the Cot θ relationship, solvent power relationships may be placed on a single functional graph as illustrated by Figure 4.6. One may choose the desired percent of asphaltenes to be precipitated from Figure 4.6 and thereby determine what Cot θ relationship is required for the system, thus adjusting the solvent content of the system to the exact specifications required to obtain the necessary Cot θ relationship in the system and thereby control the degree of precipitation of asphaltenes.

The significance of selecting Cot θ for deasphalting becomes apparent when one realizes that a solvent system exhibiting a negative value yields an excellent asphaltene precipitation. If the solvent system exhibits a very low solvent power, its respective Cot θ is liable to lie close to the X-axis and exhibit a value of minus infinity. If the system exhibits a good solvent power, it will lie closer to the Y-axis. A good solvent has a positive value by definition, the Cotangent θ having a value greater than zero.

FIGURE 4.6: PERCENT ASPHALTENES IN SYSTEM PRECIPITATED

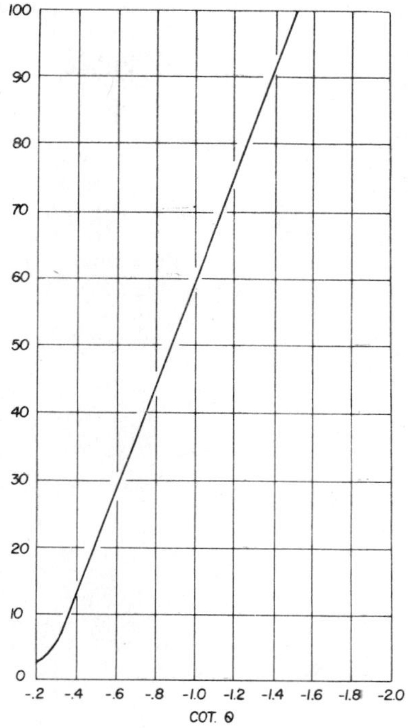

Source: D.L. Mitchell and J.G. Speight; U.S. Patent 3,779,902; December 18, 1973

Therefore, a solvent having a Cotangent θ less than zero will be a relatively poor solvent. It is preferred that the Cotangent θ of the solvent and oil system utilized in the deasphalting of heavy oil be less than minus 0.3 if a significant degree of asphaltene precipitation is desired. To exhibit the use of solvent power adjustment of the system to control the yield of precipitated asphaltenes, the following examples are presented.

Example 1: The solvent powers in terms of Cot θ of various systems of solvents and oils were determined by mixing one-to-one volumes of normally liquid or liquefied solvents with an Athabasca bitumen containing 17 weight percent asphaltenes. A measured amount of 100 g of each solvent was poured into 130 g of a water-bitumen froth, equivalent to 100 g of dry bitumen. The mixture was shaken vigorously for 5 to 10 minutes at 0.5 hr intervals and allowed to settle; this procedure was conducted for approximately 8 hr. At the end of 8 hr period the fractions were easily separated by decantation followed by filtration, with light suction of the solvent/heavy oil solution.

After removal of any residual solvent by blowing, the water content of the asphaltene fraction was determined by the standard ASTM D-9558 method for the determination of water in petroleum products. For the heavy oil fraction formed, the solvent was removed by blowing followed by evaporation and the water separated mechanically. Results of this analytical procedure are set forth in the table on the following page.

Results of Deasphalting Bitumen

Solvent	Solvent Power of System (Cot θ)	Weight Percent Asphaltenes Precipitated
Propane	−(2.5 to 3.5) est.	100
Butane	−2	100
N-Pentane	−1.5	100
Hexane Isomers	−1.2	75
Heptane Isomers	−1.1	67
Pentane/Hexane	−1.3	83
Pentane/Heptane	−1.2	75
Pentane/Cyclopentane (1:1)	−0.7	35
Pentane/Cyclohexane (1:1)	−0.7	35
Cyclopentane	−0.1	1
Cyclohexane	0	0
Heptane/Cyclopentane (1:1)	−0.5	22
Heptene/Cyclohexane (1:1)	−0.5	23

Example 2: A solution of bi-constituent solvent consisting of liquid propane and varsol was utilized in a sample of Athabasca tar bitumen in order to determine the percent asphaltenes precipitated. As liquid propane itself precipitates asphaltenes plus resins and possibly other fractions of bitumen, initial concentrations with 100% propane solvent were determined. Yields of precipitated materials ranged from between 40 to 48% with liquid propane used in large excess. The solution of the propane with varsol produced a solvent blend of increased solvent power. Consequently, the yield of precipitated material decreased.

For example, a solvent bitumen weight ratio of 3.5 to 1 and a propane-varsol concentration of 2.5 to 1 resulted in a 30.7% yield of precipitate at a Cot θ of −2.6. In a similar manner, when bitumen was present in a higher concentration, the contribution of the soluble aromatic portion was added to the solvent power, the precipitating media in the yield of precipitating material decreased as the Cot θ value of the system increased. In the second case, an approximate propane-varsol weight ratio of between 1.5 and 3.5 to 1 was blended as a solvent. The bitumen ratio was between 1.0 and 1.6 to 1 as compared to four to one in the first instance and the Cot θ was −2.5. The yield of precipitating material was 18.4% in this case, showing considerable decrease in precipitate as compared to the 30.7% as illustrated in the first case.

Therefore, it can be seen through this experimentation and Figure 4.6 that the amount of asphaltene precipitate can be varied over a wide range, from say, approximately 0.5% to 100% of the total asphaltene content of the system, dependent upon the adjusted solvent power of the precipitating medium or solvent system. The inherent advantage of this control is that subsequent crops of asphaltenes can be precipitated as essentially mineral-free material in a second deasphalting tower with the majority of the minerals precipitated with a first crop of material in a first deasphalting tower.

RECOVERY OF METAL VALUES

ALUMINUM

Soda Ash-Caustic Soda Treatment for Cell-Grade Alumina

According to a process described by *F.C. Haas; U.S. Patent 3,721,730; March 20, 1973; assigned to The Oil Shale Corporation,* dawsonite is a basic sodium aluminum carbonate having the formula $NaAlCO_3(OH)_2$. This material, until recently, has not been of great importance. However, as the world supply of the chief aluminum-bearing mineral, namely bauxite, diminishes, there is increasing emphasis on other sources of aluminum-bearing minerals. One such source of aluminum is dawsonite which is found in association with oil shale, quartz, feldspar and other matrixes.

Moreover, in view of the increasing world interest in the recovery of oil from oil shale due to the depletion and nationalization of petroleum reserves, there is now a greater demand for making the shale oil extraction process more economically attractive. To this end, attempts have been made to develop an economical method for the recovery of aluminum values present in the dawsonite-bearing oil shale, found in the western state of Colorado, and throughout the world.

Previous efforts towards the recovery of aluminum values from dawsonite have met with little success, because they have been uneconomical in view of their dependence upon the use of large concentrations of costly acidic or alkaline leach liquors, and also the use of highly elevated temperatures and pressures during the leaching operation.

One such process which has been proposed previously is described in U.S. Patent 3,389,975, where retorted dawsonite-bearing oil shale is leached with an aqueous soda ash (sodium carbonate)-caustic soda (sodium hydroxide) leach liquor having a mol ratio of soda ash to caustic soda of at least about 1.5:1 up to 8:1 and providing a sodium-to-aluminum high atomic ratio of at least about 5:1 up to 60:1 to dissolve the aluminum values which are thereafter precipitated in the form of sodium aluminum carbonate hydroxide. Such a procedure has the drawbacks of also solubilizing appreciable amounts of silica contaminant, of requiring large amounts of sodium-containing reagents and of producing the recovered aluminum values in a form which has to be further processed into the more commercially useful anhydrous alumina form from which aluminum metal may be obtained electrolytically.

Another procedure is described in U.S. Patent 3,510,255 where a dawsonite-bearing oil shale matrix is pyrolyzed at 750° to 975°F and then roasted in an oxidizing atmosphere at a higher temperature of 1000° to 1400°F to convert the dawsonite to sodium aluminate and to convert the oil shale to shale oil vapors and shale ash, i.e., spent shale from which the fixed carbon has been combusted or burned off, and subsequently the solids blend of sodium aluminate and shale ash is leached with a leach liquor which is water or aqueous soda ash solution or aqueous soda ash-caustic soda solution having a high mol ratio of soda ash to caustic soda of 1.5:1 to 7.5:1 to provide a sodium aluminate solution from which alumina is thereafter precipitated.

This process, while being highly useful, has the drawback that it requires the roasting of the spent shale product of the pyrolysis of oil shale so as to burn off or combust the fixed carbon content thereof and thereby provide a product, namely, shale ash, which can be satisfactorily treated with the leach liquor.

Moreover, such a process requires a high mol ratio of soda ash to caustic soda when caustic soda is used in the leach liquor, which, together with the alkaline material produced during roasting, has the drawback of increasing the soluble silica contaminant content of the leach liquor to high values, e.g., in the range of 0.12 to 0.22 gram of silica per liter, and adversely affecting the purity of the alumina product.

In accordance with this process, ground oil shale containing dawsonite (usually varies from 5 to 30% thereof) is pyrolyzed at a temperature of from 750° to 975°F, preferably from 850° to 950°F, to yield shale oil hydrocarbonaceous vapors and a solids blend of spent shale containing residual or fixed carbon, e.g., usually from 3 to 5% up to as high as 10% by weight, and sodium aluminate ($NaAlO_2$).

The sodium aluminate is formed from the dawsonite during its heating to temperatures of 660°F or higher. The pyrolysis of the dawsonite-bearing oil shale can be conducted during a period ranging from several minutes to several hours using conventional retort chambers or pyrolyzers such as fluid beds, rotary and stationary kilns and the like. Generally, the heat of pyrolysis or retorting is provided by hot solid heat exchange bodies such as sand, shale ash, metal or ceramic balls, and the like.

The shale oil vapors are separated from the solids blend of spent shale and sodium aluminate, for example, by withdrawing them through separate educt units of the retort or pyrolyzer. The shale oil vapors made up of gaseous and vaporous hydrocarbons are treated in a recovery unit.

The solids blend of spent shale and sodium aluminate is leached with a leach liquor at a temperature of from 70° to 212°F, and preferably at a temperature of from 120° to 170°F, for from 1 to 60 minutes, preferably for from 5 to 15 minutes. The leach liquor is a dilute aqueous soda ash-caustic soda solution having a concentration of from 5 to 20 grams per liter of soda ash (or from 0.5 to 2% by weight thereof) and from 8 to 20 grams per liter of caustic soda (or from 0.8 to 2% by weight thereof). Recycle leach liquor can also be used.

The mol ratio of soda ash to caustic soda in the leach liquor is from 0.1:1 to 0.9:1 and preferably is from 0.2:1 to 0.4:1. This mol ratio is highly significant, because it has been found that lower and higher mol ratios result in poor alumina extraction and/or in high contamination of the alumina with silica. The solids content during the leaching operation is from 5 to 50% by weight. The pH of the leach liquor during the leaching operation is maintained at an alkaline value of from 11 to 13. The leaching effects the extraction of the alkali-soluble sodium aluminate from the solids blend to provide a sodium aluminate solution which usually contains only very small amounts of contaminating solubilized silica, e.g., 0.02 gram of silica per liter or appreciably less.

The sodium aluminate solution resulting from the leaching operation is separated from the spent shale solids by any suitable means, such as by filtration, decantation, centrifugation, settling, and the like.

The sodium aluminate solution and/or the recycle leach liquor can be desilicated to remove solubilized silica by seeding the solution with sodalite or noselite, although such procedure is in most instances not necessary because of the low solubilized silica content.

The sodium aluminate solution is then treated, for example, at from 77° to 158°F, with at least a stoichiometric amount of carbon dioxide to form thereby an alumina hydrate precipitate and a supernatant liquor containing water-soluble sodium carbonate by-product. The rate of precipitation can be increased by seeding the solution with alumina trihydrate. The alumina hydrate precipitate is thereafter separated from the supernatant liquid by suitable separation means, such as those mentioned above.

The separated alumina hydrate product [$Al_2O_3 \cdot 3H_2O$ or $2\,Al(OH)_3$] can be thereafter dried and calcined, e.g., at 950°C for 30 minutes, to provide anhydrous cell-grade alumina (Al_2O_3) suitable for use in the electrolytic production of aluminum metal.

The process is highly useful in that it results in an alumina recovery or extraction of at least 70% and usually 80 to 90% or higher without significant silica contamination of the alumina. The process is illustrated by the following data and examples.

Examples 1 through 19: The process described above in detail was conducted where the mol ratio of soda ash (Na_2CO_3) to caustic soda (NaOH) in the leach liquor was varied appreciably. The amount of dissolved silica impurities in the leach liquor was determined as well as the percent alumina extracted from the pyrolyzed dawsonite-bearing oil shale. The data are reported below in Table 1 where Examples 1 through 11 are illustrative of the process and Examples 12 through 19 are comparative examples of processes outside the scope of the process because of the mol ratio of soda ash to caustic soda used in the leach liquors.

Table 2 presents the product analysis for Example 7. The pregnant leach liquor of Example 7 was carbonated with CO_2 at 70°C for 2 hours in the presence of aluminum trihydrate seeds to precipitate the aluminum as aluminum hydroxide. The precipitate was calcined to form the very pure alumina product presented in Table 2.

TABLE 1

Example	Leach Solution g Na_2CO_3/l- g NaOH/l	Mol Ratio Na_2CO_3:NaOH	% Alumina Extracted	Dissolved Impurities (g SiO_3/l
1	5 - 15	0.13:1	82	0.021
2	5 - 10	0.19:1	75	0.002
3	10 - 18	0.21:1	90	0.020
4	10 - 15	0.25:1	92	0.011
5	7 - 10	0.26:1	72	0.004
6	10 - 13	0.29:1	86	0.011
7	10 - 10	0.38:1	85	0.010
8	13 - 10	0.49:1	79	0.008
9	15 - 10	0.57:1	79	0.009
10	20 - 10	0.75:1	84	0.021
11	20 - 8.5	0.89:1	84	0.010
12	50 - 0	∞	43	0.013
13	5 - 0	∞	37	0.005
14	20 - 3.75	2.16:1	39	0.04
15	20 - 5.0	1.51:1	43	0.01
16	20 - 7.5	1.00:1	54	0.01
17	5 - 20	0.09:1	96	0.036
18	5 - 25	0.08:1	70	0.047
19	5 - 30	0.06:1	67	0.064

TABLE 2: PRODUCT ANALYSIS OF EXAMPLE 7

Al_2O_3	99.8734 (by difference)
Mg	0.0010
Mn	0.0001
Pb	0.0015
Fe	0.0300
Sn	0.0015
Cu	0.0015
Na	0.0700
B	0.0010
SiO_2	0.0200
Total	100.00

The data set forth above in Table 1 clearly demonstrate the criticality of using an aqueous soda ash-caustic soda leach liquor which has a mol ratio of soda ash to caustic soda of from 0.1:1 to 0.9:1 and preferably from 0.2:1 to 0.4:1. Thus, in Examples 1 through 11 and particularly Examples 3 through 7 of illustrating the process where such mol ratios were employed, the percent alumina extracted ranged from a satisfactory value of 72% up to a remarkable value of 92%.

Equally important is the concomitant fact that the dissolved silica contaminant in the leach liquor was present therein in an amount of only 0.02 gram per liter down to 0.002 gram per liter. However, in comparative Examples 12 through 19 where the requisite mol ratio of soda ash to caustic soda was not employed, the data obtained were not satisfactory.

Thus, in comparative Examples 12 through 16 where the mol ratio was varied from 1.0:1 to infinity and was too high because of the relatively small amount and even absence of caustic soda in the leach liquor, only a small amount of dissolved silica contaminant was usually present in the leach liquor but more significantly the percent alumina extracted was very poor and uneconomical in that it ranged from low values of only 37 to 54%.

In comparative Examples 17 through 19 the mol ratio of soda ash to caustic soda was lower (0.06:1 to 0.09:1) than the requisite values for the process (0.1:1 to 0.9:1) and although satisfactory alumina extraction was obtained, namely, from 67 to 96%, more significantly the dissolved silica contaminant in the leach liquor and hence carried forward or occluded in the alumina precipitate was excessively high and appreciably above 0.02 gram of silica per liter, namely, from 0.036 to 0.064 gram of silica per liter.

Alumina Trihydrate as Seed

According to a process described by *F.C. Haas; U.S. Patent 3,859,413; January 7, 1975; assigned to The Oil Shale Corporation* it has been found that essentially silica-free alumina product is obtained from dawsonite-bearing oil shale by a process which involves rapidly wetting and leaching the retorted residue at about 5° to about 40°C, with a dilute alkaline leach liquor followed by precipitation of the extracted alumina directly from the leach liquor at about 60°C to about 75°C in the presence of alumina seed.

The solubilized aluminum values are recovered directly from the leach liquor via precipitation in the presence of alumina trihydrate seed. The product precipitates as alumina trihydrate which is thereafter calcined to alumina. Since the leaching operation is conducted at from 5° to 40°C, the aluminum-containing liquor is withdrawn to a separate vessel and heated to between 60° and 75°C.

The precipitation is conducted over a period of from 30 minutes to 4 hours depending on the weight ratio of alumina trihydrate seeds to alumina in solution. The alumina product which precipitates under the foregoing conditions conforms to the silica specification set for the production of electrolytic cell grade alumina, i.e., 0.03 weight percent SiO_2 or less.

Example 1: This example illustrates the effect of varying the leach temperature and time upon the amount of alumina and silica leached from alumina-containing retorted oil shale residue. A high alumina-to-silica weight ratio is indicative of high alumina and low silica extraction.

A sample of dawsonite-bearing oil shale was retorted at 480°C for 5 minutes to provide oil vapors and a finely divided residue product containing aluminum values. The residue product was essentially free of benzene-soluble hydrocarbons and contained about 4 weight percent fixed organic carbon.

Portions of the sample were wetted and leached with an aqueous alkaline leach liquor containing caustic soda. The initial caustic soda concentrations were from 10 to 20 grams per liter. The leach liquor was used to first wet the residue product and then to leach the alumina. Both operations were conducted in a commercially available attrition scrubber.

Visual inspection of the wetting procedure indicated that the residue was usually completely wetted within 15 to 30 seconds. Leaching begins as the particles are wetted. Consequently, the length of the leach time is difficult to determine with certainty since some leaching occurs before the sample is entirely wetted. For this reason, it will be noted in the following Table 1 that maximum leach times are presented.

In two instances, times of less than 0.1 minute were recorded on samples obtained immediately after the solids appeared to be completely wetted, i.e., the dusty-like appearance of the solids was rendered wet. The sample portions were leached at 20 to 40% solids concentration at leach temperatures which were varied from 5° to 60°C. The percent alumina (Al_2O_3) extraction and the Al_2O_3 and SiO_2 content of the pregnant leach liquor were determined as a function of leach temperature and time. The results of 33 such tests are set forth in the following table.

TABLE 1

Test No.	Leach Temp. °C	Max. Leach Time Min.	% Solids	Leach Liquor g/l NaOH	Pregnant Liquor, g/l Al_2O_3	Pregnant Liquor, g/l SiO_2	Al_2O_3/SiO_2 Ratio	% Al_2O_3 Ext'n
G-51	60	10	40	20	18.05	0.143	126	95
H-29	60	10	20	10	8.10	0.062	131	100
G-229	60	1	40	20	18.04	0.090	200	96
G-228	60	1	40	10	15.83	0.041	386	88
G-230	40	1	40	20	19.11	0.057	335	99
G-226	25	1	40	20	15.85	0.030	528	91
G-221	25	1	40	20	16.48	0.023	717	99
H-414	25	1	20	10	8.38	0.015	516	99
H-412	5	1	20	20	7.13	0.012	600	100
71-235	25	0.5	40	20	19.08	0.030	636	92
G-248	25	0.5	40	15	17.76	0.019	935	96
G-249	25	0.5	40	10	16.96	0.016	1060	83
71-237	25	0.25	40	20	19.73	0.030	658	98
71-238	25	<0.1	40	20	20.21	0.039	518	95
19*	25	<0.1	40	20	—	—	710	95-100

*Average of 19 Tests

From the above data, it is readily apparent that the highest Al_2O_3/SiO_2 weight ratios result from conducting the leaching operation at temperatures from 5° to 25°C for 1 minute or less after the residue has been wetted.

Example 2: This example illustrates the effect of raising the precipitation temperature from that used in the leach operation to maintain silica in solution during the precipitation of alumina directly from the pregnant leach liquor.

A sample of retorted shale residue was wetted and leached with an aqueous alkaline liquor containing 20 grams per liter caustic soda at a % solids concentration. The total time required for the wetting and subsequent leaching operations was 1 minute at 25°C. Immediately after conducting the leaching operation, the pregnant liquor was withdrawn from the leach vessel and heated to varying higher temperatures and the alumina allowed to precipitate directly from the heated liquor in the presence of alumina trihydrate seeds. All precipitations were done with a 5:1 seed ratio. The percent of silica in the alumina product (calculated) as a function of the precipitation temperature and time is reported in the following table.

TABLE 2

Alumina and Silica Precipitation versus Time and Temperature

Precip. Temp.,°C	Precip. Time, hrs	Solution Analyses g Al_2O_3/l	g SiO_2/l	% Al_2O_3 Precip.	% SiO_2 Precip.	% SiO_2 in Prod. (calc.)
30	0	19.26	0.027			
	0.5	19.26	0.027	0	0	0
	1	17.56	0.025	8.8	7.4	0.12
	2	14.40	0.026	25.2	3.7	0.021
	3	12.90	0.024	33.0	11.1	0.047
40	0	19.14	0.027			
	0.5	17.84	0.027	13.0	0	0
	1	15.38	0.025	19.6	7.4	0.053
	2	12.35	0.025	35.5	7.4	0.029
	3	10.51	0.024	45.1	11.1	0.035
50	0	19.27	0.030			
	0.5	15.61	0.027	19.0	10	0.082
	1	11.41	0.027	40.8	10	0.038
	2	9.32	0.026	51.6	13.3	0.040
	3	8.87	0.026	54.0	13.3	0.038
60	0	18.90	0.027			
	0.5	12.37	0.026	34.6	3.7	0.015
	1	10.55	0.025	44.2	7.4	0.024
	2	9.08	0.025	52.0	7.4	0.020
	3	8.13	0.024	57.0	11.1	0.028
65	0	19.01	0.027			
	0.5	11.73	0.025	38.3	7.4	0.027
	1	10.00	0.025	47.4	7.4	0.022
	2	8.53	0.025	55.1	7.4	0.019
	3	7.82	0.025	58.9	7.4	0.018
70	0	19.50	0.027			
	0.5	11.65	0.024	40.3	11.1	0.038
	1	9.88	0.024	49.3	11.1	0.031
	2	8.74	0.025	55.2	7.4	0.019
	3	8.33	0.024	57.3	11.1	0.027
75	0	18.69	0.028			
	0.5	15.24	0.025	18.5	10.7	0.087
	1	10.14	0.025	45.8	10.7	0.035
	2	8.90	0.025	52.4	10.7	0.031
	3	8.68	0.025	53.6	10.7	0.030
80	0	19.27	0.030			
	0.5	11.62	0.026	39.7	13.3	0.052
	1	10.32	0.027	46.4	10.0	0.034
	2	9.79	0.026	49.2	13.3	0.042
	3	9.68	0.028	49.8	6.7	0.021
90	0	18.20	0.027			
	0.5	12.41	0.023	31.8	14.8	0.069
	1	12.06	0.023	33.7	14.8	0.064
	2	11.91	0.023	34.6	14.8	0.064
	3	11.76	0.023	35.4	14.8	0.062

From the data in the above table, it is readily apparent that the optimum conditions for alumina precipitation and meeting the silica specifications of 0.03 weight percent are 65°C for 1 to 2 hours.

The data also show that the temperature could go as low as 60°C and as high as 75°C and still produce a satisfactory product, but the silica content would be very close to specification. Below and above this temperature range, silica is above specification and the percentage of alumina precipitated is lower.

Example 3: This example illustrates the effect of varying the seed ratio on the alumina precipitation time and the silica content of the product. One test was conducted in which retorted alumina-containing oil shale residue was leached for 1 minute at 25°C with 20 grams per liter of caustic soda (initial concentration) at 40% solids concentration. Four samples of the pregnant leach liquor were withdrawn to separate precipitation vessels and heated to and maintained at 65°C in the presence of alumina trihydrate seed.

Seed ratios of 1:1, 3:1, 5:1, and 8:1 were tested. Reagent grade fine alumina trihydrate seed was used in each test. The results for precipitation times up to 3 hours are presented in the following table.

TABLE 3

		Aluminum and Silica Precipitation versus Seed Ratio and Time				
Seed Ratio	Precip. Time, hrs	Solution Analyses g Al_2O_3/l	g SiO_2/l	% Al_2O_3 Precip.	% SiO_2 Precip.	% SiO_2 in Prod. (Calc.)
1:1	0	18.26	0.030			
	0.5	18.26	0.030	0	0	0
	1	17.79	0.029	2.6	3.3	0.21
	2	12.49	0.027	31.6	10.0	0.052
	3	10.70	0.029	41.4	3.3	0.013
3:1	0	18.26	0.030			
	0.5	12.90	0.029	29.4	3.3	0.019
	1	10.55	0.028	42.2	5.7	0.026
	2	9.02	0.029	50.6	3.3	0.011
	3	8.44	0.029	53.8	3.3	0.010
5:1	0	19.01	0.027			
	0.5	11.73	0.025	38.3	7.4	0.027
	1	10.00	0.025	47.4	7.4	0.022
	2	8.53	0.025	55.1	7.4	0.019
	3	7.82	0.025	58.9	7.4	0.018
8:1	0	18.19	0.030			
	0.5	9.02	0.030	50.4	0	0
	1	8.37	0.029	54.0	3.3	0.010
	2	7.46	0.028	59.0	5.7	0.019
	3	7.19	0.030	61.4	0	0

From the above data, it is apparent that a virtually silica-free, high-yield alumina product is obtained with a high 8:1 seed ratio and a precipitation of 30 minutes. Satisfactory product yields and purities are obtainable at seed ratios as low as 3:1 for two to three hours precipitation time.

Controlled Pyrolysis to Prevent Formation of Insoluble Oxides

A process described by *M. Prats; U.S. Patent 3,502,372; March 24, 1970; assigned to Shell Oil Company* is directed to recovery of shale oil and soluble aluminum compounds such as dawsonite from underground dawsonite-containing oil shale formations by sequential treatment of such formations to effect pyrolysis of the organic matter, preferably by in situ combustion thereby recovering oil and subsequently subjecting the formation to solution mining to recover the dawsonite by means of aqueous solutions capable of dissolving the dawsonite and effecting above ground the separation of the dawsonite from the solution.

To effect and facilitate this recovery process of both oil and soluble aluminum compounds such as dawsonite, the formation containing these materials should be fragmented, for example, by means well known in the art, prior to treating or injecting sequentially into the resulting fracture-permeated zone of the formation a kerogen-pyrolyzing fluid and an aqueous solution capable of dissolving therein the soluble aluminum compounds such as dawsonite.

In order to accomplish the recovery of both the oil and dawsonite from dawsonite-containing oil shale formations most efficiently, it has been found that it is essential to control the pyrolyzing fluid temperature and/or composition within specified limits so as to effect desired oil recovery while preventing the decomposition of the dawsonite or other soluble aluminum compounds into an insoluble oxide. To accomplish this the fracture-permeated zone of the oil shale formation is treated under controlled conditions by:

(1) Flowing into the recovery zone hot oil-shale-pyrolyzing fluid having a pH of less than about 8 and a temperature above the formation temperature (and preferably above 500°F) through portions of a fracture-permeated oil shale that contains an aqueous fluid soluble aluminum compound, such as dawsonite, so that organic components of pyrolyzed oil shale are entrained and displaced by the oil-shale-pyrolyzed fluid;

(2) Flowing aqueous liquid into contact with oil shale from which organic components have been removed by the oil-shale-pyrolyzing fluid, so that aluminum-containing mineral components, such as dawsonite, are dissolved in the aqueous liquid; and

(3) Recovering oil and aluminum from fluids produced from the fluid-contacted oil shale.

The oil-shale-pyrolyzing fluid can comprise the combustion products of substantially any type of in situ combustion in which the combustion-supporting fluid is sufficiently free of alkaline material to form nonalkaline combustion products, hot hydrocarbons such as benzene, volatile compounds of oil shale, steam or hot aqueous nonalkaline fluids, phenolic materials, mixtures of hydrocarbons with phenols, and polyacids.

Leaching Technique

M.N. Papadopoulos and R.C. Ueber; U.S. Patent 3,700,280; October 24, 1972; assigned to Shell Oil Company describe a method for recovering hydrocarbons and dawsonite decomposition products from underground oil-shale formations containing nahcolite-dawsonite deposits.

The process is particularly applicable to various subsurface oil shale formations, such as in the Green River formation in the Colorado area of the United States, containing rich deposits of nahcolite and dawsonite. The nahcolite and dawsonite are apt to occur within an oil shale formation in the form of beds, lenses, nodules, loads, veins, or the like, having sizes ranging from microscopic particles to layers that are many feet in thickness and many miles in extent.

Dawsonite is particularly apt to occur in the form of microscopic particles in amounts of up to about 10 or 12% by weight of the oil shale while the nahcolite can vary in amounts of 5 to 40%. The procedural steps in carrying out the process are essentially as follows:

(1) establishing at least one well borehole into a nahcolite-dawsonite-rich oil-shale formation;

(2) forming a cavern by leaching a portion of the nahcolite with an aqueous fluid such as water which may be cool or hot, preferably the latter;

(3) circulating a hot fluid such as hot water and/or steam in the upper region and preferably along the cavern roof to effect decomposition of the nahcolite and dawsonite to carbon dioxide and water-insoluble products of dawsonite such as insoluble oxide and silicate complexes of aluminum such as analcite to effect enlargement of the cavern by upward migration of the cavern roof as well as fracturization and/or rubbling;

(4) circulating sequentially or simultaneously a cooling fluid such as hydrogen along the cavern roof and a kerogen-pyrolyzing fluid such as steam and/or solvents such as benzene, phenol, etc. through the rubbled oil shale to effect hydrocarbon recovery; and

(5) solution mining to recover the aluminum complexes from dawsonite.

Example: Into an oil-shale formation rich in nahcolite and dawsonite a well was completed at below about 2,000 feet and a portion of the nahcolite bed was water-leached to form a cavern. Steam was injected along the cavern roof to decompose the nahcolite and dawsonite to form carbon dioxide thereby building up pressure and causing upward migration of the cavern roof and oil shale rubbling. The dawsonite is essentially converted to water-insoluble complexes of aluminum such as analcite. A cooling fluid hydrogen is thereafter circulated along the cavern roof to prevent further upward roof migration while steam is injected into the oil shale rubble to effect hydrocarbon recovery.

After the hydrocarbons such as oil have been recovered a basic solution containing a chelating agent such as an alkali metal salt of ethylenediaminetetraacetic acid, e.g., Na_4EDTA (the sodium salt of ethylenediaminetetraacetic acid) or Na salt of 1,2-diaminocyclohexane-tetraacetic acid is injected into the treated oil recovery zone of the shale formation preferably at a temperature of from 100° to 500°F to effect recovery of analcite.

Pyrolytic Extractive Fluid and Chelate

A process described by *C.C. Templeton; U.S. Patent 3,572,838; March 30, 1971; assigned to Shell Oil Company* is directed to recovery of shale oil and soluble aluminum compounds such as dawsonite from a subterranean or underground dawsonite-containing oil shale formation by fracturing and/or rubblizing the formation so as to form a permeable zone and circulating alternating slugs of a pyrolytically extractive fluid and an aqueous alkaline fluid containing an effective amount, sufficient to prevent insoluble product formation, of an acid-insoluble chelating agent and thereafter recovering and processing the outflowing fluid so as to effect recovery of the oil phase from the aqueous phase and separately processing the oil phase to recover oil and aluminum compounds, e.g., dawsonite and the chelating agent, from the water-phase components.

In certain instances, depending upon the properties of the oil shale formation being treated and/or the pyrolytically extractive fluid being used, additional portions of aqueous-fluid-solubilized shale oil can be recovered from the water-phase components of the outflowing fluid.

Figure 5.1 is an illustration of a preferred form of this process where **30** is the oil shale formation of which zone **31** is fractured and fragmented and from which oil and dawsonite are to be recovered. The formation is penetrated by an injection well **32** and a production well **33** each of which have tubing strings **34** and **35**, respectively, in which through tubing string **34** is alternatively injected a pyrolytic extracting fluid from **36** and an alkaline aqueous alkaline fluid containing a chelating agent from vessel **37**.

The injection via tubing string **34** of fluids **36** and **37** is alternatively carried out as required to effect oil pyrolysis and dawsonite solubilization to be recovered as outflowing fluids via well **33** through tubing string **35**.

At the surface the outflowing fluid from **35** is processed for separation by flowing into separator **38** via lines **39a** and **39b** between which is a surface processing unit **39** and the oil phase is separated in **38** and directed via line **40** to vessel **41**. The aqueous phase is directed to through line **42** into vessel **43** where it is acidified in line **44** and the chelating agent precipitated in vessel **45** separated and returned to vessel **37** via line **46**. The pH adjusted aqueous solution of dawsonite is directed via line **47** into vessel **48** where the solution is treated with a base and on precipitation the precipitated aluminum oxide product is recovered in vessel **49** and the aqueous waste discarded via line **50**.

In a preferred process for conducting the process the pyrolytically extractive fluid is formed in situ by initiating an underground combustion within the fracture-permeated zone in the oil shale formation and injecting a combustion-supporting fluid to produce a pyrolytically

FIGURE 5.1: RECOVERY OF ALUMINUM COMPOUNDS AND OIL

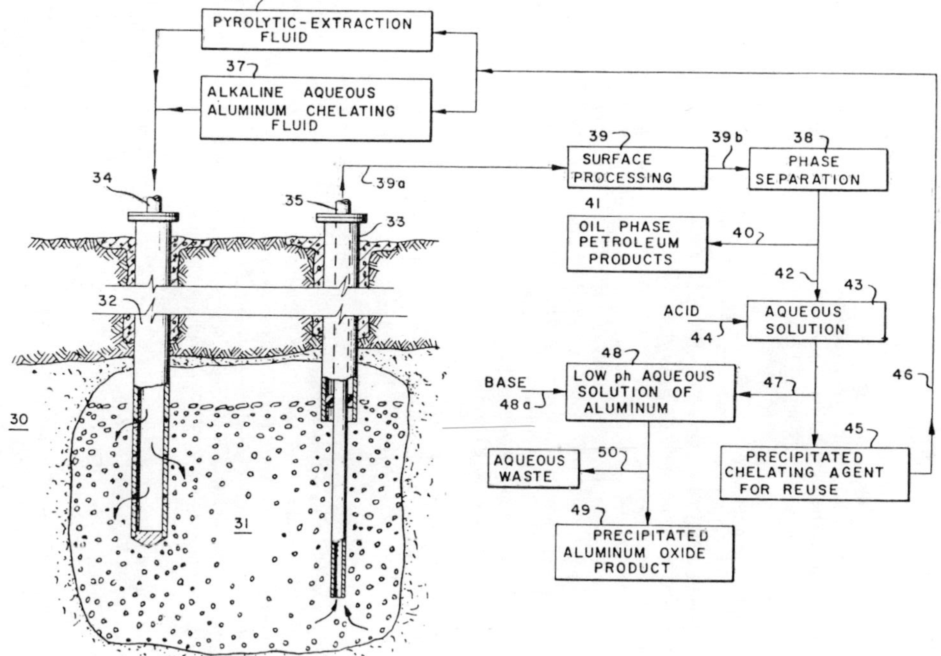

Source: C.C. Templeton; U.S. Patent 3,572,838; March 30, 1971

extractive fluid resulting from the products of combustion. The combustion-supporting fluid may contain components, such as an aqueous liquid, to control the temperature and heat-transport properties of the combustion, foam to control the sweep efficiency of fluids moving through interstices within the fracture-permeated zone, fuel to supplement that available in the exposed oil shale surfaces, etc.

In the preferred process, a slug of an aqueous alkaline solution containing a chelating agent is preferably injected prior to the initiation of the underground combustion such as an aminocarboxylic acid, e.g., ethylenediaminetetraacetic acid (EDTA) and salts thereof (e.g., Na_4EDTA) via vessel **37** as shown in Figure 5.1. The inflowing aqueous solution is heated, preferably by means of surface-located or borehole-located heating device, so that it preheats the oil shale to a temperature approaching the ignition temperature of the oil shale.

The oil shale fragments are preheated with an aqueous liquid to cause a reduction in their particle size and improve the distribution of permeabilities and surface area-to-volume ratios within the cavern prior to the initiation of underground combustion.

Such an aqueous-fluid preheating is particularly advantageous in that it induces the exfoliation of portions of the fracture-permeated oil shale into small pieces and thus improves both the distribution of effective permeabilities within the interstices between the pieces of oil shale and the ratio of surface area of the oil shale pieces to the volume of the oil shale pieces.

In general, each slug of aqueous alkaline solution which contains a chelating agent, e.g., Na_4 EDTA, should have a volume at least exceeding the solid-free void space of both the fracture-permeated zone within the oil shale formation and the inflow and outflow conduits. If desired, the outflowing portions of a relatively small size slug can be reheated and recycled through the fracture-permeated zone.

Where the fracture-permeated portion of the oil shale has not been preheated, the inflowing portions of the chelating-agent solution are preferably heated, continuously or incrementally, to increasingly higher temperatures which are significantly greater than the normal temperature of the oil shale formation.

The pressure of the inflowing chelating-agent solution is preferably increased, by maintaining a back pressure on outflowing portions, if necessary, to the extent required to maintain substantially all of the solution in the liquid phase. When the chelating-agent solution is being injected prior to the initiation of an underground combustion, the injecting and heating are preferably conducted so that, at least near the point of inflow into the fracture-permeated zone of oil shale, the oil shale is heated to a temperature of at least several hundred degrees F for a period of at least several days with temperature of at least about 400°F and times of at least one week being preferred.

In general, the alternations of slugs of chelating-agent solution and pyrolytically extractive fluid are preferably repeated by interrupting the circulation of the pyrolytically extractive fluid (e.g., by maintaining an underground combustion), initiating circulation of chelating-agent solution, interrupting the circulation of chelating-agent solution, reinitiating the circulation of pyrolytically extractive fluid, etc.

Where the regional tectonics and/or the well assembly equipment make it undesirable to inflow an aqueous fluid at a pressure sufficient to keep it substantially liquid at the temperatures created by a pyrolytic extraction, e.g., those created by an underground combustion, the inflowing of the chelating-agent solution is preferably preceded by the inflowing of a slug of water that is softened to an extent preventing scaling at the temperature existing within the fracture-permeated zone. The time and volume of such a water circulation are preferably adjusted to lower the temperature, at least near the point of inflow into the oil shale, to a temperature at which inflowing portions of a chelating-agent solution can be kept substantially liquid under pressures which are desirable for employment.

Example: After a rubblized or fractured subterranean oil shale formation has been subjected to a dry in situ combustion, an aqueous solution containing an alkaline agent such as NaOH or Na_2CO_3 and a chelating agent such as Na_4EDTA is injected through a well into the partly cooled recovery zone of the shale formation at a temperature of from 200° to 225°C, with the aluminum remaining soluble as a mixture of Al-EDTA and aluminate ion AlO_2^-. The solution returning from a production well to the surface containing some EDTA is in the form of metal complexes (Al·$EDTA^-$, Ca·$EDTA^=$, or Mg·$EDTA^=$ expected from leaching spent oil shale), since the metal-EDTA form is more stable at higher temperatures than the free EDTA.

The produced solution is cooled in a heat exchange scheme to less than 100°C to recover the EDTA as precipitated H_4EDTA by acidifying the produced solution with an acid such as hydrochloric acid or sulfuric acid to a pH of about 1.6. The solid H_4EDTA is separated from the solution by either filtration or decantation, the solution is then alkalized by $NaHCO_3$ or Na_2CO_3, or a mixture of both, so that at a pH of 6.5 to 7.5 the $Al(OH)_3 \cdot xH_2O$ when filtered off may contain enough impurities to warrant its solution in acid, and a second alkalization with Na_2CO_3 to obtain high-purity $Al(OH)_3 \cdot xH_2O$.

Finally, the $Al(OH)_3 \cdot xH_2O$ is again filtered off, and calcined to Al_2O_3. Such Al_2O_3 can then be sent to conventional electrolytic plants for production of aluminum metal. The separated H_4EDTA is dissolved in a NaOH solution, which can be blended with fresh natural brine, to form an aqueous phase suitable to be sent to the injection wells.

Sodium Aluminate

According to a process described by *J.R. Dyni; U.S. Patent 3,642,433; February 15, 1972; assigned to U.S. Secretary of the Interior* dawsonite and dawsonitic oil shales are heated to decompose the dawsonite into sodium aluminate. Water is then employed to dissolve and form an aluminate solution from which alumina can be recovered by known methods. The following examples illustrate the effectiveness of the process on dawsonitic oil shale.

Example 1: An approximately 6.6 g sample (labeled Sample A) of Piceance Creek Basin (Colorado) dawsonitic oil shale was fired to 720°C, held at this temperature for 1 hour, cooled to room temperature and leached with 350 ml of water at 100°C for a period of 10 to 15 minutes. Thereafter a small stream of CO_2 was bubbled through the solution for 45 minutes to precipitate out aluminum hydroxide.

A second approximately 6.6 g sample (Sample B) of Piceance Creek dawsonitic oil shale was treated in the same manner as Sample A except that 600 ml of leaching water at 28°C was employed. Analyses of the precipitates from both samples were as follows.

Sample	Weight of Unfired Sample, g	Total Aluminum in Unfired Oil Shale Sample, wt %	Total Weight of Precipitate, g	Aluminum in Precipitate, wt %
A	6.55	6.8	0.03	41.2
B	6.59	6.3	0.34	32.0

As can be seen from these results as much as 26% of the total aluminum in the dawsonite-bearing oil shale can be removed by the process.

The following example shows the relative effectiveness of different leaching mediums on Piceance Creek Basin oil shale.

Example 2: Three different dawsonite-bearing oil shale samples were each leached at different temperatures with various solvents and the following results were obtained.

Sample C – [Total Al content of shale sample = 4.20 weight percent; Al present in shale sample as dawsonite (x-ray analysis) = 3.8 weight percent].

Extracted Aluminum (wt % of shale)

Restoring Temperature, °C	Hot Distilled Water Leach	2% Hot NaOH Solvent Leach	2% Hot H_2SO_4 Solvent Leach
400	1.4	2.25	3.4
500	1.45	1.85	2.2
600	1.75	3.0	2.5
700	1.85	2.4	3.0
800	1.7	1.5	3.2
900	1.55	2.25	3.2
1000	1.5	2.4	3.0

Sample D – (Total Al content of shale sample = 4.50 weight percent; Al present in shale sample as dawsonite = 2.9 weight percent).

Extracted Aluminum (wt % of shale)

Temperature, °C	Hot Distilled Water Leach	2% Hot NaOH Solvent Leach	2% Hot H_2SO_4 Solvent Leach
400	0.5	1.5	2.2
500	0.7	0.35	3.2
600	0.75	1.4	2.75
700	0.65	1.25	2.4
800	0.45	1.5	2.5
900	0.35	1.55	2.2
1000	0.5	1.8	2.0

Sample E – (Total Al content in shale sample = 6.82 weight percent, Al present in shale sample as dawsonite = 2.5 weight percent).

Extracted Aluminum (wt % of shale)

Temperature, °C	Hot Distilled Water Leach	2% Hot NaOH Solvent Leach	2% Hot H_2SO_4 Solvent Leach
400	0.8	3.7	4.3
500	0.65	3.8	4.2
600	0.5	3.2	3.9
700	0.55	3.1	4.0
800	0.1	3.4	3.25
900	0.1	2.2	2.6
1000	0.1	0.25	2.0

As can be seen from the above results, dilute NaOH and H_2SO_4 not only extract that portion of the dawsonite which has been decomposed to water-soluble aluminate, but also extract some of the aluminum which is originally present in the shale in other forms such as possible amorphous dawsonite, organically associated aluminum, or aluminum hydroxide.

In addition to the substantial aluminum extraction obtained by the process, the use of water or dilute acid or base leaches inhibits dissolution of alumino-silicate or silica minerals such as feldspar and quartz which materials could contribute silica, a deleterious substance in alumina recovery processes.

Sodium Aluminate–Critical Roasting Temperature

R.N. Hall and F.C. Haas; U.S. Patent 3,510,255; May 5, 1970; assigned to The Oil Shale Corporation describe a process for recovering a water-soluble sodium aluminate from a dawsonite-bearing matrix which involves roasting the dawsonite-bearing matrix at an elevated temperature of about 1000° to 1400°F to form a blend of sodium aluminate and spent residue; and leaching the blend at low temperature and atmospheric pressure with a leach liquor, which is water, or an aqueous soda ash solution, to extract the sodium aluminate.

Example 1: This example illustrates the effect of varying roasting temperatures upon the amount of aluminum recovered. A sample of dawsonite, $NaAlCO_3(OH)_2$, was prepared synthetically by precipitating the salt with carbon dioxide from a sodium aluminate solution. The x-ray diffraction pattern of the synthetic dawsonite is very similar to natural dawsonite.

The sample of dawsonite was then heated in an oxidizing atmosphere at temperatures from 900° to 1400°F and then leached with hot water for about 15 minutes at atmospheric pressure and at a solids content of 4%. The percent of water-soluble aluminum and percent of theoretical aluminum available as a function of roasting temperature is reported in the following table.

TABLE 1

Roasting Temperature, °F	% Soluble Aluminum	% of Theoretical	pH of Leach
900	4.7	25.1	9.2
1000	5.0	26.7	9.8
1050	6.3	33.7	-
1100	9.0	48.1	-
1150	14.2	75.9	-
1200	16.1	86.1	10.3
1300	15.5	82.8	-
1400	14.0	74.8	11.1

From the above data, it is readily apparent that the highest yield of water-soluble sodium aluminate results from conducting the roasting operation at the intermediate temperatures of about 1150° to 1300°F.

Example 2: Two samples (designated TG-1 core, Nos. 403 and 450) of a Colorado oil shale containing dawsonite, one high in nahcolite (18% sodium bicarbonate content) and one low in nahcolite (3% sodium bicarbonate content), were roasted at temperatures from 1000° to 1400°F for one hour, and then leached at atmospheric pressure with hot water for 15 minutes at a solids content of 4% to extract the water-soluble aluminate therefrom. The results, which illustrate the effect of varying the roasting temperature, are set forth on the following table.

TABLE 2

Examples	Roasting Temperature, °F	% Water-Soluble Aluminum
High-nahcolite-content oil shale	1000	1.0
High-nahcolite-content oil shale	1100	1.0
High-nahcolite-content oil shale	1200	1.3
High-nahcolite-content oil shale	1300	1.5
High-nahcolite-content oil shale	1400	1.2
Low-nahcolite-content oil shale	1000	0.5
Low-nahcolite-content oil shale	1100	0.7
Low-nahcolite-content oil shale	1200	1.0
Low-nahcolite-content oil shale	1300	0.4
Low-nahcolite-content oil shale	1400	0.1

From the above data, it is readily apparent that the highest yield of water-soluble sodium aluminate results from conducting the roasting operation at the intermediate temperatures of about 1100° to 1300°F. The data in Table 2 further indicate that the presence of nahcolite (which decomposes to soda ash during the roasting) in association with the dawsonite-bearing oil shale matrix aids in the recovery of water-soluble sodium aluminate therefrom. This is further shown by the fact that when the nahcolite was removed from the big-nahcolite-content oil shale and the oil shale roasted at 1250°F, the amount of water-soluble aluminum recovered fell from 1.5 to 0.5%.

Example 3: This example illustrates the effect of leaching time using the different leach liquors. A composite sample (designated TG-1 core, Composite No. 2) of a Colorado oil shale containing dawsonite was roasted at 1250°F until all the carbon was burned off and then leached at atmospheric pressure and at ambient temperature with different leach liquors, namely, 4% aqueous soda ash and water, at a solids content of 43%. As the leaching progressed, samples were removed and the filtrate analyzed for soluble aluminum. The results are tabulated in the following table.

TABLE 3

Leaching Time, hours	4% Soda Ash Leach, g Al/l	Water Leach, g Al/l
0.017	3.25	0.00
0.05	3.40	0.50
0.08	3.50	0.45
0.25	3.79	0.36
0.50	3.79	0.31
1.0	3.63	0.29
24.0	3.13	$<$0.05

It will be noted from the data set forth in the above table, that the amount of soluble aluminum recovered by the leaching operation decreases as the leaching time increases and that favorable leaching times are, in general, from about 5 minutes to about 60 minutes.

Recovery of Alumina

R.A. Van Nordstrand; U.S. Patent 3,516,787; June 23, 1970; assigned to Sinclair Research, Inc. describes a process for separating oil and aluminum values from oil shale containing kerogens, sodium aluminum carbonate hydroxide, quartz and dolomite by retorting the shale at 500° to 1200°F to separate the oil, leaching the resulting spent shale with an alkaline solution at a temperature of up to 220°F to dissolve the aluminum values from the shale without substantial precipitation of SiO_2, and recovering the aluminum values as hydrous alumina from the alkaline solution. The following examples illustrate the process.

Example 1: An oil shale from Section 21, Township 1S.98W., Rio Blanco County, Colo., taken over the depth interval 2,186 to 2,189 was crushed to a ⅛ to ¼-inch size. In addition to about 30% kerogens, this shale contains about 20% dawsonite [$NaAlCO_3(OH)_2$] with the remainder being primarily quartz (SiO_2) and dolomite [$CaMg(CO_3)_2$]. Twenty samples of 80 g each were separated. Each sample was placed in an oil shale assay retort constructed so that both the oil and water produced from each sample is recovered and measured.

The retort operates with limited access of air, so the pyrolysis of the shale is carried out in the effective absence of oxygen. Pyrolysis of this sample was carried out for two hours at 900°F. Each of the 20 portions produced about the same amount of oil, the average being 13.9 cc for the 80 g sample, corresponding to 41.7 gallons of oil per ton of shale. The weight loss of the shale is estimated at 25%.

The retort was opened while the sample was still hot, about 500°F, and the retorted oil shale was then cooled in air. The retorted oil shale was a black, porous, friable substance. A small portion of this retorted sample was crushed and analyzed by x-ray diffraction. The dawsonite was no longer present in its original crystal form and there was no crystalline compound which could be identified by x-ray diffraction as the product of the dawsonite decomposition. The quartz appears to be present in unchanged amount. The dolomite was partially lost. The expected amount of calcite was observed in the retorted shale. The dawsonite was apparently decomposed to a sodium aluminate ($NaAlO_2$).

A portion of the retorted sample was then treated by a leaching operation to obtain alumina. The leaching solution consisted of sodium hydroxide and distilled water, specifically, 50 g of NaOH per liter of solution. A portion of the retorted shale described above was pulverized, and 5 g were placed in a 250-ml Pyrex beaker. Thirty-nine ml of the leaching solution was added. The mixture was heated to about 180°F and maintained at this temperature for 15 minutes, with frequent stirring.

The mixture was diluted with water and filtered. The filtrate was acidified with hydrochloric acid and heated (driving off carbon dioxide and hydrogen sulfide). The solution was then made slightly alkaline with ammonium hydroxide to precipitate aluminum hydroxide. The latter was filtered off, ignited to Al_2O_3 and weighed. The five-gram sample of retorted oil shale produced 0.2283 g alumina, or a 4.57% recovery. This value when corrected for the 25% weight loss on retorting gives a value of 3.4% alumina based on the raw shale or 68 lb alumina per ton of raw shale.

Example 2: The table below illustrates the oil and alumina yield obtained from a second group of 20 samples from a second oil shale obtained in the vicinity of the first shale and having essentially the same composition. After retorting at 900°F for 2 hours, the twenty (3-gram) samples were leached according to the procedure described in Example 1 using three-fifths of the amount of leaching solution. The alumina yield from samples obtained at depths of 2,050 to 2,530 ft average 4.1% alumina based on the retorted shale which is about 3.1% based on raw shale or 62 lb alumina per ton of raw shale. The alumina yield is clearly related to the dawsonite content as measured by x-ray diffraction analysis of the individual oil shale samples.

Sample (foot depth)	Oil Produced	Alumina Produced (percent on retorted shale)	Oil Yield (previous assay) (gal./ton)	Dawsonite Content (by x-ray diffraction)
850	Yes	0.11	14	0.00
1,041	Yes	0.17	9	0.00
1,257	Yes	0.59	10	0.00
1,371	Yes	0.06	28	0.00
1,474	Yes	0.03	26	0.00
1,567	Yes	0.00	13	0.00
1,666	Yes	0.00	31	0.00
1,820	Yes	2.07	11	0.27
2,039	Yes	0.23	11	0.03
2,050	Yes	4.94	17	0.61
2,110	Yes	2.05	24	0.47
2,170	Yes	2.41	11	0.25
2,230	Yes	5.50	41	1.00
2,290	Yes	5.22	25	0.49
2,350	Yes	6.62	55	0.80
2,410	Yes	3.38	33	0.65
2,470	Yes	3.40	48	0.63
2,530	Yes	3.62	33	0.75
2,588	Yes	0.15	9	0.00
2,650	Yes	0.03	45	0.00

VANADIUM

Lime Treatment Prior to Leaching

B.A. Hardwick and J.H. Patterson; U.S. Patent 3,826,809; July 30, 1974; assigned to The Oil Shale Corporation, Australia describe a method of improving the leaching of vanadium and/or molybdenum values from the ash obtained by roasting low-grade ores containing a high content of calcium carbonate.

The process owes its origin to difficulties encountered when seeking to provide an economic method of recovering vanadium and molybdenum from the deposits of vanadiferous shale at Julia Creek in Queensland.

These shale deposits comprise low contents of vanadium and molybdenum (about 0.3% V and about 0.03% Mo), together with a high content of calcium carbonate (up to 50%), kerogen (up to 20%), quartz (up to 25%), pyrite (up to 5%), and traces of micas and clays. The content of kerogen is relatively much higher in the underground shale deposits—the case of oil shale—than in the weathered surface shale deposits (where in fact there may be no kerogen), and the underground deposits therefore constitute a valuable source of fuel oil.

The vanadium is contained predominantly in a mixed layer montmorillonite/mica type of clay mineral; and the molybdenum is contained in separate phases associated with iron, namely pyrite (in the case of oil shale) or geothite (in the case of weathered shale).

According to standard practice for recovering commercially valuable fuel oil from oil shale, the latter is retorted at a temperature between about 400° and about 500°C, under non-oxidative conditions, thus volatilizing the oil from its precursor (kerogen), and leaving a residue of spent shale. In the case of Julia Creek oil shale, the vanadium and molybdenum values after retorting are found to have remained in the spent shale together with inter alia the calcium carbonate.

Methods have been proposed previously for recovering vanadium from vanadiferous ores generally in which the ore is crushed, roasted under oxidative conditions at a temperature of the order of 800°C, and then leached with sulfuric acid. Standard procedures, for example including the step of solvent extraction with di-2-ethylhexyl phosphoric acid in kerosene, are then available for treating the leach liquors to obtain therefrom the vanadium (typically recovered as V_2O_5).

Similarly, it has been found possible to recover vanadium and molybdenum from Julia Creek shale deposits by methods including roasting the shale (possibly after a retorting step to recover fuel oil), and then leaching the roasted shale with a suitable reagent.

When a shale containing calcium carbonate is roasted at temperatures above about 600°C, and particularly at temperatures, approaching 800°C, it will be appreciated that the process is inherently liable to transform the calcium carbonate into free lime with evolution of carbon dioxide. As for the case of recovering vanadium from vanadiferous ores, generally, sulfuric acid is a possible reagent for leaching vanadium and molybdenum values from Julia Creek shale ash. However, because of the combination in this ash of quite low vanadium and molybdenum contents and an unusually high total content of acid reactive material (calcium carbonate and derivative-free lime), it has been found that the requirement of sulfuric acid for a given recovery of vanadium and molybdenum is so high as to render the total process uneconomic.

In the course of experiments leading up to the process, there has been an extensive investigation into the various parameters relevant both to the preparation by roasting of Julia Creek shale and to the process of leaching the resulting shale ashes with different reagent systems. As a result of this investigation, it has been found possible to achieve satisfactory recoveries of vanadium and molybdenum (for example, 65%) by leaching Julia Creek shale ashes with aqueous reagents selected from the group consisting of carbonates and bicarbonates, all of which tend to be incompletely reactive with calcium carbonate.

Typical members within this group are (1) alkaline reagents such as sodium carbonate, sodium bicarbonate, and mixtures thereof; ammonium carbonate, ammonium bicarbonate, and mixtures thereof; and the product of passing carbon dioxide into aqueous ammonia (referred to hereinafter as "carbonated ammonia"); and (2) the weakly acidic reagent carbon dioxide/water. While satisfactory recoveries of vanadium and molybdenum can be achieved with carbonate reagents, it is a disadvantage that the unrecovered fraction of these metal values can still be high.

It is accordingly an object of this process to provide a method of improving the recovery of vanadium and/or molybdenum values when using a carbonate reagent to leach an ash comprising such values and having a high content of calcium carbonate.

The process is predicated on the discovery that, when shale ashes were prepared from Julia Creek shale by roasting at a temperature above about 600°C, the subsequent recovery of vanadium and molybdenum therefrom by leaching with a carbonate reagent was dramatically improved by first conditioning the ashes in a nonleaching medium comprising water.

Flowing from this work, a method has broadly been provided of treating an ore to recover therefrom metal values selected from the group consisting of vanadium and molybdenum, the method comprising the steps of:

(1) roasting the ore to an ash;
(2) forming an aqueous slurry of the ash; and
(3) leaching the values from the ash by introducing into the slurry, a water-soluble reagent selected from the group consisting of a carbonate and a bicarbonate.

The method is further characterized in step (2) by maintaining the slurry in the presence of free lime for a sufficient period at a selected temperature to condition the ash and thereby improve the recovery of the values in step (3). As explained subsequently, this requirement regarding free lime is met inherently in the case of an ash prepared from Julia Creek shale by roasting at a temperature above about 600°C.

By way of general illustration, when a vanadiferous weathered shale is roasted to a temperature above about 600°C, it is found possible typically to obtain a recovery of vanadium of only about 25% when the ash is leached immediately at ambient temperature with carbonated ammonia; however, when the same ash is first conditioned by immersion for 1 day in water at 80°C, a recovery of vanadium typically of about 60% is obtained when leaching is then carried out with the same reagent under the same conditions. Significant improvements in the leaching of molybdenum values are also achieved by first conditioning this ash in the same way.

Example 1: This example relates to samples of ash which were prepared from Julia Creek weathered shale by roasting in a muffle furnace, conditioned (if at all) in water, and then leached at ambient temperature (20°C) with carbonated ammonia. The example explores those variations in the recoveries of vanadium and molybdenum which could be attributed to variations in (1) the temperature of roasting the shale, and (2) the period of conditioning.

A quantity of weathered shale was divided into four samples. Each sample was roasted at a different temperature, and each resulting ash was divided into control and experimental ashes. The control ash in each case was submitted immediately to leaching, while the experimental ashes before leaching were first conditioned for different periods in water at 60°C.

Recoveries (percent) of vanadium and molybdenum in the various cases are given in Table 1. It is apparent from the results (1) that higher recoveries of vanadium and molybdenum are generally associated with roasting at higher temperatures; (2) that conditioning invariably leads to higher recoveries of both these values; and (3) that, in the case of vanadium, the improvement in recovery is greater for ashes prepared at lower temperatures.

TABLE 1

Roasting temperature (° C.)	Vanadium recovery (percent) for periods of conditioning—				Molybdenum recovery (percent) for periods of conditioning—			
	0 (a), hr.	24 hrs.	48 hrs.	72 hrs.	0 (a), hr.	24 hrs.	48 hrs.	72 hrs.
750	43	54	69	73	57	61	66	63
850	53	60	67	70	56	69	71	67
950	75	81	80	85	69	80	80	------
1,050	77	86	86	85	69	82	85	84

NOTE.—(a) denotes control ash.

Example 2: This example relates to samples of ash which were prepared from spent Julia Creek oil shale by roasting in a fluidized bed furnace, conditioned (if at all) in water, and then leached at ambient temperature (20°C) with carbonated ammonia. The example explores those variations in the recoveries of vanadium which could be attributed to variations in (1) the temperature of conditioning, and (2) the period of conditioning.

Two different samples of spent shale were roasted at a temperature of about 790°C and the resulting ashes, X and Y, were each divided into control and experimental categories. The control ashes were submitted immediately to leaching, while the experimental ashes before leaching were first conditioned for different periods in water at temperatures selected from: ambient temperature (20°C), 58.5°C, and 80°C. Recoveries (percent) of vanadium in the various cases are given in Table 2.

It is apparent from the results given in the table that higher recoveries of vanadium are obtained by conditioning the ash for the same periods at higher temperatures, and that substantially maximum recoveries can be obtained by conditioning the ash for as little as about one-half day at 80°C.

TABLE 2

Ash sample	Conditioning temperature (° C.)	Conditioning period (hr.)	Vanadium recovery (percent)
X	20	(a) 0	60
		24	64
		72	70
		168	76
		264	77
		336	80
X	58.5	(a) 0	60
		24	75
		48	83
		72	83
Y	80	(a) 0	65
		6	66
		8	74
		10	75
		12	77
		16	82
		48	83

NOTE.—(a) denotes control ash.

Example 3: A quantity of ash was prepared from spent Julia Creek oil shale by roasting in a fluidized bed furnace at 790°C. The ash was divided into portions A and B and the treatment of portion B described below illustrates the process.

Portion A was submitted immediately to leaching at 100°C (atmospheric pressure) with a reagent consisting of an aqueous mixture of sodium carbonate and sodium bicarbonate. Portion B was first conditioned for 20 hours in water at 80°C before leaching under the same conditions with the same reagent. The recovery of vanadium from portion B was 61% compared with 26% from portion A.

Hydrometallurgical Treatment and Anion Exchange Resin

J.F. Nutter, F.C. Haas and D.L. Thompson; U.S. Patent 3,709,974; January 9, 1973; assigned to The Oil Shale Corporation describe a hydrometallurgical process for recovering vanadium values from a vanadium-bearing calcium-rich ore. The process involves treating an aqueous slurry of the ore with lime under elevated pressure and temperature conditions. The vanadium values are thereafter solubilized and recovered.

The process is particularly useful for recovering vanadium values from low-grade ores which contain relatively large concentrations up to 75 weight percent or more of lime-producing calcium compounds such as calcite, dolomite, and the like. Such lime-producing compounds may constitute a major portion of the ore's matrix, although the process is also applicable to calcium-deficient vanadium-bearing ores to which calcium compounds may be added to form highly alkaline aqueous slurries. A calcium-deficient ore is one which contains little or no lime-producing compounds.

According to the process, finely divided vanadium-bearing calcitic shale ore is slurried in water and mixed with from about 150 to about 225 pounds of calcium oxide per ton of ore. The slurry is placed in an autoclave and heated to a temperature between 175° and 225°C. Such temperatures result in pressures which vary from 129 psig at 175°C to 370 psig at 225°C. The autoclave is maintained at the desired temperature for a period of from 0.5 to 2 hours. The solids content of the slurry is between 2 and 50% by weight. The slurry is then cooled, the pH lowered by the addition of carbon dioxide to solubilize the vanadium values which are then recovered.

In certain situations where the ore body is located in a remote region, it may be convenient to utilize the calcitic content of the ore to provide the lime (calcium oxide) requirement. Under such circumstances, the lime may be obtained by calcining a portion of the ore.

It is known that the particular vanadium species formed according to this process has a limited water solubility at pH's above about 10.5 and is highly water-soluble at pH's between 5 and 9 with the best recoveries being obtained at pH's between 6 and 7.5. These lower pH values are difficult to measure accurately because as the pressure is released by opening the vessel, carbon dioxide is evolved thereby raising the pH.

The initial pH of the slurry prior to the addition of carbon dioxide is usually above 10.5. Carbon dioxide is added to lower the pH to between 5 and 9, although the optimum recoveries have been obtained at pH's between 6 and 7.5 and at a solids content of 20% by weight. The addition of carbon dioxide to the slurry not only solubilizes the vanadium values, it also reacts with the lime to precipitate calcium carbonate. The amount of carbon dioxide required is ideally that amount which is formed during calcite decomposition upon roasting a portion of the calcitic ore and returned to the slurry to convert the lime back to calcite.

While the carbonation solubilizing technique yields varying amounts of soluble vanadium values, vanadium recoveries may be significantly improved by shifting the solubility equilibrium. The improved recovery is obtained by mixing a strong basic anion exchange resin with the slurry in an amount of from 0.25 cubic foot to 0.75 cubic foot of resin/ton of ore.

In a continuous process, it is desirable to maintain the pH of the slurry within the desired range through the constant addition of carbon dioxide thereto. The carbonation step may be carried out prior to or concurrent with the addition of resin to the slurry. As the soluble vanadium anions are extracted or adsorbed by the resin, the concurrent introduction of carbon dioxide maintains the desired pH to effect additional vanadium value solubilization and extraction. The adsorbed vanadium is thereafter stripped or eluted from the resin and recovered by conventional techniques. The use of such anion exchange resins makes it feasible to adsorb up to 99% of the extractable or solubilized vanadium for overall yields from the ore in the order of at least 50 to 60%. Optimum vanadium extraction is accomplished through the use of multiple countercurrent adsorption and elution stages.

Among the suitable anion exchange resins are the strong base quaternary amine anion exchange resins. Some of the resins found to be useful for extracting vanadium values from the pulp are the resins Amberlite IRA—400, Amberlite IRA—425, and Dowex—21K. Amberlite IRA—400 and Amberlite IRA—425 are strongly basic anion exchange resins containing a quaternary ammonium group prepared by reacting a tertiary amine with a chloromethylated copolymer of styrene and divinylbenzene as described in U.S. Patent 2,591,573. These two resins are the same except that Amberlite IRA—425 is specifically made for resin-in-pulp (RIP) processes. Dowex—21K is a strong base quaternary amine resin manufactured from a copolymer of styrene and divinylbenzene which has been chloromethylated and then aminated with trimethylamine. The Dowex—21K resin is also specifically made for resin-in-pulp operations.

Examples 1 through 19: These examples demonstrate the effect of lime (calcium oxide) concentration, autoclave temperature and pressure, and treatment time. Portions of a composite calcitic shale ore sample containing 0.72% vanadium pentoxide were mixed with varying amounts of reagent grade lime and slurried in water at a solids content of 20% by weight. The slurries were placed in an autoclave and heated to various temperatures and pressures. The autoclave was maintained at the desired temperature for various time periods.

After cooling the autoclave to room temperature, the slurry having a pH of about 10.5 or higher was removed and carbonated by bubbling carbon dioxide into the slurry for about one to two minutes to lower the pH to 6.5.

The carbonated slurries were mixed with 0.5 cubic feet of various forms of a strong base anion exchange resin (Amberlite IRA—425) per ton of ore and extracted for about 15 minutes at ambient temperature to load the resin with vanadium anions. Carbon dioxide was continuously added to the slurry during the extraction operation. The resin was screened from the slurry, thereafter stripped or eluted of the vanadium anions with a sodium chloride eluting solution, the vanadium precipitated as ammonium metavanadate with ammonium chloride and the precipitate roasted to form V_2O_5. The results are presented in the following table.

TABLE 1: CO_2-RIP LEACH OF REAGENT GRADE LIME-TREATED CALCITE SHALE

Example number	Lbs. CaO added per ton of ore	Treatment: Temp. (° C.)	Pressure (p.s.i.g.)	Time (hrs.)	Percent V_2O_5 extracted
Control	0	225	370	1	0
1	42	200	225	1	24
2	42	200	225	2	22
3	42	225	370	1	23
4	128	225	370	0.5	34
5	128	225	370	1	39
6	150	225	370	1	49
7	170	150	69	0.5	14
8	170	150	69	1	26
9	170	175	129	0.5	28
10	170	175	129	1	32
11	170	200	225	0.5	29
12	170	200	225	1	39
13	170	225	370	1	54
14	170	250	577	1	42
15	212	225	370	1	48
16	255	225	370	0.5	35
17	255	225	370	1	37
18	255	225	370	2	35
19	425	250	577	1	42

From the comparative data in the above table, it can be seen that optimum vanadium recoveries as high as 54% V_2O_5 (Example 13) are achieved at autoclave temperatures and pressures in the order of 225°C and 370 psig using 170 pounds of lime per ton of ore for a treatment period of one hour. When lime was omitted during the pressure hydrometallurgical treatment of the ore (Control Example), no vanadium was recovered.

Data for Examples 3, 5, 6, 13, 15 and 17 indicate that at constant temperature, pressure and time the amount of vanadium recovered increases to a maximum as the amount of lime used is increased and then decreases as the amount of lime used is further increased. Hence, excessive lime is not warranted. Data for Examples 8, 10, 12, 13 and 14 indicate that at constant lime and time the amount of vanadium recovered increases to a maximum as the temperature and pressure are increased and then decreases as the temperature and pressure are further increased. Accordingly, the use of excessive temperatures and pressures is not warranted.

Examples 20 through 22: These examples illustrate an alternate procedure for providing the source of calcium oxide (lime) for use in the pressure hydrometallurgical treatment procedure.

A portion of a vanadium-bearing calcitic shale ore sample containing about 75% calcium carbonate was calcined at about 800°C for one hour to convert the calcium carbonate, i.e., calcite, into lime ash by driving off the carbon dioxide. Portions of the uncalcined ore were mixed with the lime ash in various amounts equivalent to the calcium oxide requirement. The mixture was then slurried in water at a solids content of 20% by weight and heated in an autoclave at about 225°C and 370 psig for one hour.

The product of the autoclave treatment was thereafter carbonated with carbon dioxide to a pH of 6.5 and mixed with 0.5 cubic foot of Amberlite IRA—425, a strong basic anion exchange resin, per ton of ore and extracted or leached for about 15 minutes at 60°C while maintaining the desired pH of 6.5 of the slurry through the continuous introduction of carbon dioxide thereto. After screening of the loaded resin from the slurry, the vanadium values were stripped or eluted from the resin, precipitated and calcined to V_2O_5 as set forth hereinabove. (The carbon dioxide product of the lime ash-producing calcination procedure may advantageously be utilized to satisfy the various carbon dioxide process requirements.) The results of this procedure are presented in the following table.

TABLE 2: CO_2-RIP LEACH OF LIME ASH-TREATED CALCITIC SHALE

Example number	Lbs. 800° C. lime ash/ton of ore	Equiv. CaO	Percent V_2O_5 extracted
20	400	216	45
21	300	162	47
22 (3-stage resin leach)	300	162	52

From the data for Examples 20 and 21 in the above Table 2 vis-a-vis the data for Examples 15 and 13, respectively in the above Table 1, it is readily apparent that a portion of the ore can be used (calcined) to provide the source of lime, i.e., equivalent CaO, without adversely affecting the vanadium recovery (i.e., V_2O_5 or 45% vs 48 and 47% vs 54%). The data for Examples 21 and 22 indicate that a multistage resin leaching or extracting procedure improves the recovery of the vanadium values.

Carbonation Technique to Control pH

F.C. Haas; U.S. Patent 3,656,936; April 18, 1972; assigned to The Oil Shale Corporation describes a process for recovering vanadium values from highly alkaline, calcium-rich calcines, which are usually the roasted product of low-grade vanadium-bearing calcitic ores. The process involves slurrying the calcine in water and carbonating the slurry to a pH of 5.0 to 9.0 to extract the soluble vanadium species from the slurry. A basic anion exchange resin is added to the carbonated slurry thereby increasing the vanadium recovery by shifting the extraction equilibrium.

Example 1: This example demonstrates the effect of pH range on the amount of vanadium extracted. Portions of a composite sample containing 1.07% vanadium pentoxide were roasted at 600°C in a muffle furnace for one hour and quenched in water. The slurries thus formed had a pH of about 13. Carbon dioxide was added to the slurries to adjust the pH to between 7.0 and 12.5.

The carbonated slurries were mixed with various forms of a strong anion base exchange resin (Amberlite IRA—400) and leached for about 15 minutes at ambient temperature to load the resins with vanadium ions. Atmospheric pressure was used in all but one of the tests. The resins were stripped of the vanadium ions and the vanadium precipitated. The results of aqueous leaching at various pH's are presented in the following table.

TABLE 1

Test No.	% Solids	Pressure psig	pH	Resin Form	% V_2O_5 Extracted
S-553	40	Ambient	7	HSO_4^-	50
S-555	40	65	7	HSO_4^-	55
S-556	40	Ambient	7	OH^-	50
S-557	40	Ambient	7	Cl^-	53
S-558	40	Ambient	8.9	Cl^-	54
70-31B	25	Ambient	10	Cl^-	4
70-31A	25	Ambient	11	Cl^-	0
S-561	25	Ambient	12.5	Cl^-	1
S-560	25	Ambient	7	Cl^-	53
S-565	25	Ambient	7	HCO_3^-	55

From the data in the above table, it can be seen that vanadium extraction decreases sharply at pH's above 9.0. It also is apparent that the use of pressure during the leach operation aids in increasing vanadium extraction and that virtually any form of the resin may be used.

Example 2: This example illustrates the equilibrium shift caused by adding a basic anion exchange resin to the slurry which results in increased vanadium extraction. Two synthetically prepared samples and two samples of vanadium-bearing shale ash serve to illustrate the effect of resin addition.

The synthetic sample was prepared by blending 50 g of calcium carbonate with 0.75 g of vanadium pentoxide and roasting the mixture at 1000°C for one hour. The vanadium-bearing shale was also roasted at 1000°C for one hour. The calcines were quenched in water to form aqueous alkaline slurries having a solids content of about 25%. A strong basic anion exchange resin (Amberlite IRA—425) was added to two slurries containing the synthetically prepared sample and the shale ash sample, respectively. Carbon dioxide was then introduced into all four slurries to lower the pH from 13.0 to 6.5 to 7.0 and then they were leached for 15 minutes at ambient conditions.

The resins were thereafter screened from the slurries and stripped of the vanadium values. The liquor from the two slurries, which did not contain the resin, was analyzed for the percent V_2O_5 extracted. The results of the four tests are presented in the following table.

TABLE 2

Sample Type	Resin in Slurry	% V_2O_5 Extracted
Synthetic Blend	No	66
Synthetic Blend	Yes	88
Shale Ash	No	5
Shale Ash	Yes	55

From the data in the above Table 2, it is evident that the equilibrium shift resulting from the addition of a basic anion exchange resin results in the recovery of a substantially greater portion of the vanadium values present in the calcined material and partially solubilized in the aqueous carbonated slurry.

Example 3: This example shows the effect of varying roasting temperatures and times upon the solubility of a synthetically prepared blend of calcium carbonate and vanadium pentoxide that is slurried in water and leached with carbon dioxide.

The same was prepared by blending reagent grade calcium carbonate and vanadium pentoxide in the ratio of 100:1 and roasting portions of the mixture at various temperatures for different time periods. The roasted calcines were then slurried in water at a solids content of about 25%. The slurries were carbonated by adding carbon dioxide until the slurry pH was reduced from 13.0 to 6.5 to 7.0. This also precipitated calcium carbonate. The slurries containing the water-soluble calcium vanadate species and calcium carbonate solids were leached for 15 minutes. The solids were filtered and the solution analyzed for the percent of vanadium pentoxide extracted. For all but three of the tests, an oxidizing atmosphere was maintained during the roasting operation.

The percent of vanadium pentoxide (V_2O_5) extracted as a function of roasting time, temperature and atmosphere is presented in the following table.

TABLE 3

Test No.	Temp, °C	Time, min	Air Addition	% V_2O_5 Extraction
1	600	5	Yes	26
2	600	15	Yes	27
3	600	30	Yes	22
4	600	60	Yes	20
5	800	5	Yes	37
6	800	15	Yes	52
7	800	30	Yes	11
8	800	60	Yes	15
9	1000	5	Yes	13
10	1000	15	Yes	15
11	1000	30	Yes	15
12	1000	60	Yes	44
13	600	5	No	44
14	600	15	No	33
15	600	30	No	35

From the above data in Table 3, it is readily apparent that the optimum condition is 800°C in an oxidizing atmosphere for 15 minutes. It is also apparent that the absence of air is also beneficial since very good extractions are attainable at temperatures as low as 600°C.

Sulfur Dioxide

B.A. Hardwick and J.H. Patterson; U.S. Patent 3,840,637; October 8, 1974; assigned to The Oil Shale Corporation, Australia describe a method of recovering acid leachable vanadium values from vanadiferous ores comprising a high content of calcium carbonate. The ore is first roasted to an ash under conditions selected to minimize the decomposition of the calcium carbonate component to free lime, and vanadium values are then leached from an aqueous slurry of the ash by means of an acid gas. The method is characterized particularly by using an acid gas comprising sulfur dioxide to effect pH reduction in the range below about 6, and various parameters (for example, concentration of sulfur dioxide in the acid gas and temperature of slurry) are controlled whereby to favor incomplete reaction with the calcium carbonate component of the ash. The final leaching system (of pH preferably between about 4 and about 2) is not at equilibrium, and is filtered immediately for maximum recovery of vanadium.

Example 1: A sample of spent Julia Creek oil shale was roasted in a fluidized bed at 800°C and the resulting ash was divided into portions A to D. Each portion of ash was formed into a corresponding aqueous slurry (20% solids by weight) and the slurry was then leached at 55°C to a selected final pH below 6 by introducing thereto a selected acid gas.

In the case of slurries A to C, the pH was first lowered to 6.5 (first-stage leaching) by introducing thereto a gas comprising carbon dioxide as the only acidic component. The pH was then lowered to the selected final value (second-stage leaching) by means of a gas of composition as specified in the table below.

In the case of slurry D, the pH was lowered to the selected final value (single-stage leaching) by means of the gas of composition specified in the table. At the termination of leaching, the slurry in each case was filtered immediately and washed (uniform procedures being followed); and the filtrate was analyzed for vanadium by standard procedures. The table records the recovery (percent) of vanadium and the sulfur dioxide input in each case.

Slurry	Gas composition (percent by volume)				Final pH	Percent vanadium extracted	SO_2 input (lb./100 lb. ash)
	SO_2	CO_2	O_2	N_2			
A	3.8	11.4	6.3	78.5	3.0	71	13
B	4.6	----	----	95.4	3.1	38	17
C	4.8	----	95.2	----	3.5	33	17
D	7.5	92.5	----	----	2.9	69	13

It is apparent from these results that acceptably high vanadium recoveries can be obtained with reasonably low sulfur dioxide inputs when the acid gas used is sulfur dioxide in mixture either with carbon dioxide per se (single-stage leaching: Case D) or with carbon dioxide, oxygen and nitrogen (two-stage leaching: case A). However, the vanadium recoveries are reduced, and the sulfur dioxide inputs are at the same time increased, when the final pH is achieved by means of an acid gas comprising sulfur dioxide in mixture either with oxygen or nitrogen alone.

ZIRCONIUM

According to a process described by *S. Penzes; U.S. Patent 3,656,938; April 18, 1972; assigned to Canada Cities Service, Ltd., Imperial Oil Limited; Atlantic Richfield Corporation; Royalite Oil Company, Limited, Canada* heavy metals, especially zirconium, may be recovered from bituminous sands.

While the presence of zirconium and titanium in very minor quantities in coarse grains of bituminous sands has been previously recognized, these metals have not appeared to be present in economically recoverable concentrations.

It has been found that when bituminous sand containing minor quantities of heavy metals, especially metals of Group IVB of the periodic table of elements (titanium, zirconium and hafnium), and particularly zirconium, is treated for recovery of bitumen by processes such as the hot and cold water methods, the heavy metals concentrate in the bituminous froth recovered from the body of hot water to an extent such that recovery of the metals from the froth becomes economical. The concentration of metals in such froth has been found to be at least 10 and frequently as much as 40 or 50 times that of the corresponding metals in the raw bituminous sand.

In the case of titanium, the concentration in the froth is usually about 20 times that in the raw sand while zirconium is usually found in the froth in about forty times the concentrations found in the original bituminous sand.

Preferably, the bituminous froth containing concentrations of heavy metals is produced from bituminous sand by the hot water method. In this process, raw bituminous sand is mixed with warm water in a mixer such as a conventional rotating cylindrical tumbler having perforations for the slurry to drop out and provision for rejecting oversized rocks and lumps of sand which cannot be broken down. The slurry produced in the mixer is preferably at a temperature between about 140° and about 180°F. The slurry temperature may be controlled by control of the temperature of the water added to the mixer or by direct addition of steam to the mixer or by other suitable means.

Average residence time of bituminous sand in the mixer is preferably between about 0.5 and about 5 minutes. Sufficient water is normally added in the mixer to form a slurry containing between about 20 and about 40 weight percent water. After leaving the mixer the slurry is usually flooded with additional water to bring its water content to between about 40 and about 60 weight percent and the flooded slurry is then introduced into a body of warm water maintained at a temperature between about 140° and about 200°F.

If desired, aeration gas may be added to the hot water to assist in separation of bitumen from sand particles. Air entrained in the slurry during the mixing step as well as any introduced into the separation zone causes bitumen to rise to the upper surface of the water to form a froth while sand is allowed to settle to the bottom. Sand tailings may be withdrawn from the bottom of the separation zone while froth containing bitumen, heavy metals and other solids is recovered from the upper portion of the separation zone.

Following recovery from the hot water separation zone, the froth containing bitumen, heavy metals and other solids is treated for recovery of bitumen and metals. In one example, the froth is subjected to thermal dehydration treatment to evaporate water and is then subjected to gravity treatment as in a cyclone separator to obtain an overhead stream containing the majority of the bitumen and an underflow stream containing most of the solids and substantially all of the heavy metals. If desired, the underflow stream may be filtered to further separate liquid and solids. The solids from the underflow stream are then preferably dried as in a rotary kiln.

Solids recovered from bituminous sand froth, as described above, are generally in the size range between about 20 and about 44 microns and contain concentrated amounts of heavy metals. In the process, such solids are further treated with alkali metal hydroxide followed by flotation with aeration gas to further concentrate the heavy metals and especially the zirconium present in the form of zircon.

The alkali metal hydroxide is preferably in the form of an aqueous solution and, while the strength of the solution is not considered especially important, concentrations of between about 5 and about 30 mol percent of the alkali metal hydroxide are commonly used. When treating the solids with the alkali metal hydroxide, enough liquid should be used to completely wet the solids and maintain a continuous liquid phase. This may usually be accomplished with the use of at least 3 volumes of liquid per volume of solids with liquid to solids ratios in the range of about 3 to about 20 being most usual. While any alkali metal hydroxide may be used, sodium hydroxide has been found to be especially useful.

Once the heavy metal solids have been concentrated in the lower portion of the flotation zone, as described above, they may be recovered and the heavy metal recovered in accordance with standard metal recovery techniques.

One suitable method for recovering zirconium (with or without hafnium enrichment) and titanium from such solids involves first passing the dried solids through magnetic separation equipment to recover ilmenite therefrom. The tailings from the magnetic separation step are then passed across a conventional shaking table to remove light solids, leaving zircon and rutile. These may be treated in conventional flotation equipment with a soap collector such as sodium oleate and a frothing additive to obtain separation of zircon from rutile. It should be understood that the zircon may be enriched with minor amounts of hafnium such as up to about 5% by weight of the zircon.

Example: A typical example of recovery of heavy metals from tar sands involves recovery of such metals from bituminous sand having the following composition:

Ingredients	Weight Percent
Solids	86.12
Bitumen	10.43
Water	3.45
Hafnium-Enriched Zirconium	
as Zr	0.05
as Zircon	0.10
Titanium	
as Ti	0.21
as Rutile	0.35

This tar sand is mixed with water to form a flooded slurry containing 60 weight percent water and is then introduced into a hot water separation zone containing a body of hot water maintained at a temperature of 159°F. Froth recovered from the upper portion of the separation zone contains 22.14 weight percent water, 68.50 weight percent bitumen, and 9.36 weight percent solids. Zirconium is present in the solids in this froth to the extent of 1.95 weight percent based on Zr and 3.92 weight percent based on zircon.

Titanium is present in the solids in the froth to the extent of 4.04 weight percent on Ti and 6.74 weight percent based on rutile. Following thermal dehydration and cycloning, the underflow solids from the cyclone contain 70 to 80 weight percent of the solids present in the froth except that about 90% of the rutile, ilmenite and hafnium-enriched zircon is present in the solids recovered from the cyclone underflow stream due to the relatively high gravity of these materials.

The underflow solids from the cyclone are then treated with 5 volumes of 10 mol percent aqueous solution of sodium hydroxide for a period of one-half hour at a temperature of 190°F and are subsequently contacted with upflowing air in a flotation zone for one-half hour to preferentially float solids other than heavy metals. This results in a concentrated layer of heavy metal solids in the bottom of the flotation zone and the solids are then removed and heavy metals recovered from the metal recovery techniques described above.

COMPANY INDEX

The company names listed below are given exactly as they appear in the patents, despite name changes, mergers and acquisitions which have, at times, resulted in the revision of a company name.

INVENTOR INDEX

U.S. PATENT NUMBER INDEX

NOTICE

POLLUTION CONTROL IN THE PETROLEUM INDUSTRY 1973

by H. R. Jones

Pollution Technology Review No. 4

The petroleum industry can logically be divided into three major divisions: (1) Production, (2) refining, (3) subdividing and distributing. Production includes the operations involved in locating and drilling oil fields, removing oil from the ground, pretreatment at the well site, and transporting the crude to the refinery. The extent of pretreatment depends on the type of crude oil, and usually comprises removal of gas and brine. Refining includes the operations necessary to convert the crude into saleable products. Subdividing and distribution involves handling the finished petroleum products.

All of these activities are or can be sources of emissions into the air and water. This book shows you practical technology for avoiding or reducing these harmful emissions, and how to clean up after unavoidable damage has been done. There is also a separate chapter on the costs and economics of pollution control.

Chapter headings are given here, followed by examples of important subtitles.

ISBN 0-8155-0494-2 **322 pages**

OFFSHORE DRILLING TECHNOLOGY 1975

by Frank R. Carmichael

Energy Technology Review No. 5
Ocean Technology Review No. 3

The pecuniary interests, impacts, and high stakes involved in offshore drilling for petroleum are clearly revealed in the patent applications of the last few years.

Indeed, many of the concepts and design parameters now being implemented by the major construction companies around the world for use in the North Sea and other offshore locations are fully described and illustrated in the U.S. patent literature.

This book presents over 190 different processes and equipment designs for all phases of modern offshore drilling techniques including descriptions and uses of drilling ships, platforms, wellhead completions and a considerable number of very complex subsea facilities. Many of the processes in this area are described as complete systems, thus consolidating multiple functions in this emerging technology. However, for continuity, these processes are presented in the sections deemed most appropriate for their overall contributions to offshore exploration and exploitation.

A partial and condensed table of contents follows here. Numbers in parentheses indicate a plurality of processes per topic. Chapter headings and some of the more important subheadings and subtitles are given here.

ISBN 0-8155-0566-3

392 pages